2012 INTERNATIONAL RESIDENTIAL CODE®
STUDY COMPANION

INTERNATIONAL
CODE COUNCIL®

2012 International Residential Code
Study Companion

ISBN: 978-1-60983-157-8

Cover Design:	Ricky Razo
Publications Manager:	Mary Lou Luif
Project Editor:	Roger Mensink
Illustrator/Interior Design:	Mike Tamai
Manager of Development:	Doug Thornburg

COPYRIGHT 2011

INTERNATIONAL
CODE COUNCIL®

TABLE OF CONTENTS

INTRODUCTION

This study companion provides practical learning assignments for independent study of the provisions of the 2012 *International Residential Code (*IRC®). The independent study format affords a method for the student to complete the program in an unregulated time period. Progressing through the workbook, the learner can measure his or her level of knowledge by using the exercises and quizzes provided for each study session.

The workbook is also valuable for instructor-led programs. In jurisdictional training sessions, community college classes, vocational training programs and other structured educational offerings, the study guide and the IRC can be the basis for classroom instruction.

All study sessions begin with a general learning objective specific to the session, the specific code sections or chapter under consideration and a list of questions summarizing the key points of study. Each session addresses selected topics from the IRC and includes code text, a commentary on the code provisions, illustrations representing the provisions under discussion and multiple choice questions that can be used to evaluate the student's knowledge. Before beginning the quizzes, the student should thoroughly review the IRC, focusing on the key points identified at the beginning of each study session.

The workbook is structured so that after every question the student has an opportunity to record his or her response and the corresponding code reference. The correct answers are found in the back of the workbook in the answer key.

This study companion was developed by Douglas W. Thornburg, AIA, C.B.O., Technical Director of Product Development for the International Code Council. In addition to authoring numerous educational texts and resource materials, he instructs seminars nationally on both the IBC and IRC. Doug has more than 30 years experience in the application and enforcement of building codes.

The information presented in this publication is believed to be accurate; however, it is provided for informational purposes only and is intended for use only as a guide. As there is a limited discussion of selected code provisions, the code itself should always be referenced for more complete information. In addition, the commentary set forth may not necessarily represent the views of any enforcing agency, as such agencies have the sole authority to render interpretations of the IRC.

Questions or comments concerning this study companion are encouraged. Please direct your comments to ICC at *studycompanion@iccsafe.org*.

About the International Code Council

The International Code Council® (ICC®) is a member-focused association dedicated to helping the building safety community and construction industry provide safe, sustainable and affordable construction through the development of codes and standards used in the design, build and compliance processes. Most U.S. communities and many global markets choose the International Codes®. ICC Evaluation Service (ICC-ES), a subsidiary of the International Code Council, has been the industry leader in performing technical evaluations for code compliance fostering safe and sustainable design and construction.

Headquarters: 500 New Jersey Avenue, NW, 6th Floor, Washington, DC 20001-2070

District offices: Birmingham, AL; Chicago, IL; Los Angeles, CA

1-888-422-7233

www. iccsafe.org

Enhance Your Study Experience

ICC's bonus online quiz is a new practice tool just right for you.

Your 2012 *International Residential Code Study Companion* includes a helpful online practice quiz developed to help you retain information and prepare for exams. The 60-question online practice quiz includes features such as:

- Many new questions to supplement those in the study companion;
- Format similar to computer based exams, with point and click convenience;
- Review of the quiz available upon completion, including code references;
- Questions covering topics from all study sessions of the study companion; and
- Ability to repeat the quiz multiple times.

2012 International Residential Code

Study Companion Online Quiz

The study companion online quiz is beneficial when studying for the following 2012 Certification exams:

- Residential Inspector
- Residential Plans Examiner
- Residential Contractor

To take your online quiz, visit www.iccsafe.org/Quiz2012 **and enter coupon code:**

SM12007

2012 IRC Chapters 1 and 44
Administration and Referenced Standards

OBJECTIVE: To obtain an understanding of the administrative provisions of the *International Residential Code®* (IRC®), including the scope and purpose of the code, duties of the building official, issuance of permits, submission of construction documents, inspection procedures, certificate of occupancy and referenced standards.

REFERENCE: Chapters 1 and 44, 2012 *International Residential Code*

KEY POINTS:
- What is the purpose and scope of the *International Residential Code*?
- When materials, methods of construction or other requirements are specified differently in separate provisions, which requirement shall govern?
- When there is a conflict between a general requirement and a specific requirement, which provision shall be applicable?
- When do the provisions of the appendix apply?
- How are existing buildings to be addressed?
- What term is applied to the building department? How are deputies appointed?
- What are the powers and duties of the building official in regard to the application and interpretation of the code? In regard to right of entry?
- What degree of liability does the building official have in regard to performance of their duties?
- Under what conditions may the building official grant modifications to the code?
- How may alternative materials, designs and methods of construction be approved?
- When is a permit required? What types of work are exempted from permits?
- Is work exempted from a permit required to comply with the provisions of the code?
- What is the process outlined for obtaining a permit?
- What information is required on the permit application?
- What conditions or circumstances would bring the validity of a permit into question?
- When does a permit expire? What must occur when a permit expires prior to completion of a building?
- When are submittal documents required? When must plans be prepared by a registered design professional?

KEY POINTS: • How long must approved construction documents be retained by the building department?
(Cont'd) • How are temporary buildings addressed? To what levels of conformance must they comply?

• What fees are set forth in the code? How are fees to be determined?

• What types of inspections are specifically required by the code? When are inspections required?

• When is a certificate of occupancy required? For what reasons is revocation permitted?

• When may a temporary certificate of occupancy be issued?

• What is the purpose of a board of appeals? Who shall serve on the board?

• What limitations are placed on the authority of the board of appeals?

• When should a stop-work order be issued?

• What are referenced standards? How are they applied to the IRC®?

Topic: Dwelling Units

Category: Administration

Reference: IBC 101.2, Exception

Subject: General Requirements

Code Text: *Detached one- and two-family dwellings and multiple single-family dwellings (townhouses) not more than three stories above grade plane in height with a separate means of egress and their accessory structures shall comply with the* International Residential Code.

Discussion and Commentary: Many residential structures are exempt from the requirements of the *International Building Code®* (IBC®) and are regulated instead by a separate and distinct document, the *International Residential Code®* (IRC®). The IRC contains prescriptive requirements for the construction of detached single-family dwellings, detached duplexes, townhouses and all structures accessory to such buildings. Limited to three stories in height above grade plane with individual egress facilities, these residential buildings are fully regulated by the IRC for building, plumbing, mechanical, electrical and energy provisions.

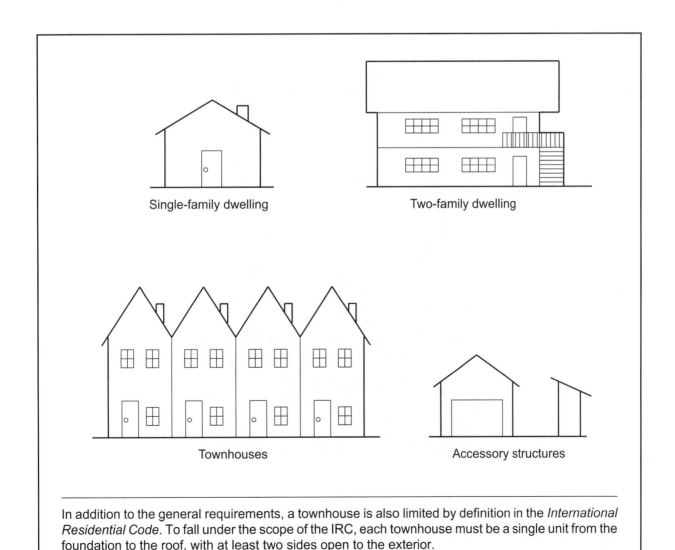

Single-family dwelling

Two-family dwelling

Townhouses

Accessory structures

In addition to the general requirements, a townhouse is also limited by definition in the *International Residential Code*. To fall under the scope of the IRC, each townhouse must be a single unit from the foundation to the roof, with at least two sides open to the exterior.

Code Text: *The provisions of the* International Residential Code for One- and Two-family Dwellings *shall apply to the construction, alteration, movement, enlargement, replacement, repair, equipment, use and occupancy, location, removal and demolition of detached one- and two-family dwellings and townhouses not more than three stories above grade plane in height with a separate means of egress and their accessory structures.* See exceptions for 1) live/work units complying with the requirements of Section 419 of the IBC and 2) owner-occupied lodging houses with no more than five guest rooms.

Discussion and Commentary: The *International Residential Code* is intended to regulate the broad spectrum of construction activities associated with residential buildings and structures. The provisions address building planning and construction, as well as energy efficiency, mechanical, plumbing and electrical aspects.

Accessory buildings—such as private garages, carports, sheds, playhouses and similar structures incidental to the major dwelling or townhouse—are also regulated by the IRC, provided each accessory structure is limited to 3,000 square feet in floor area and two stories in height.

Code Text: *Provisions in the appendices shall not apply unless specifically referenced in the adopting ordinance.*

Discussion and Commentary: The appendix chapters of the IRC address subjects that are inappropriate as a mandatory portion of the code. Rather, the appendices are optional, with each jurisdiction adopting all, some or none of the appendix chapters, depending on its needs for enforcement in any given area. There are various reasons why certain issues are placed in the appendix. Often, the provisions are limited in application or interest. Some appendix chapters are merely extensions of requirements set forth in the body of the code. Others address issues that are often thought of as outside of the scope of a traditional building code. For whatever reason, no appendix chapter is applicable unless specifically adopted.

IRC Appendix Chapters

Appendix A	Sizing and Capacities of Gas Piping
Appendix B	Sizing of Venting Systems Serving Appliances Equipped with Draft Hoods, Category I Appliances, and Appliances Listed for Use and Type B Vents
Appendix C	Exit Terminals of Mechanical Draft and Direct-Vent Venting Systems
Appendix D	Recommended Procedure for Safety Inspection of an Existing Appliance Installation
Appendix E	Manufactured Housing Used as Dwellings
Appendix F	Radon Control Methods
Appendix G	Swimming Pools, Spas and Hot Tubs
Appendix H	Patio Covers
Appendix I	Private Sewage Disposal
Appendix J	Existing Buildings and Structures
Appendix K	Sound Transmission
Appendix L	Permit Fees
Appendix M	Home Day Care – R-3 Occupancy
Appendix N	Venting Methods
Appendix O	Automatic Vehicular Gates
Appendix P	Sizing of Water Piping System
Appendix Q	ICC *International Residential Code* Electrical Provisions/National Electrical Code Cross Reference

---

Appendix chapters not adopted as a portion of a jurisdiction's building code may still be of value in application of the code. Provisions in the appendices might provide some degree of assistance in evaluating proposed alternative designs, methods or materials of construction.

Code Text: *The department of building safety is hereby created and the official in charge thereof shall be known as the building official. The building official shall be appointed by the chief appointing authority of the jurisdiction. In accordance with the prescribed procedures of this jurisdiction and with the concurrence of the appointing authority, the building official shall have the authority to appoint a deputy building official, the related technical officers, inspectors, plan examiners and other employees. Such employees shall have powers as delegated by the building official.*

Discussion and Commentary: The building official is an appointed officer of the jurisdiction and charged with the administrative responsibilities of the department of building safety. It is not uncommon for the jurisdiction to use a different position title to identify the building official, such as Chief Building Inspector, Superintendent of Central Inspection, or Director of Code Enforcement. Regardless of the jurisdictional title, the code recognizes the individual in charge as the building official.

Inspectors, plan reviewers and other technical staff members are typically given some degree of authority to act for the building official in the decision-making process, including the making of appropriate interpretations on various provisions of the code.

Code Text: *The building official is hereby authorized and directed to enforce the provisions of the IRC. The building official shall have the authority to render interpretations of the IRC and to adopt policies and procedures in order to clarify the application of its provisions. Such interpretations, policies and procedures shall be in conformance with the intent and purpose of the IRC. Such policies and procedures shall not have the effect of waiving requirements specifically provided for in the IRC.*

Discussion and Commentary: The building official must be knowledgeable to the extent that he or she can rule on those issues that are not directly addressed or are unclear in the code. The basis for such a determination is the intent and purpose of the *International Residential Code*, which often takes some research to discover.

City of (Jurisdiction)

Department of Building Safety

Photo

_____
Name of individual

_____
Job function

The individual identified on the badge is a duly authorized employee of (the Jurisdiction) and is a designated representative of the Department of Building Safety.

Valid
through   _____   _____
            Date   Building Official

Proper Identification Mandated by Section R104.5 when Inspecting Structures or Premises

Although the IRC gives broad authority to the building official in interpreting the code, this authority also comes with great responsibility. The building official must restrict all decisions to the intent and purpose of the code; the waiving of any requirements is strictly prohibited.

Code Text: *The building official, member of the board of appeals or employee charged with the enforcement of the IRC, while acting for the jurisdiction in good faith and without malice in the discharge of the duties required by the IRC or other pertinent law or ordinance, shall not thereby be rendered liable personally and is hereby relieved from personal liability for any damage accruing to persons or property as a result of any act or by reason of an act or omission in the discharge of official duties.*

Discussion and Commentary: The protection afforded by the code regarding employee liability is limited to only those acts that occur in good faith. Absolute immunity from all tort liability is not provided where an employee acts maliciously. An employee is not relieved from personal liability where malice can be shown, and it is probable that the jurisdiction will not provide for the employee's defense.

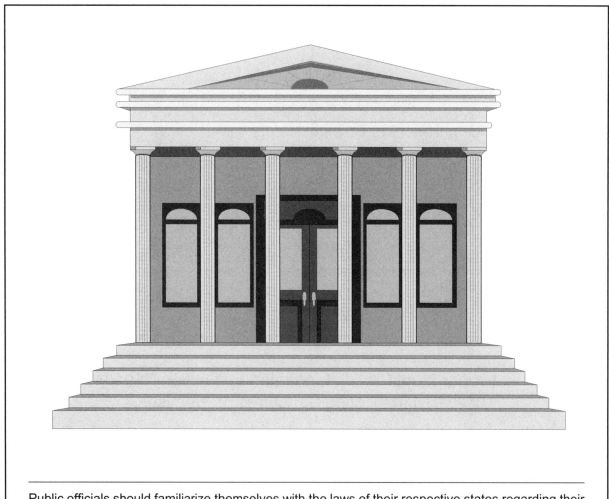

Public officials should familiarize themselves with the laws of their respective states regarding their exposure to tort liability.

Topic: Alternative Materials	**Category:** Administration
Reference: IRC R104.11	**Subject:** Duties and Powers of Building Official

Code Text: *The provisions of the IRC are not intended to prevent the installation of any material or to prohibit any design or method of construction not specifically prescribed by the IRC, provided than any such alternative has been approved.*

Discussion and Commentary: The building official is granted broad authority in the acceptance of alternative materials, designs and methods of construction. Arguably the most important provision of the IRC, the intent is to implement the adoption of new technologies. Furthermore, it gives the code even more of the character of a performance code. The provisions encourage state-of-the-art concepts in construction, design and materials as long as they meet the performance level intended by the IRC. In a portion of the provision not quoted above, reference is also made to the use of any performance-based provisions of the other *International Codes®* as acceptable alternates.

Building official may approve alternative materials, design and methods of construction, if:

- Proposed alternative is satisfactory

- Proposed alternative complies with intent of code

- Material, method or work is equivalent in:

  (1) Quality

  (2) Strength

  (3) Effectiveness

  (4) Fire Resistance

  (5) Durability

  (6) Safety

Advisable for building official to:

- Require sufficient evidence or proof

- Record any action granting approval

- Enter information into files

The building official should ensure that any necessary substantiating data or other evidence be submitted to show that the alternative is in fact equivalent in performance. Moreover, where tests are performed, reports of such tests must be retained by the building official.

Topic: Permits Required and Exempted   **Category:** Administration
Reference: IRC R105.1, R105.2   **Subject:** Permits

Code Text: *Any owner or authorized agent who intends to construct, enlarge, alter, repair, move, demolish, or change the occupancy of a building or structure . . . shall first make application to the building official and obtain the required permit.* See multiple exceptions where a permit is not required. *Exemption from permit requirements of the IRC shall not be deemed to grant authorization for any work to be done in any manner in violation of the provisions of the IRC or any other laws or ordinances of the jurisdiction.*

Discussion and Commentary: Except in those few instances specifically listed, such as minor repairs, maintenance, finish work and limited equipment replacement, all construction-related work requires a permit and is subject to subsequent inspections.

Building work exempt from building permit:

- One-story detached accessory buildings such as tool and storage sheds, limited to 200 square feet in floor area

- Fences not over 7 feet high

- Retaining walls not over 4 feet high

- Water tanks supported directly on grade, if the capacity does not exceed 5,000 gallons and the height to diameter ratio does not exceed 2 to 1

- Sidewalks and driveways

- Painting, papering, tiling carpeting, cabinets, counter tops and similar finish work

- Prefabricated swimming pools less than 24 inches deep

- Swings and other playground equipment

- Window awnings supported by an exterior wall that do not project more than 54 inches from the exterior wall

- Decks, if 1) 200 square feet maximum), 2) limited to 30 inches above grade, 3) not attached to a dwelling and 4) not serving the required exit door

---

Whether or not a building permit is required by the code, all work must be done in accordance with the code requirements. It is important that the owner be responsible for proper and safe construction.

Topic: Submittal Documents	**Category:** Administration
Reference: IRC R106	**Subject:** Construction Documents

Code Text: *Submittal documents consisting of construction documents and other data shall be submitted in two or more sets with each application for a permit.* See exception that authorizes building official to waive the submission of construction documents. *Manufacturer's installation instructions, as required by* the IRC, *shall be available on the job site at the time of inspection. The construction documents submitted with the application for permit shall be accompanied by a site plan showing the size and location of new construction and existing structures on the site and distances from lot lines.*

Discussion and Commentary: After the construction documents and other submittal documents have been submitted, they are to be reviewed by the building department for compliance with the adopted codes and other jurisdictional requirements. Once the construction documents have been approved, a permit is to be issued.

As a part of the necessary record keeping process, the retention by the building department of at least one set of approved construction documents is required. Such documents must be retained for a minimum of 180 days after the completion date of the work, or as otherwise mandated by state or local laws.

Topic: Payment, Schedule and Valuations   **Category:** Administration
Reference: IRC R108   **Subject:** Fees

Code Text: *A permit shall not be valid until the fees prescribed by law have been paid. On buildings, structures, electrical, gas, mechanical and plumbing systems or alterations requiring a permit, a fee for each permit shall be paid as required, in accordance with the schedule as established by the applicable governing authority. Building permit valuation shall include total value of the work for which a permit is being issued, such as electrical, gas, mechanical, plumbing equipment and other permanent systems, including materials and labor.*

Discussion and Commentary: Fees are typically established at a level that will provide enough funds to adequately pay for the costs of operating the various building department functions, including administration, plan review and inspection. It is common that the fee schedule be based on the projected construction cost (valuation) of the work. Although there are several different methods for determining the appropriate valuation, it is important that a realistic and consistent approach be taken so that permit fees are applied fairly and accurately.

APPENDIX L
PERMIT FEES

TOTAL VALUATION	FEE
$1 to $500	$24
$501 to $2,000	$24 for the first $500; plus $3 for each additional $100 or fraction thereof, to and including $2,000
$2,001 to $40,000	$69 for the first $2,000; plus $11 for each additional $1,000 or fraction thereof, to and including $40,000
$40,001 to $100,000	$487 for the first $40,000; plus $9 for each additional $1,000 or fraction thereof, to and including $100,000
$100,001 to $500,000	$1,027 for the first $100,000; plus $7 for each additional $1,000 or fraction thereof, to and including $500,000
$500,001 to $1,000,000	$3,827 for the first $500,000; plus $5 for each additional $1,000 or fraction thereof, to and including $1,000,000
$1,000,001 to $5,000,000	$6,327 for the first $1,000,000; plus $3 for each additional $1,000 or fraction thereof, to and including $5,000,000
$5,000,001 and over	$18,327 for the first $5,000,000; plus $1 for each additional $1,000 or fraction thereof

---

A fee schedule is set forth in Appendix L of the IRC. Like the other appendix provisions, the fee schedule is only applicable where specifically adopted by the jurisdiction.

Code Text: *For onsite construction, from time to time the building official, upon notification from the permit holder or his agent, shall make or cause to be made any necessary inspections and shall either approve that portion of the construction as completed or shall notify the permit holder or his or her agent wherein the same fails to comply with* the IRC.

Discussion and Commentary: The inspection function is possibly the most critical activity in the entire code enforcement process. At the various stages of construction, an inspector often provides the final check of the building for safety-related compliance. If necessary, the building official may accept inspection reports from approved agencies verifying compliance with the applicable provisions of the code. It is important that such agencies be highly qualified and reliable.

Required inspections (where applicable):

- Foundation
- Plumbing, mechanical, gas and electrical
- Floodplain
- Frame and masonry
- Fire-resistance-rated construction
- Others as required by the building official
- Final

The permit holder or authorized agent must notify the building official that the work is ready for inspection. Furthermore, the responsibility to make the work accessible and available for inspection rests with the person requesting the inspection.

Code Text: *Work shall not be done beyond the point indicated in each successive inspection without first obtaining the approval of the building official. The building official, upon notification, shall make the requested inspections and shall either indicate the portion of the construction that is satisfactory as completed, or shall notify the permit holder or an agent of the permit holder wherein the same fails to comply with the IRC. Any portions that do not comply shall be corrected and such portion shall not be covered or concealed until authorized by the building official.*

Discussion and Commentary: Each successive inspection must be approved prior to further work. This practice helps to control the concealment of any work that must be inspected, which would result in the removal of materials that block access.

The building official must establish a consistent and convenient procedure for notification of the appropriate persons when an inspection approval is not granted. The notification method should encourage an efficient and effective resolution to the issues under consideration.

Code Text: *No building or structure shall be used or occupied, and no change in the existing occupancy classification of a building or structure or portion thereof shall be made until the building official has issued a certificate of occupancy therefor as provided herein.* See exceptions for 1) work exempt from permits, and 2) accessory structures. *Issuance of a certificate of occupancy shall not be construed as an approval of a violation of the provisions of the IRC or of other ordinances of the jurisdiction.*

Discussion and Commentary: The certificate of occupancy is the tool by which the building official regulates and controls the uses and occupancies of the various buildings and structures within the jurisdiction. The code makes it unlawful to use or occupy a building unless a certificate of occupancy has been issued for that specific use. A temporary certificate of occupancy may be issued prior to completion of all of the work, but only in those portions of the building that can be safely occupied.

Certificate of Occupancy

(Address of Structure)

This (applicable portion of structure) has been inspected for compliance with the laws and ordinances of (jurisdiction) and is hereby issued a Certificate of Occupancy

Building permit number _____

Applicable edition of code _____

Sprinkler system provided/ required _____

Special conditions _____

_____

_____

_____

Building Official _____

Name and address of owner _____

_____

The building official is permitted to suspend or revoke a certificate of occupancy based on any of the following reasons: 1) when the certificate is issued in error, 2) when incorrect information is supplied or 3) when the building is in violation of the code.

Code Text: *In order to hear and decide appeals of orders, decisions or determinations made by the building official relative to the application and interpretation of the IRC, there shall be and is hereby created a board of appeals. An application for appeal shall be based on a claim that the true intent of the code or the rules legally adopted thereunder have been incorrectly interpreted, the provisions of the code do not fully apply, or an equally good or better form of construction is proposed. The board shall have no authority to waive requirements of the IRC.*

Discussion and Commentary: Any aggrieved party with a material interest in the decision of the building official may appeal such a decision before a board of appeals. This provides a forum other than the court system in which the building official's action can be reviewed.

Members of the board of appeals shall be qualified by experience and training to rule on issues pertaining to building construction. Appendix B of the *International Building Code* provides greater detail on administrative procedures and the necessary qualifications for board service.

Code Text: *This chapter lists the standards that are referenced in various sections of the IRC. The standards are listed herein by the promulgating agency of the standard, the standard identification, the effective date and title, and the section or sections of the IRC that reference the standard. The application of the referenced standard shall be as specified in Section R102.4.*

Discussion and Commentary: Limited in their scope, standards define more precisely the general provisions in the code. The *International Residential Code* references several hundred standards, each addressing a specific aspect of building design or construction. In general terms, the standards referenced by the IRC are primarily materials, testing and installation standards.

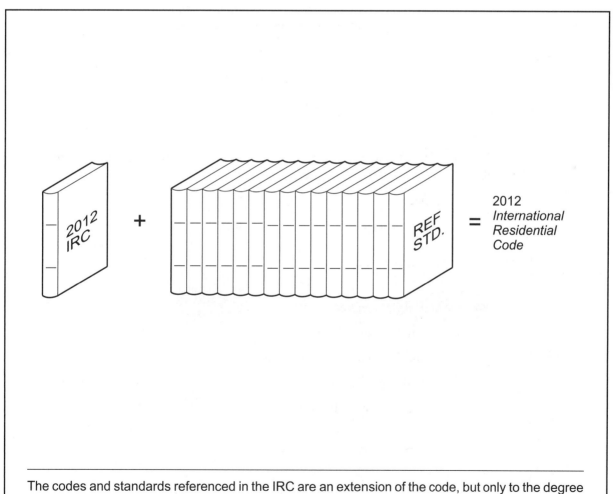

The codes and standards referenced in the IRC are an extension of the code, but only to the degree prescribed by the IRC. Where there is a conflict between the provisions in the IRC and any referenced code or standard, the provisions of the IRC shall apply.

Quiz

Study Session 1
IRC Chapters 1 and 44

1. The *International Residential Code* is applicable to single-family dwellings a maximum of _____ stories in above-grade-plane height.

    a. one

    b. two

    c. three

    d. four

    Reference _____

2. If there is a conflict in the code between a general requirement and a specific requirement, the _____ requirement shall apply.

    a. general

    b. specific

    c. least restrictive

    d. most restrictive

    Reference _____

3. Provisions of the appendices do not apply unless _____ .

    a. specified in the code

    b. applicable to unique conditions

    c. specifically adopted

    d. relevant to fire or life safety

    Reference _____

4. _____ is the term used in the IRC to describe the individual in charge of the department of building safety.

    a. Building official   b. Code official

    c. Code administrator   d. Chief building inspector

Reference _____

5. The building official has the authority to _____ the provisions of the code.

    a. ignore   b. waive

    c. violate   d. interpret

Reference _____

6. Used materials may be utilized under which one of the following conditions?

    a. they meet the requirements for new materials

    b. when approved by the building official

    c. used materials may never be used in new construction

    d. a representative sampling is tested for compliance

Reference _____

7. The building official has the authority to grant modifications to the code _____ .

    a. for only those issues not affecting life safety or fire safety

    b. for individual cases where the strict letter of the code is impractical

    c. where the intent and purpose of the code cannot be met

    d. related only to administrative functions

Reference _____

8. The building official shall not grant modifications to any provision related to areas prone to flooding without the _____ .

    a. granting of a variance by the board of appeals

    b. approval of the jurisdiction's governing body

    c. presentation of an approved design certificate

    d. submission by a registered design professional

Reference _____

9. Tests performed by _____ may be required by the building official where there is insufficient evidence of code compliance.

    a. the owner   b. the contractor

    c. an approved agency   d. a design professional

Reference _____

10. A permit is not required for the construction of a one-story detached accessory structure when it has a maximum floor area of _____ square feet.

    a. 100   b. 120

    c. 150   d. 200

Reference _____

11. Prefabricated swimming pools are subject to a building permit where they have a minimum depth of _____ .

    a. 12 inches   b. 24 inches

    c. 30 inches   d. 36 inches

Reference _____

12. Unless an extension is authorized, a permit becomes invalid when work does not commence within _____ after permit issuance.

    a. 90 days   b. 180 days

    c. one year   d. two years

Reference _____

13. The building permit, or a copy of the permit, shall be kept _____ until completion of the project.

    a. at the job site   b. by the permit applicant

    c. by the contractor   d. by the owner

Reference _____

14. When a building permit is issued, the construction documents shall be approved
_____ .

     a. in writing or by stamp

     b. and two sets returned to the applicant

     c. and stamped as "Accepted as Reviewed"

     d. pending payment of the plan review fee

Reference _____

15. Unless otherwise mandated by state or local laws, the approved construction documents shall be retained by the building official for a minimum of _____ from the date of completion of the permitted work.

     a. 90 days   b. 180 days

     c. one year   d. two years

Reference _____

16. Where applicable, which one of the following inspections is not specifically identified by the *International Residential Code* as a required inspection?

     a. foundation inspection

     b. frame inspection

     c. fire-resistance-construction inspection

     d. energy efficiency inspection

Reference _____

17. Whose duty is it to provide access to work in need of inspection?

     a. the permit holder or agent

     b. the owner or owner's agent

     c. the contractor

     d. the person requesting the inspection

Reference _____

18. The certificate of occupancy shall contain all of the following information except:

    a. the name and address of the owner

    b. the name of the building official

    c. the edition of the code under which the permit was issued

    d. the maximum occupant load

Reference _____

19. A temporary certificate of occupancy is valid for what period of time?

    a. 30 days

    b. 60 days

    c. 180 days

    d. a period set by the building official

Reference _____

20. The board of appeals is not authorized to rule on an appeal based on a claim that _____.

    a. the provisions of the code do not fully apply

    b. a code requirement should be waived

    c. the rules have been incorrectly interpreted

    d. a better form of construction is provided

Reference _____

21. The membership of the board of appeals _____

    a. shall include a jurisdictional member

    b. must consist of at least five members

    c. shall be knowledgeable of building construction

    d. must include an engineer or an architect

Reference _____

22. A permit is not required for construction of a fence with a maximum height of
_____feet.

      a. 5   b. 6

      c. 7   d. 8

Reference _____

23. When a stop work order is issued, it shall be given to any of the following individuals except for the _____ .

      a. owner   b. owner's agent

      c. permit applicant   d. person doing the work

Reference _____

24. Which of the following standards is applicable for factory-built fireplaces?

      a. ASTM A 951–06   b. CPSC 16 CFR Part 1404

      c. NFPA 259-08   d. UL 127–08

Reference _____

25. ACI 318-11 is a reference standard addressing _____ .

      a. structural concrete   b. wood construction

      c. structural steel buildings   d. gypsum board

Reference _____

26. All of the following issues are specifically identified for achieving the purpose of the *International Residential Code*, except for _____ .

      a. affordability   b. structural strength

      c. energy conservation   d. usability and accessibility

Reference _____

27. A permit is not required for the installation of a window awning provided the awning projects a maximum of _____ inches from the exterior wall and does not require additional support.

      a. 30   b. 36

      c. 48   d. 54

Reference _____

28. Where a self-contained refrigeration system contains a maximum of _____ pound(s) of refrigerant, a permit is not required.

    a. 1   b. 2

    c. 5   d. 10

Reference _____

29. It is the duty of the _____ or their agent to notify the building official that work is ready for inspection.

    a. permit holder   b. owner

    c. contractor   d. design professional

Reference _____

30. What is the role of the building official in relationship to the board of appeals?

    a. advisor only   b. ex officio member

    c. full voting member   d. procedural reviewer

Reference _____

31. Where the enforcement of a code provision would violate the conditions of an appliance's listing, the conditions of the listing _____ .

    a. are no longer valid   b. may be disregarded

    c. shall apply   d. are optional

Reference _____

32. Inspection reports shall be retained in the official records for what minimum period?

    a. until after the certificate of occupancy is issued

    b. 180 days after the report is issued

    c. the time required for the retention of public records

    d. 180 days after the certificate of occupancy is issued

Reference _____

33. Which of the following reasons is not specifically identified by the code as authority to suspend or revoke a permit?

    a. the permit was issued in error

    b. required inspections have not been performed

    c. it was issued in violation of a jurisdictional ordinance

    d. inaccurate information was provided at the time of issuance

Reference _____

34. A permit for a temporary structure is limited as to time of service, but shall not be allowed for more than _____ days.

    a. 30   b. 60

    c. 90   d. 180

Reference _____

35. The issuance of a certificate of occupancy is not required for what maximum size of accessory building?

    a. 120 square feet

    b. 150 square feet

    c. 200 square feet

    d. all accessory buildings are exempt

Reference _____

2012 IRC Sections R301 and R302
Building Planning I

OBJECTIVE: To gain an understanding of the design criteria to be used in the application of the prescriptive provisions of the *International Residential Code*, including wind and seismic limitations, snow loads and live loads, as well as the methods for addressing buildings requiring some degree of fire-resistance-rated construction.

REFERENCE: Sections R301 and R302, 2012 *International Residential Code*

KEY POINTS:
- What is the definition of light-framed construction? Under what conditions must such construction be designed in accordance with accepted engineering practice?
- What design standards are identified as acceptable alternatives to the prescriptive structural provisions of the IRC?
- At what wind speed must the structural system be designed based on a source other than the IRC? What design standards are acceptable for residential wind design?
- What is the most common design wind speed in the United States? Where are the hurricane-prone regions?
- How is the appropriate wind exposure category determined? Which category is assumed unless the site meets the definition of another category?
- What are the two methods of establishing wind speed? How do they compare to each other? Which method is used in the IRC?
- Buildings constructed in which seismic design categories are subject to the seismic provisions of the IRC? Which category requires design to the requirements of the IBC?
- In which seismic design categories are irregular buildings required to be designed? How is an irregular building determined?
- What alternate methods are permitted in the determination of a site's seismic design category?
- What is the maximum snow load permitted when using the conventional construction provisions of the IRC?
- What are the limitations on story height for the various types of construction methods?
- What is a dead load? A live load?

- What is the design live load to be used for sleeping rooms? For rooms other than sleeping rooms?
- How is an uninhabitable attic defined? A habitable attic? How does the minimum live load for attic areas differ based on roof slope? Based upon the attic's use?
- What is the maximum deflection permitted for floor structural members? For roof rafters with a ceiling attached directly to the bottom of the rafters?
- What is fire separation distance? How is it measured?
- For an nonsprinklered building, at what fire separation distance from an interior lot line do exterior walls not need a fire-resistance rating? For a sprinklered building? If a rated wall is required, what is the minimum required rating of the wall?
- How far must a projection be located from an interior lot line?
- At what fire separation distance are openings permitted in exterior walls? Where is the amount of openings limited?
- Under what condition may the exterior wall of an accessory structure located on the property line have no fire-resistance rating?
- How are penetrations of exterior walls required to have a fire-resistance rating to be addressed?
- What minimum required fire-resistance-rated separation is mandated between townhouses? Under what conditions are parapets required? When is structural independence of the fire separation wall mandated?
- In a two-family dwelling, how must the two units be separated? Where the separation is horizontal, how is the supporting construction regulated?
- How are penetrations of dwelling unit separations required to be protected?
- How must a private garage be separated from the residence? Under what conditions are openings permitted? What type of door is required? How are duct penetrations to be addressed?
- What type of protection is required for enclosed accessible spaces under stairways?
- How are interior walls and ceilings regulated for flame spread? For smoke development? What special requirements apply to insulation?
- Where is fireblocking required? What materials are acceptable for fireblocking purposes?
- Where is draftstopping required? What materials can be used for draftstopping?
- What clearance is mandated between combustible insulation and recessed luminaires?

Topic: Application

Reference: IRC R301.1

Category: Building Planning

Subject: Design Criteria

Code Text: *Buildings and structures, and all parts thereof, shall be constructed to safely support all loads, including dead loads, live loads, roof loads, flood loads, snow loads, wind loads and seismic loads as prescribed by the IRC. The construction of buildings and structures in accordance with the provisions of* the IRC *shall result in a system that provides a complete load path that meets all requirements for the transfer of all loads from their point of origin through the load-resisting elements to the foundation. Buildings and structures constructed as prescribed by* the IRC *are deemed to comply with the requirements of* Section R301.1.

Discussion and Commentary: In structural design, loads are generally divided into two categories: gravity loads, which act vertically; and lateral loads, which act horizontally. All structures should be built to support these loads and provide a complete load path capable of transferring these loads from their point of origin through the appropriate load-resisting elements and foundation and, ultimately, to the supporting soil.

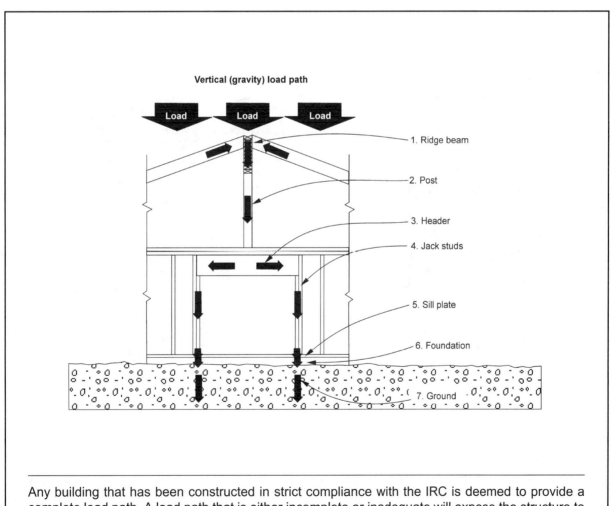

Any building that has been constructed in strict compliance with the IRC is deemed to provide a complete load path. A load path that is either incomplete or inadequate will expose the structure to damage just as surely as will an undersized structural member.

Code Text: *As an alternative to the requirements in Section R301.1 the following standards are permitted subject to the limitations of* the IRC *and the limitations therein. Where engineered design is used in conjunction with these standards, the design shall comply with the* International Building Code. *1) American Forest and Paper Association (AF&PA) Wood Frame Construction Manual (WFCM); 2) American Iron and Steel Institute (AISI), Standard for Cold-Formed Steel Framing—Prescriptive Method for One-and Two-Family Dwellings (AISI S230) and 3) ICC Standard on the Design and Construction of Log Structures (ICC 400).*

Discussion and Commentary: Although the *International Residential Code* is intended to address most structural conditions encountered in residential construction, it may be advantageous to utilize more comprehensive prescriptive design methods that are set forth in the various industry standards.

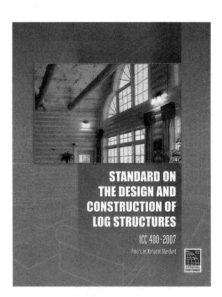

Where utilizing engineering design along with the *Wood Frame Construction Manual, the Cold-Formed Steel Design Manual* or ICC-400, the design criteria must be in accordance with the IBC. The IBC becomes the design basis in those cases where the prescriptive provisions are not used.

Topic: Design
Reference: IRC R301.1.2, R301.1.3

Category: Building Planning
Subject: Design Criteria

Code Text: *The requirements of the IRC are based on platform and balloon-frame construction for light-frame buildings. The requirements for concrete and masonry buildings are based on a balloon framing system. When a building of otherwise conventional construction contains structural elements exceeding the limits of Section R301 or otherwise not conforming to the IRC, these elements shall be designed in accordance with accepted engineering practice.*

Discussion and Commentary: Light-framed construction is a type of construction whose vertical and horizontal structural elements are primarily formed by a system of repetitive wood or light-gage steel framing members. Platform construction is defined as a method of construction by which floor framing bears on load-bearing walls that are not continuous through the story levels or floor framing. In balloon framing, the wall studs extend beyond the floor line.

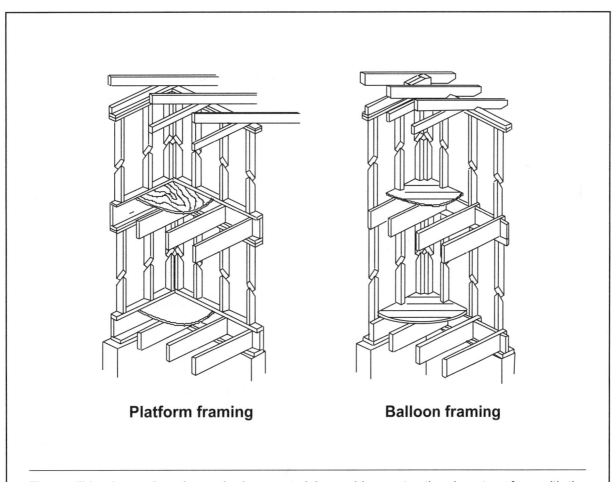

Platform framing   **Balloon framing**

There will be times when the methods or materials used in construction do not conform with the prescriptive structural provisions of the IRC. In such situations, it is acceptable to use an engineered solution to satisfy the requirements of the code.

Code Text: *Buildings shall be constructed in accordance with the provisions of the IRC as limited by the provisions of Section R301. Additional criteria shall be established by the local jurisdiction and set forth in Table R301.2(1).*

Discussion and Commentary: Table R301.2(1) is designed so that jurisdictions recognize and use certain climatic and geographic design criteria that vary from location to location. Communities are directed to complete the table with a variety of factors. The code references the table in many of its provisions, mandating completion of the table in order to understand the code requirements. Footnotes to the table provide guidance on how and where the determination of the appropriate information is obtained.

TABLE R301.2(1)
CLIMATIC AND GEOGRAPHIC DESIGN CRITERIA

| GROUND SNOW LOAD | WIND DESIGN | | SEISMIC DESIGN CATEGORY[f] | SUBJECT TO DAMAGE FROM | | | WINTER DESIGN TEMP[e] | ICE BARRIER UNDERLAYMENT REQUIRED[h] | FLOOD HAZARDS[g] | AIR FREEZING INDEX[i] | MEAN ANNUAL TEMP[j] |
	Speed[d] (mph)	Topographic effects[k]		Weathering[a]	Frost line depth[b]	Termite[c]					

For SI: 1 pound per square foot = 0.0479 kPa, 1 mile per hour = 0.447 m/s.

a. Weathering may require a higher strength concrete or grade of masonry than necessary to satisfy the structural requirements of this code. The weathering column shall be filled in with the weathering index (i.e., "negligible," "moderate" or "severe") for concrete as determined from the Weathering Probability Map [Figure R301.2(3)]. The grade of masonry units shall be determined from ASTM C 34, C 55, C 62, C 73, C 90, C 129, C 145, C 216 or C 652.

b. The frost line depth may require deeper footings than indicated in Figure R403.1(1). The jurisdiction shall fill in the frost line depth column with the minimum depth of footing below finish grade.

c. The jurisdiction shall fill in this part of the table to indicate the need for protection depending on whether there has been a history of local subterranean termite damage.

d. The jurisdiction shall fill in this part of the table with the wind speed from the basic wind speed map [Figure R301.2(4)]. Wind exposure category shall be determined on a site-specific basis in accordance with Section R301.2.1.4.

e. The outdoor design dry-bulb temperature shall be selected from the columns of $97^1/_2$-percent values for winter from Appendix D of the *International Plumbing Code*. Deviations from the Appendix D temperatures shall be permitted to reflect local climates or local weather experience as determined by the building official.

f. The jurisdiction shall fill in this part of the table with the seismic design category determined from Section R301.2.2.1.

g. The jurisdiction shall fill in this part of the table with (a) the date of the jurisdiction's entry into the National Flood Insurance Program (date of adoption of the first code or ordinance for management of flood hazard areas), (b) the date(s) of the Flood Insurance Study and (c) the panel numbers and dates of all currently effective FIRMs and FBFMs or other flood hazard map adopted by the authority having jurisdiction, as amended.

h. In accordance with Sections R905.2.7.1, R905.4.3.1, R905.5.3.1, R905.6.3.1, R905.7.3.1 and R905.8.3.1, where there has been a history of local damage from the effects of ice damming, the jurisdiction shall fill in this part of the table with "YES." Otherwise, the jurisdiction shall fill in this part of the table with "NO."

i. The jurisdiction shall fill in this part of the table with the 100-year return period air freezing index (BF-days) from Figure R403.3(2) or from the 100-year (99%) value on the National Climatic Data Center data table "Air Freezing Index- USA Method (Base 32°)" at www.ncdc.noaa.gov/fpsf.html.

j. The jurisdiction shall fill in this part of the table with the mean annual temperature from the National Climatic Data Center data table "Air Freezing Index-USA Method (Base 32°F)" at www.ncdc.noaa.gov/fpsf.html.

k. In accordance with Section R301.2.1.5, where there is local historical data documenting structural damage to buildings due to topographic wind speed-up effects, the jurisdiction shall fill in this part of the table with "YES." Otherwise, the jurisdiction shall indicate "NO" in this part of the table.

The frost line depth should be established by the local jurisdiction based on past practice or available information. Since frost penetration is dependent on soil conditions and elevations as well as expected winter temperatures, it may vary greatly within a limited geographical area.

| **Topic:** Wind Limitations | **Category:** Building Planning |
| **Reference:** IRC R301.2.1, R301.2.1.1 | **Subject:** Design Criteria |

Code Text: *Buildings and portions thereof, shall be constructed in accordance with the wind provisions of the* IRC *using the basic wind speed in Table R301.2(1) as determined from Figure R301.2(4)A. The structural provisions of the* IRC *for wind loads are not permitted where wind design is required as specified in Section R301.2.1.1. The wind provisions of the* IRC *shall not apply to the design of buildings where wind design is required in accordance with Figure R301.2(4)B or where the basic wind speed from Figure R301.2(4)A equals or exceeds 110 miles per hour (49 m/s).* See exceptions for concrete construction and structural insulated panel construction.

Discussion and Commentary: The prescriptive wind provisions of the IRC are applicable in most geographical areas as depicted in Figure R301.2(4)A. However, wind design is required where at least one of two conditions occur: 1) where Figure R301.2(4)B identifies special wind regions, or 2) where the basic wind speed equals or exceeds 110 miles per hour. In such high-wind areas, buildings must be designed for wind loads in accordance with one or more of five identified methods.

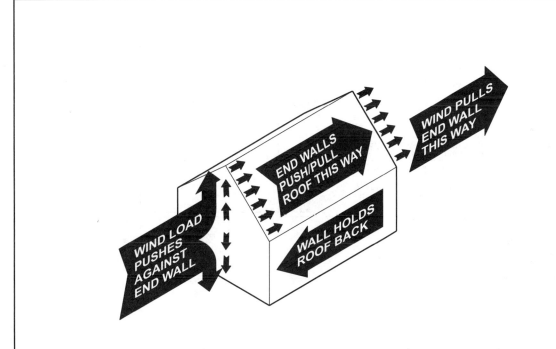

Wind loads are a major consideration in designing a structure's lateral force-resisting system. In addition, other building elements are affected as evidenced by provisions for roof tie-downs, for example. Windows, skylights and exterior doors must withstand the component and cladding pressures.

Topic: Protection of Openings   **Category:** Building Planning
Reference: IRC R301.2.1.2   **Subject:** Design Criteria

Code Text: *Exterior glazing in buildings located in windborne debris regions shall be protected from windborne debris.* See exception for use of precut wood structural panels with complying attachment hardware.

Discussion and Commentary: Opening protection in windborne debris regions is usually in the form of permanent shutters or laminated glass that meet the requirements of the Large Missile Test of ASTM E 1996 and E 1886. The exception provides a prescriptive approach, which is only permitted in one- and two-story buildings. This method of opening protection uses wood structural panels with a maximum span of 8 feet along with complying fasteners. The builder must precut these panels to fit each glazed opening and provide the necessary attachment hardware.

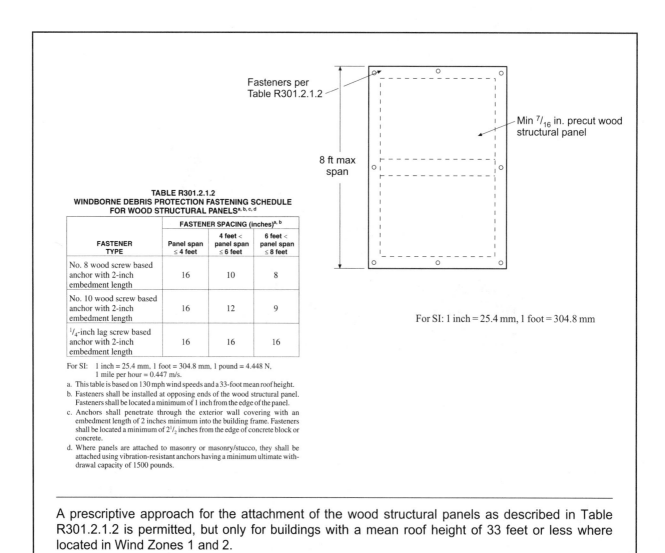

Fasteners per Table R301.2.1.2

Min $^7/_{16}$ in. precut wood structural panel

8 ft max span

TABLE R301.2.1.2
WINDBORNE DEBRIS PROTECTION FASTENING SCHEDULE
FOR WOOD STRUCTURAL PANELS[a, b, c, d]

FASTENER TYPE	FASTENER SPACING (inches)[a, b]		
	Panel span ≤ 4 feet	4 feet < panel span ≤ 6 feet	6 feet < panel span ≤ 8 feet
No. 8 wood screw based anchor with 2-inch embedment length	16	10	8
No. 10 wood screw based anchor with 2-inch embedment length	16	12	9
$^1/_4$-inch lag screw based anchor with 2-inch embedment length	16	16	16

For SI: 1 inch = 25.4 mm, 1 foot = 304.8 mm, 1 pound = 4.448 N, 1 mile per hour = 0.447 m/s.

a. This table is based on 130 mph wind speeds and a 33-foot mean roof height.

b. Fasteners shall be installed at opposing ends of the wood structural panel. Fasteners shall be located a minimum of 1 inch from the edge of the panel.

c. Anchors shall penetrate through the exterior wall covering with an embedment length of 2 inches minimum into the building frame. Fasteners shall be located a minimum of 2$^1/_2$ inches from the edge of concrete block or concrete.

d. Where panels are attached to masonry or masonry/stucco, they shall be attached using vibration-resistant anchors having a minimum ultimate withdrawal capacity of 1500 pounds.

For SI: 1 inch = 25.4 mm, 1 foot = 304.8 mm

A prescriptive approach for the attachment of the wood structural panels as described in Table R301.2.1.2 is permitted, but only for buildings with a mean roof height of 33 feet or less where located in Wind Zones 1 and 2.

Code Text: *For each wind direction considered, an exposure category that adequately reflects the characteristics of ground surface irregularities shall be determined for the site at which the building or structure is to be constructed. For any given wind direction, the exposure in which a specific building or other structure is sited shall be assessed as being in Exposure Category A, B, C or D.*

Discussion and Commentary: In addition to wind speed, wind loading on structures is a function of the site's exposure category. The exposure category reflects the characteristics of ground surface irregularities and accounts for variations in ground surface roughness that arise from natural topography and vegetation as well as from constructed features. Additional requirements apply for those local areas where historical data indicate that significant wind speed-up has occurred at isolated hills, ridges and escarpments.

Exposure B should be used unless the site meets the definition for another category of exposure. Exposure B includes urban and suburban areas, wooded areas and other terrain with numerous closely-spaced obstructions having the size of single-family dwellings or larger.

Code Text: *The seismic provisions of the IRC shall apply to 1) townhouses in seismic design categories C, D_0, D_1 and D_2, and 2) detached one- and two-family dwelling in seismic design categories, D_0, D_1 and D_2. Buildings shall be assigned a Seismic Design Category in accordance with Figure R301.2(2).*

Discussion and Commentary: Earthquakes generate internal forces in a structure due to inertia. These forces can cause a building to be distorted and severely damaged. The objective of earthquake resistant construction is to resist these forces and the resulting distortions. Buildings are deemed to be located in one of several seismic design categories, ranging from Category A to Category E. Buildings in Categories A and B as well as detached one- and two-family dwellings in Category C perform satisfactorily when constructed in accordance with the basic prescriptive provisions of the IRC. On the other hand, Category E buildings must typically be designed in accordance with the seismic provisions of the *International Building Code*.

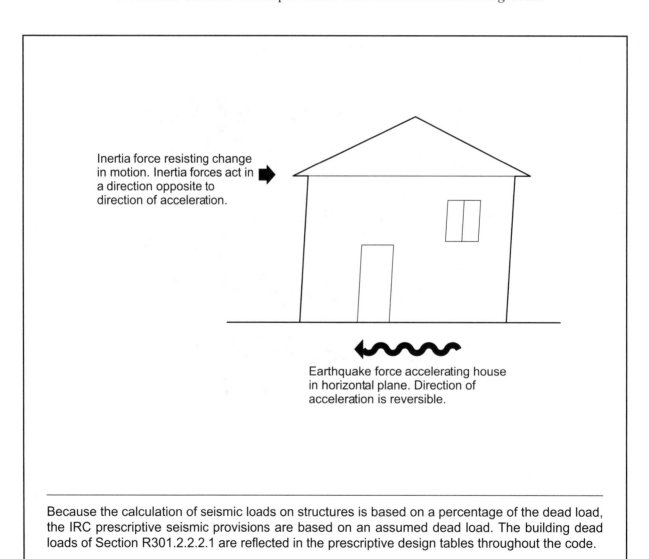

Inertia force resisting change in motion. Inertia forces act in a direction opposite to direction of acceleration.

Earthquake force accelerating house in horizontal plane. Direction of acceleration is reversible.

Because the calculation of seismic loads on structures is based on a percentage of the dead load, the IRC prescriptive seismic provisions are based on an assumed dead load. The building dead loads of Section R301.2.2.2.1 are reflected in the prescriptive design tables throughout the code.

Topic: Irregular Buildings **Category:** Building Planning
Reference: IRC R301.2.2.2.5 **Subject:** Design Criteria

Code Text: *The seismic provisions of the* IRC *shall not be used for irregular structures located in Seismic Design Categories C, D_0, D_1 and D_2. Irregular portions of structures shall be designed in accordance with accepted engineering practice to the extent the irregular features affect the performance of the remaining structural system. A building or a portion of a building shall be considered to be irregular when one or more of seven specific conditions occur. See seven possible conditions.*

Discussion and Commentary: Conventional light-frame construction typically allows cantilevers and offsets within certain limits in order to accommodate common design features and options. These features are not well suited to resisting earthquake loads. This becomes more of a concern in areas of higher seismic hazard.

- Portions of buildings are considered irregular where one or more of seven listed conditions exist:

  - Exterior shear wall lines or braced wall panels not in one plane vertically from foundation to uppermost story where required (exception for limited cantilevers and setbacks).

  - Section of floor or roof not laterally supported by shear walls or braced wall lines on all edges (exception for limited extension beyond shear wall or braced wall line).

  - End of braced wall panel occurs over an opening below and ends more than 1 foot from edge of opening (exception for opening with complying header).

  - Opening in a floor or roof exceeds 12 feet or exceeds 50 percent of least floor or roof dimension.

  - Portions of floor level vertically offset (exceptions for framing directly supported by continuous foundations or where floor framing lapped or tied together).

  - Shear walls and braced wall lines do not occur in two perpendicular directions.

  - Stories above grade plane braced by wood wall framing include masonry or concrete construction.

The IRC, consistent with the IBC, identifies building features that are considered irregular and are not permitted under conventional light-frame construction provisions for the higher seismic design categories C, D_0, D_1 and D_2.

Code Text: *Wood-framed construction, cold-formed steel-framed construction and masonry and concrete construction, and structural insulated panel construction in regions with ground snow loads 70 psf (3.35 kPa) or less, shall be in accordance with Chapters 5, 6 and 8. Buildings in regions with ground snow loads greater than 70 psf (3.35 kPa) shall be designed in accordance with accepted engineering practice. The roof shall be designed for the live load indicated in Table R301.6 or the snow load indicated in Table R301.2(1), whichever is greater.*

Discussion and Commentary: The prescriptive tables for floor, wall and roof construction are limited to buildings constructed in areas having a maximum 70 psf snow load. Structures in regions exceeding this limitation require an engineered design for all elements that carry snow loads. The basic 20 psf live load may be reduced based on the tributary area supported by any structural member of the roof, but no reduction is permitted for snow loads.

TABLE R301.6
MINIMUM ROOF LIVE LOADS IN POUNDS-FORCE
PER SQUARE FOOT OF HORIZONTAL PROJECTION

	TRIBUTARY LOADED AREA IN SQUARE FEET FOR ANY STRUCTURAL MEMBER		
ROOF SLOPE	**0 to 200**	**201 to 600**	**Over 600**
Flat or rise less than 4 inches per foot (1:3)	20	16	12
Rise 4 inches per foot (1:3) to less than 12 inches per foot (1:1)	16	14	12
Rise 12 inches per foot (1:1) and greater	12	12	12

For SI: 1 square foot = 0.0929 m², 1 pound per square foot = 0.0479 kPa,
             1 inch per foot = 83.3 mm/m.

For roofs, both the live load and snow load must be considered, with the more restrictive loading condition used. The live load takes into account that individuals and materials may be present on the roof during roofing operations.

Topic: Floodplain Construction	**Category:** Building Planning
Reference: IRC R301.2.4	**Subject:** Design Criteria

Code Text: *Buildings and structures constructed in whole or in part in flood hazard areas (including A or V Zones) as established in Table R301.2(1) shall be designed and constructed in accordance with Section R322. Buildings and structures located in whole or in part in identified floodways shall be designed and constructed in accordance with ASCE 24.*

Discussion and Commentary: Application of the flood provisions of the code cannot prevent or eliminate all future flood damage. Rather, the provisions represent a reasonable balance of the knowledge and awareness of flood hazards, methods to guide development to less hazard-prone locations, methods of design and construction intended to resist flood damage, and each community's and landowner's reasonable expectations of land use.

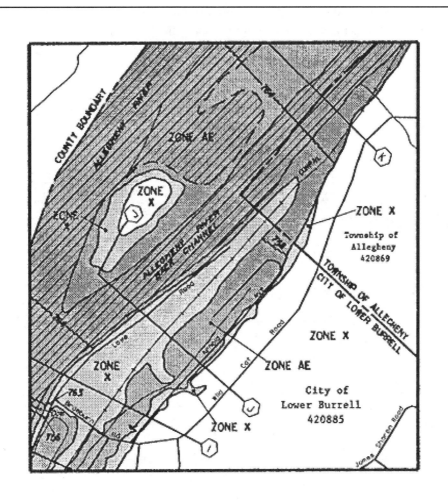

The flood hazard areas shown on FEMA maps are determined using the base flood, which is defined as have a 1-percent chance of occurring in any given year. The maps do not show the worst case flood nor the "flood of record," which usually is the most severe in history.

Topic: Story Height

Category: Building Planning

Reference: IRC R301.3

Subject: Design Criteria

Code Text: *The wind and seismic provisions of the IRC shall apply to buildings with story heights not exceeding the following: 1) for wood wall framing, the laterally unsupported bearing wall stud height permitted by Table R602.3(5) plus a height of floor framing not to exceed 16 inches (see exception for braced walls); 2) for steel wall framing, a stud height of 10 feet, plus a height of floor framing not to exceed 16 inches; 3) for masonry walls, a maximum bearing wall clear height of 12 feet plus a height of floor framing not to exceed 16 inches (see exception for gable end walls); 4) for insulating concrete form walls, the maximum bearing wall height per story as permitted by Section 611 tables plus a height of floor framing not to exceed 16 inches and 5) for structural insulated panel (SIP) walls, the maximum bearing height per story as permitted by Section 613 tables shall not exceed 10 feet plus a height of floor framing not to exceed 16 inches.*

Discussion and Commentary: This section identifies the maximum story heights for use of the IRC's prescriptive provisions, with an allowance for a floor system up to 16 inches in depth.

TABLE R602.3(5)
SIZE, HEIGHT AND SPACING OF WOOD STUDS[a]

	BEARING WALLS					NONBEARING WALLS	
STUD SIZE (inches)	Laterally unsupported stud height[a] (feet)	Maximum spacing when supporting a roof-ceiling assembly or a habitable attic assembly, only (inches)	Maximum spacing when supporting one floor, plus a roof-ceiling assembly or a habitable attic assembly (inches)	Maximum spacing when supporting two floors, plus a roof-ceiling assembly or a habitable attic assembly (inches)	Maximum spacing when supporting one floor height[a] (feet)	Laterally unsupported stud height[a] (feet)	Maximum spacing (inches)
2 × 3[b]	—	—	—	—	—	10	16
2 × 4	10	24[c]	16[c]	—	24	14	24
3 × 4	10	24	24	16	24	14	24
2 × 5	10	24	24	—	24	16	24
2 × 6	10	24	24	16	24	20	24

For SI: 1 inch = 25.4 mm, 1 foot = 304.8 mm, 1 square foot = 0.093 m².

a. Listed heights are distances between points of lateral support placed perpendicular to the plane of the wall. Increases in unsupported height are permitted where justified by analysis.

b. Shall not be used in exterior walls.

c. A habitable attic assembly supported by 2 × 4 studs is limited to a roof span of 32 feet. Where the roof span exceeds 32 feet, the wall studs shall be increased to 2 × 6 or the studs shall be designed in accordance with accepted engineering practice.

These requirements are height-limiting solely due to the prescriptive structural limitations of the code. Such story heights may be increased through the engineered design provisions in the *International Building Code.*

Topic: Live Loads	**Category:** Building Planning
Reference: IRC R301.5	**Subject:** Design Criteria

Code Text: *The minimum uniformly distributed live load shall be as provided in Table R301.5.*

Discussion and Commentary: Table R301.5 provides the minimum loads based on the use of a particular area or portion of a structure. These loads must be considered for the design of corresponding structural elements of any residence constructed under the code. For instance, bedrooms (sleeping rooms) require the use of a minimum 30-psf live load, whereas all other rooms must be designed for at least a 40-psf uniform live load. Exterior balconies and decks are to be designed for a minimum 40-psf uniform live load. Guardrails and handrails must be designed for a single 200-pound concentrated load applied in any direction along the top of the rail. In addition, the in-fill components of a guard system, such as intermediate rails, must be designed to withstand a minimum 50-pound load applied on a one-square-foot area.

TABLE R301.5
MINIMUM UNIFORMLY DISTRIBUTED LIVE LOADS
(in pounds per square foot)

USE	LIVE LOAD
Uninhabitable attics without storage[b]	10
Uninhabitable attics with limited storage[b, g]	20
Habitable attics and attics served with fixed stairs	30
Balconies (exterior) and decks[e]	40
Fire escapes	40
Guardrails and handrails[d]	200[h]
Guardrail in-fill components[f]	50[h]
Passenger vehicle garages[a]	50[a]
Rooms other than sleeping room	40
Sleeping rooms	30
Stairs	40[c]

For SI: 1 pound per square foot = 0.0479 kPa, 1 square inch = 645 mm², 1 pound = 4.45 N.

a. Elevated garage floors shall be capable of supporting a 2,000-pound load applied over a 20-square-inch area.
b. Uninhabitable attics without storage are those where the maximum clear height between joists and rafters is less than 42 inches, or where there are not two or more adjacent trusses with web configurations capable of accommodating an assumed rectangle 42 inches high by 24 inches in width, or greater, within the plane of the trusses. This live load need not be assumed to act concurrently with any other live load requirements.
c. Individual stair treads shall be designed for the uniformly distributed live load or a 300-pound concentrated load acting over an area of 4 square inches, whichever produces the greater stresses.
d. A single concentrated load applied in any direction at any point along the top.
e. See Section R502.2.2 for decks attached to exterior walls.
f. Guard in-fill components (all those except the handrail), balusters and panel fillers shall be designed to withstand a horizontally applied normal load of 50 pounds on an area equal to 1 square foot. This load need not be assumed to act concurrently with any other live load requirement.
g. Uninhabitable attics with limited storage are those where the maximum clear height between joists and rafters is 42 inches or greater, or where there are two or more adjacent trusses with web configurations capable of accommodating an assumed rectangle 42 inches in height by 24 inches in width, or greater, within the plane of the trusses.
   The live load need only be applied to those portions of the joists or truss bottom chords where all of the following conditions are met:
   1. The attic area is accessible from an opening not less than 20 inches in width by 30 inches in length that is located where the clear height in the attic is a minimum of 30 inches.
   2. The slopes of the joists or truss bottom chords are no greater than 2 inches vertical to 12 units horizontal.
   3. Required insulation depth is less than the joist or truss bottom chord member depth.
   The remaining portions of the joists or truss bottom chords shall be designed for a uniformly distributed concurrent live load of not less than 10 lb/ft².
h. Glazing used in handrail assemblies and guards shall be designed with a safety factor of 4. The safety factor shall be applied to each of the concentrated loads applied to the top of the rail, and to the load on the in-fill components. These loads shall be determined independent of one another, and loads are assumed not to occur with any other live load.

Attic design loads vary based on a variety of factors. Where it is assumed that there is insufficient clearance or headroom to accumulate significant storage, only 10 psf is mandated. Where storage is more probable due to a sizable attic height, a 20 psf dead load is to be used. A 30-pound minimum uniform live load is required in those attics considered habitable by Section 202, or if attic access is provided by fixed stairs.

Code Text: *Construction, projections, openings, and penetrations of exterior walls of dwellings and accessory buildings shall comply with Table R302.1(1); or dwellings equipped throughout with an automatic sprinkler system installed in accordance with Section P2904 shall comply with Table R302.1(2).* See exceptions for 1) those elements perpendicular to the measurement of fire separation distance, 2) opposing walls of two buildings on the same lot, 3) detached sheds and similar structures that are exempt from permits, 4) detached garages located within 2 feet of a lot line, and 5) foundation vents.

Discussion and Commentary: Where an exterior wall of a sprinklered building has a fire separation distance of less than 3 feet, it is mandated that the wall be of minimum 1-hour fire-resistance-rated construction. The fire-resistive rating must be obtained from both sides of the wall and comply with either ASTM E 119 or UL 263. In those situations where the building is not sprinklered, a fire-resistance-rated exterior wall is mandated where the building has a fire separation distance of less than 5 feet. In all cases, exterior walls, projections, openings and penetrations in such walls located perpendicular to the line used to determine the fire separation distance are not regulated.

TABLE R302.1(2)
EXTERIOR WALLS—DWELLINGS WITH FIRE SPRINKLERS

EXTERIOR WALL ELEMENT		MINIMUM FIRE-RESISTANCE RATING	MINIMUM FIRE SEPARATION DISTANCE
Walls	Fire-resistance rated	1 hour—tested in accordance with ASTM E 119 or UL 263 with exposure from the outside	0 feet
	Not fire-resistance rated	0 hours	3 feet[a]
Projections	Fire-resistance rated	1 hour on the underside	2 feet[a]
	Not fire-resistance rated	0 hours	3 feet
Openings in walls	Not allowed	N/A	< 3 feet
	Unlimited	0 hours	3 feet[a]
Penetrations	All	Comply with Section R302.4	< 3 feet
		None required	3 feet[a]

For SI: 1 foot = 304.8 mm.

N/A = Not Applicable

a. For residential subdivisions where all dwellings are equipped throughout with an automatic sprinkler systems installed in accordance with Section P2904, the fire separation distance for nonrated exterior walls and rated projections shall be permitted to be reduced to 0 feet, and unlimited unprotected openings and penetrations shall be permitted, where the adjoining lot provides an open setback yard that is 6 feet or more in width on the opposite side of the property line.

Regardless of their location, storage sheds and similar accessory buildings that do not exceed 200 square feet in floor area are not required to be provided with fire-resistance-rated exterior walls.

Topic: Townhouses	**Category:** Building Planning
Reference: IRC 302.2	**Subject:** Fire-Resistant Construction

Code Text: *Each townhouse shall be considered a separate building and shall be separated by fire-resistance-rated wall assemblies meeting the requirements of Section R302.1 for exterior walls. See exception for permitted use of common 1-hour fire-resistance-rated wall. The fire-resistance-rated wall or assembly separating townhouses shall be continuous from the foundation to the underside of the roof sheathing, deck or slab. The fire-resistance-rating shall extend the full length of the wall assembly, including wall extensions through and separating attached enclosed accessory structures.*

Discussion and Commentary: The application of townhouse provisions has its base in the exterior wall requirements of Section R302.1, which deal with the building's location on the lot. In general, because the "exterior wall" of the townhouse is essentially being constructed with no fire separation distance where one townhouse adjoins another, that wall must have a minimum 1-hour fire-resistance rating. The adjacent townhouse would have the same requirement; thus, two separate one-hour walls would be constructed side by side where one townhouse adjoins the other.

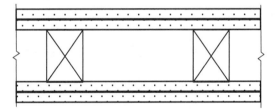

Construction type: gypsum wallboard, studs

Base layer $^5/_8$ in. Type X gypsum wallboard or veneer base applied at right angles to each side of 2 x 4 wood studs 24 in. on center with 6d coated nails, $1^7/_8$ in. long, 0.085 in. shank, $^1/_4$ in. heads, 24 in. on center. Face layer $^5/_8$ in. Type X gypsum wallboard or veneer base applied at right angles to studs over base layer with 8d coated nails, $2^3/_8$ in. long, 0.100 in. shank, $^1/_4$ in. heads, 8 in. on center. Stagger joints 24 in. on center each layer and side. Sound tested with studs 16 in. on center with nails for base layer spaced 6 in. on center (load-bearing).

2-hour wall assembly

(Fire Resistance/Sound Control Design Manual, 19th Edition, Gypsum Association)

For SI: 1 inch = 25.4 mm.

As an option, the construction of a single 1-hour fire-resistant wall is permitted in lieu of two separate 1-hour walls. Where a single 1-hour wall is constructed, no plumbing or mechanical elements are permitted within the wall.

Topic: Two-Family Dwellings

Category: Building Planning

Reference: IRC R302.3, R302.3.1

Subject: Fire-Resistant Construction

Code Text: *Dwelling units in two-family dwellings shall be separated from each other by wall and/or floor assemblies of not less than 1-hour fire-resistance rating when tested in accordance with ASTM E 119 or UL 263.* See exceptions for 1) a reduction in rating to $^1/_2$-hour for fully sprinklered buildings, and 2) a revised method of protection at the ceiling. *When floor assemblies are required to be fire-resistance-rated by Section R302.3, the supporting construction of such assemblies shall have an equal or greater fire-resistance rating.*

Discussion and Commentary: The code mandates a limited degree of fire separation to protect the occupants of one dwelling unit from the actions of their neighbor. Where the units are side by side, the separation wall must extend vertically to the underside of the roof deck. As an option to continuance of the 1-hour wall through the attic space, the ceiling can be protected with a minimum of one layer of $^5/_8$-inch Type X gypsum board, and the attic space need only be separated with a complying draftstop. Where one unit is located over the other, the fire-rated floor/ceiling assembly must extend to and be tight against the exterior walls. The required fire-resistance rating of one hour shall be verified by compliance with ASTM E 119 or UL 263, mandating an assembly constructed in accordance with its listing.

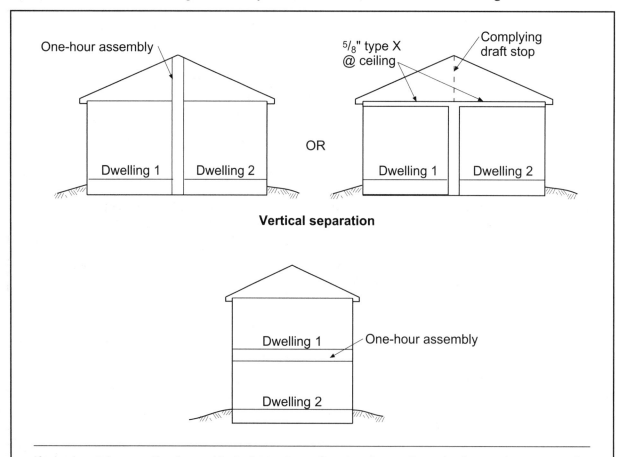

Vertical separation

If a horizontal separation is provided, elements such as bearing walls and columns that support the separation must also have a 1-hour fire-resistance rating. It is of little value to provide a complying floor/ceiling assembly that quickly fails structurally because of unprotected supports.

Code Text: *Openings from a private garage directly into a room used for sleeping purposes shall not be permitted. Other openings between the garage and residence shall be equipped with solid wood doors not less than 1³/₈ inches (35 mm) in thickness, solid or honeycomb core steel doors not less than 1³/₈ inches (35 mm) thick, or 20-minute fire-rated doors, equipped with a self-closing device.*

Discussion and Commentary: Although gypsum board or other approved material provides an adequate fire separation at the walls and/or ceiling between the garage and the dwelling unit, it is important that any other openings penetrating the separation be appropriately protected. Therefore, the type of door construction or fire rating of the door is mandated. Because the code covers only the door itself, it is not necessary to have a complete fire-rated door assembly, but it is a requirement to provide a self-closing device. The level of separation provided by the door is consistent with that provided by the gypsum board installed on the garage side.

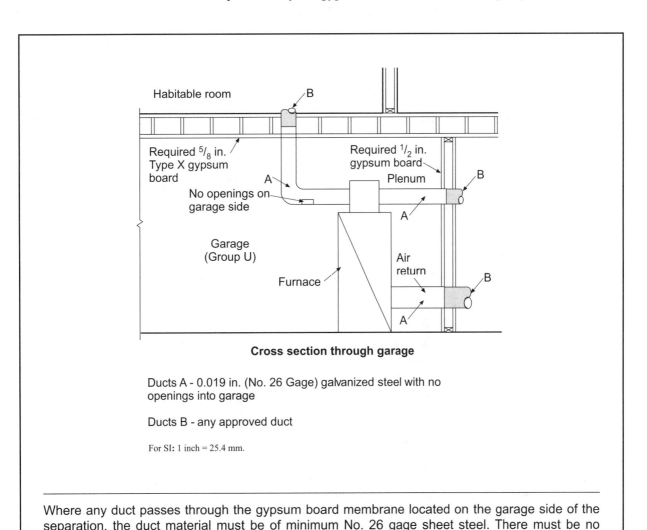

Cross section through garage

Ducts A - 0.019 in. (No. 26 Gage) galvanized steel with no openings into garage

Ducts B - any approved duct

For SI: 1 inch = 25.4 mm.

Where any duct passes through the gypsum board membrane located on the garage side of the separation, the duct material must be of minimum No. 26 gage sheet steel. There must be no openings in the duct within the garage area.

Topic: Dwelling/Garage Separation
Reference: IRC R302.6

Category: Building Planning
Subject: Fire-Resistant Construction

Code Text: *The garage shall be separated as required by Table R302.6. Openings in garage walls shall comply with Section R302.5. This provision does not apply to garage walls that are perpendicular to the adjacent dwelling unit wall.*

Discussion and Commentary: It is not uncommon for a fire to start in a private garage attached directly to a dwelling unit. In many cases, the fire may grow and go unnoticed for a period of time, becoming a distinct hazard to the residence and its occupants. The code mandates a minimum level of fire separation from the garage to the dwelling unit to allow the residents time for escape. In addition, the separation may be adequate to restrict the spread of fire to the dwelling unit until the fire can be controlled and extinguished.

TABLE R302.6
DWELLING/GARAGE SEPARATION

SEPARATION	MATERIAL
From the residence and attics	Not less than $1/2$-inch gypsum board or equivalent applied to the garage side
From all habitable rooms above the garage	Not less than $5/8$-inch Type X gypsum board or equivalent
Structure(s) supporting floor/ceiling assemblies used for separation required by this section	Not less than $1/2$-inch gypsum board or equivalent
Garages located less than 3 feet from a dwelling unit on the same lot	Not less than $1/2$-inch gypsum board or equivalent applied to the interior side of exterior walls that are within this area

For SI: 1 inch = 25.4 mm, 1 foot = 304.8 mm.

Section A-A

For SI: 1 inch = 25.4 mm.

Section B-B

The provisions regulating garages are applicable where the space is enclosed on at least three sides. Where enclosed on two sides or fewer, the parking area is considered a carport, and the separation requirements do not apply.

Code Text: *Wall and ceiling finishes shall have a flame-spread classification of not greater than 200.* See exceptions for trim, doors and windows, and very thin materials cemented to a wall or ceiling surface. *Wall and ceiling finishes shall have a smoke-developed index of not greater than 450. Tests shall be made in accordance with ASTM E 84 or UL 723.*

Discussion and Commentary: The control of interior finish materials is an important aspect of fire protection. The dangers of unregulated interior finish include both the rapid spread of fire and the contribution of additional fuel to the fire. Rapid fire spread presents a threat to the building occupants by limiting or denying their exit travel. This can be caused by either the rapid spread of the fire itself or by the production of large quantities of dense, black smoke that obscures the exit path or makes movement difficult.

Flame-spread ratings of typical construction materials

Material	Flame Spread
Ceilings:	
Glass-fiber sound-absorbing blankets	15 to 30
Mineral-fiber sound -absorbing panels	10 to 25
Sprayed cellulose fibers (treated)	20
Walls:	
Brick or concrete block	0
Cork	175
Gypsum board (with paper surface on both sides)	10 to 25
Southern pine (untreated)	130 to 190
Plywood paneling (untreated)	75 to 275
Red oak (untreated)	100

The installation of materials that will not significantly contribute to fire conditions is permitted. Trim, door and window frames, and baseboards are not regulated, because of their limited quantity. Thin materials such as wallpaper are allowed, as they tend to perform similar to their backing.

Quiz

Study Session 2
IRC Sections R301 and R302

1. What is the weathering potential for all dwelling sites located in the state of Missouri?

    a. severe   b. moderate

    c. negligible   d. none

    Reference _____

2. The basic wind speed to be used for a majority of the United States is _____ .

    a. 70 mph   b. 75 mph

    c. 85 mph   d. 90 mph

    Reference _____

3. Based on Figure R301.2(5), what is the ground snow load for most of western Colorado?

    a. 30 psf   b. 40 psf

    c. 50 psf   d. site-specific due to extreme local variations

    Reference _____

4. Unless local conditions warrant otherwise, what is the probability for termite infestation for dwellings constructed in most of Idaho?

    a. very heavy   b. moderate to heavy

    c. slight to moderate   d. none to slight

    Reference _____

5. For the purpose of determining the component and cladding loads on the roof surface of a building, the area at the ridge of a gable roof sloped at 20 degrees shall be considered as Pressure Zone _____ at other than the eaves.

   a. 0   b. 1

   c. 2   d. 3

   Reference _____

6. Where wood structural panels are used to protect windows in buildings located in windborne debris regions, #8 wood screws shall be located at a maximum of _____ on center to fasten those panels that span 8 feet.

   a. 8 inches   b. 9 inches

   c. 12 inches   d. 16 inches

   Reference _____

7. A "3-second gust" velocity of 90 miles per hour in the IRC is to be converted to a "fastest mile" wind speed of _____ for use in reference documents using the "fastest mile" criteria.

   a. 76 miles per hour   b. 85 miles per hour

   c. 90 miles per hour   d. 95 miles per hour

   Reference _____

8. What wind exposure category is appropriate for a dwelling located in a residential development in a suburban area?

   a. Exposure A   b. Exposure B

   c. Exposure C   d. Exposure D

   Reference _____

9. The seismic design category for a site having a calculated S_{DS} of 0.63g is _____ .

   a. B   b. D_0

   c. D_1 d. E

   Reference _____

10. Habitable attics shall be designed with a minimum uniformly distributed live load of _____ psf.

    a. 10   b. 20

    c. 30   d. 40

Reference _____

11. For a dwelling assigned to Seismic Design Category D_2 and constructed under the conventional provisions of the IRC, what is the maximum dead load permitted for 8-inch-thick masonry walls?

    a. 40 psf   b. 65 psf

    c. 80 psf   d. 85 psf

Reference _____

12. A portion of a building is considered irregular for seismic purposes when a floor opening, such as for a stairway, exceeds the lesser of 12 feet or _____ of the least floor dimension.

    a. 15 percent   b. 25 percent

    c. $33^1/_3$ percent   d. 50 percent

Reference _____

13. Buildings constructed in regions where the ground snow load exceeds a minimum of _____ must be designed in accordance with accepted engineering practice.

    a. 30 psf   b. 50 psf

    c. 70 psf   d. 90 psf

Reference _____

14. For rooms other than sleeping rooms, what is the minimum uniformly distributed live load that is to be used for the design of the floor system?

    a. 20 psf   b. 30 psf

    c. 40 psf   d. 50 psf

Reference _____

15. A minimum uniformly distributed live load of 10 psf is to be used for the design of uninhabitable attic areas having a maximum clear height of _____ inches.

    a. 30   b. 36

    c. 42   d. 60

    Reference _____

16. A minimum uniformly distributed live load of _____ shall be used for the design of sleeping rooms.

    a. 10 psf   b. 20 psf

    c. 30 psf   d. 40 psf

    Reference _____

17. Where no snow load is present, what is the minimum roof design live load for a 240 square foot tributary-loaded roof area having a slope of 8:12?

    a. 20 psf   b. 16 psf

    c. 14 psf   d. 12 psf

    Reference _____

18. What is the maximum allowable deflection for floor joists?

    a. L/120   b. L/180

    c. L/240   d. L/360

    Reference _____

19. The maximum allowable deflection permitted for 7:12-sloped rafters with no finished ceiling attached to the rafters is _____ .

    a. L/120   b. L/180

    c. L/240   d. L/360

    Reference _____

20. A fire-resistance rating is not required for exterior walls of nonsprinklered dwellings having a minimum fire separation distance of _____ .

    a. 3 feet   b. 4 feet

    c. 5 feet   d. 10 feet

Reference _____

21. A roof projection on a dwelling shall be located a minimum of _____ from an interior lot line.

    a. 0 feet (it may extend to the lot line)

    b. 12 inches

    c. 2 feet

    d. 4 feet

Reference _____

22. Tool and storage sheds, playhouses and similar accessory structures having a maximum floor area of _____ are not required to have exterior wall protection based on location on the lot.

    a. 100 square feet   b. 120 square feet

    c. 150 square feet   d. 200 square feet

Reference _____

23. In a dwelling, exterior wall openings located less than 3 feet from an interior lot line shall be protected with what minimum fire protection rating?

    a. 0 hours; no fire protection rating is required

    b. 20 minutes

    c. 45 minutes

    d. openings are not permitted

Reference _____

24. Where located in an exterior wall having a minimum required fire-resistance rating of one hour, through penetrations protected with an approved penetration firestop system shall have a minimum fire-resistance rating of _____ .

    a. 20 minutes   b. 30 minutes

    c. 45 minutes   d. 1 hour

Reference _____

25. Where a detached garage is located within 2 feet of a lot line, the maximum eave projection is limited to _____ inches.

    a. 4   b. 6

    c. 8   d. 12

Reference _____

26. For wind design purposes, a building located along the shoreline in the Great Lakes region is categorized as Exposure _____ where exposed to wind coming from over the water.

    a. A   b. B

    c. C   d. D

Reference _____

27. Which of the following buildings is exempt from the seismic requirements of the code?

    a.   a townhouse in Seismic Design Category C

    b.   a one-family dwelling in Seismic Design Category C

    c.   a townhouse in Seismic Design Category D_0

    d.   a two-family dwelling in Seismic Design Category D_1

Reference _____

28. Where a solid wood door is installed as a permitted opening between a garage and a residence, the minimum thickness of the door shall be _____ inches.

    a. $1^3/_8$ b. $1^1/_2$

    c. $1^5/_8$ d. $1^3/_4$

Reference _____

29. When a common wall is used to separate townhouses, it shall have a minimum _____ fire-resistance rating.

    a. 1-hour   b. 2-hour

    c. 3-hour   d. 4-hour

Reference _____

30. An in-fill panel for a guard shall be designed to withstand a minimum load of _____ applied horizontally on an area of 1 square foot.

    a. 50 pounds   b. 80 pounds

    c. 100 pounds   d. 200 pounds

    Reference _____

31. Where structural wood panels are used to provide windborne debris protection for glazed openings, the fasteners shall be long enough to penetrate through the sheathing and into wood wall framing a minimum of _____ inch(es).

    a. 1   b. $1^{1}/_{2}$

    c. 2   d. $2^{1}/_{2}$

    Reference _____

32. In the determination of the allowable deflection for cantilevered structural members, "*L*" shall be taken as _____ length of the cantilever.

    a. the actual   b. $1^{1}/_{2}$ times the

    c. twice the   d. 3 times the

    Reference _____

33. Attic spaces served by a fixed stair shall be designed to support a minimum live load of _____ .

    a. 10   b. 20

    c. 30   d. 40

    Reference _____

34. Where the roof projection of a sprinklered dwelling extends to a point two feet from an interior lot line, what minimum level of fire-resistance is required for the portion of the projection that is within 3 feet of the lot line?

    a. 30 minutes from both sides

    b. 1 hour on the underside

    c. no fire-resistance rating is required

    d. the projection is not permitted

    Reference _____

35. The aggregate area of openings in an exterior wall of a nonsprinklered dwelling located with a fire separation distance of four feet is limited to a maximum of _____ percent of the exterior wall area.

    a. 10   b. 25

    c. 50   d. 100

    Reference _____

2012 IRC Sections R303 – R310
Building Planning II

OBJECTIVE: To develop an understanding of the health and safety criteria of the code, including light and ventilation; minimum room areas and ceiling height; sanitation; toilet, bath, and shower spaces; glazing, including safety glazing; carports and garages; and emergency escape and rescue openings.

REFERENCE: Sections R303 through R310, 2012 *International Residential Code*

- Where natural light is used to satisfy the minimum illumination requirements, how is the minimum required amount of glazing determined? Where artificial light is used, what illumination level is mandated?
- Under what conditions is a whole-house mechanical ventilation system required?
- How must mechanical and gravity outside air intake openings be located in relationship to vents, chimneys, parking lots and other potential areas of a hazardous or noxious contaminant?
- Where must illumination be located in relationship to interior stairways? Exterior stairways?
- In what climatic areas must a heating system be provided? What performance level is mandated for the system?
- What is the minimum required size of the largest habitable room in a dwelling unit?
- What is the minimum dimension permitted for a habitable room other than a kitchen?
- What is the minimum ceiling height permitted for a living room or bedroom? A hallway? Bathroom? Basement? Where can a reduction in such heights be acceptable?
- How much clear floor space is required in front of a water closet? In front of a shower opening? What is the minimum distance needed between the centerline of a water closet and the nearest adjoining obstruction such as a wall or shower compartment?
- In what manner must safety glazing be identified? Multipane assemblies?
- What test standards are applicable to safety glazing materials? Which test standard is acceptable for glazing installed in any hazardous location?

KEY POINTS:
(Cont'd)

- What specific locations in and adjacent to doors are subject to human impact and require safety glazing? In tub and shower areas? In guards and railings? At stairways and stairway landings?

- When is sloped glazing considered a skylight? What glazing materials are permitted in skylights? When must a screen be installed below a skylight?

- How does a carport differ from a garage? What limitations are placed on carports?

- Where are escape and rescue openings required? What is the minimum size of such openings? Maximum sill height? What limitations are placed on the operation of the opening?

- When a window well serves an escape and rescue opening, what is its minimum size?

- How may a bulkhead enclosure be utilized as an escape and rescue opening?

- Under what conditions may an emergency escape window be located under a deck or porch?

Topic: Habitable Rooms
Reference: IRC R303.1

Category: Building Planning
Subject: Light, Ventilation and Heating

Code Text: *All habitable rooms shall have an aggregate glazing area of not less than 8 percent of the floor area of such rooms. Natural ventilation shall be through windows, doors, louvers or other approved openings to the outdoor air. The minimum openable area to the outdoors shall be 4 percent of the floor area being ventilated.* Exceptions allow the use of artificial light and mechanical ventilation.

Discussion and Commentary: A usable and sanitary interior environment depends on the inclusion of adequate light and ventilation for the habitable spaces within the dwelling unit. Traditionally, the use of natural light and, to some degree, natural ventilation has been mandated as the means for achieving such an environment. It has become increasingly more common to use artificial lighting and a mechanical ventilation system. These methods create additional design flexibility and functionality while maintaining a pleasant and sanitary living environment.

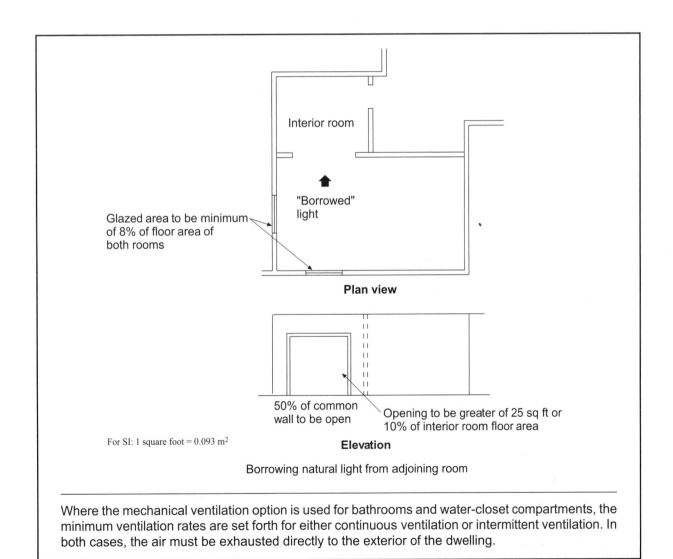

Interior room

"Borrowed" light

Glazed area to be minimum of 8% of floor area of both rooms

Plan view

50% of common wall to be open

Opening to be greater of 25 sq ft or 10% of interior room floor area

For SI: 1 square foot = 0.093 m²

Elevation

Borrowing natural light from adjoining room

Where the mechanical ventilation option is used for bathrooms and water-closet compartments, the minimum ventilation rates are set forth for either continuous ventilation or intermittent ventilation. In both cases, the air must be exhausted directly to the exterior of the dwelling.

Code Text: *Mechanical and gravity outdoor air intake openings shall be located a minimum of 10 feet (3048 mm) from any hazardous or noxious contaminant, such as vents, chimneys, plumbing vents, streets, alleys, parking lots and loading docks, except as otherwise specified in the IRC. Where a source of contaminant is located within 10 feet (3048 mm) of an intake opening, such opening shall be located a minimum of 3 feet (914 mm) below the contaminant source. Outside exhaust openings shall be located so as not to create a nuisance. Exhaust air shall not be directed onto walkways.*

Discussion and Commentary: In the context of this section, intake openings include windows, doors, combustion air intakes and similar openings that naturally or mechanically draw in air from the building exterior. The alternative to the 10-foot separation requirement, a 2-foot vertical separation distance, will allow noxious gases and contaminants to disperse into the atmosphere before they can be drawn into an air intake opening.

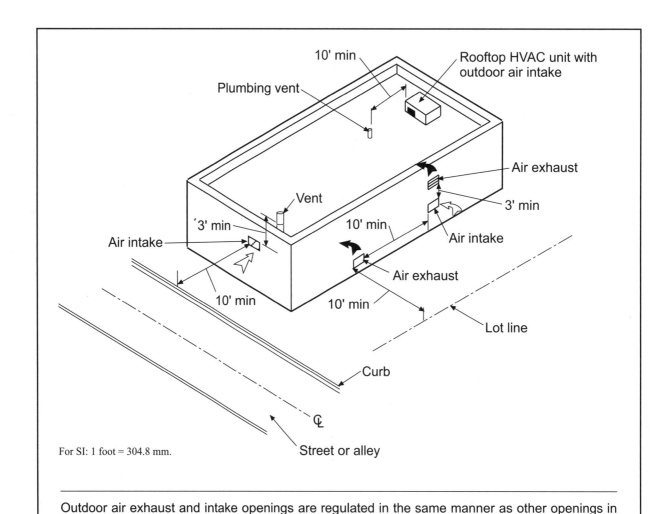

For SI: 1 foot = 304.8 mm.

Outdoor air exhaust and intake openings are regulated in the same manner as other openings in exterior walls. As such, they are not permitted in walls having a fire separation distance of less than 3 feet, except in exterior walls that are perpendicular to the lot line.

Topic: Stairway Illumination
Reference: IRC R303.7, R303.7.1

Category: Building Planning
Subject: Light, Ventilation and Heating

Code Text: *Interior stairways shall be provided with an artificial light source located in the immediate vicinity of each landing of the stairway. See exception where light source is located over each stairway section. Exterior stairways shall be provided with an artificial light source located in the immediate vicinity of the top landing of the stairway. Where lighting outlets are installed in interior stairways, there shall be a wall switch at each floor level to control the lighting outlet where the stairway has six or more risers. See exception for lights that are continuously illuminated or automatically controlled.*

Discussion and Commentary: A stairway is one of the most hazardous areas of a dwelling unit. As such, the code highly regulates the design and construction of all stairways. In addition, adequate lighting must be provided to enable the stairway user to see the treads, their nosings and any obstructions that may be present. Stairway landings must also be adequately lighted.

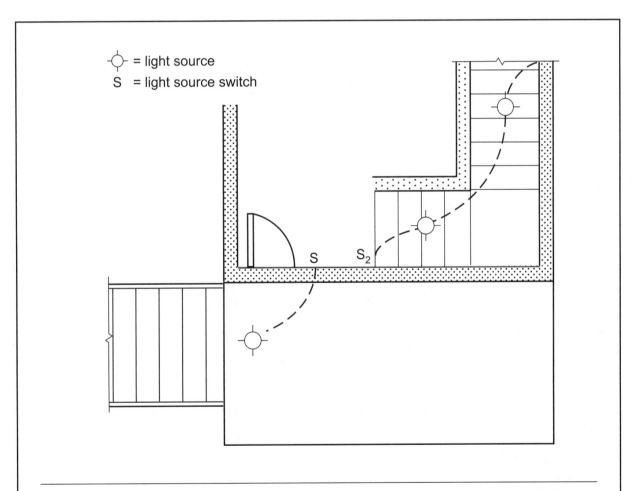

Unless the light sources are on continuously or automatically activated, interior stairway lights must be controlled from both the top and bottom of each stairway consisting of six or more risers. For exterior stairway lighting, the control switch is to be located within the dwelling unit.

Code Text: *Every dwelling unit shall have at least one habitable room that shall have not less than 120 square feet (11 m²) of gross floor area. Other habitable rooms shall have a floor area of not less than 70 square feet (6.5 m²). Habitable rooms shall not be less than 7 feet (2134 mm) in any horizontal dimension. See exceptions for kitchens regarding minimum size and horizontal dimensions.*

Discussion and Commentary: Acceptable sizes for habitable rooms have been established. Because habitable rooms are expected to be those spaces within a dwelling unit where most activities take place, they are the only areas regulated. Most habitable rooms need be only 7 feet by 10 feet to comply with the provisions; however, at least one larger room must be provided. It is seldom that any habitable room in today's typical dwelling unit would be designed with such a small floor area.

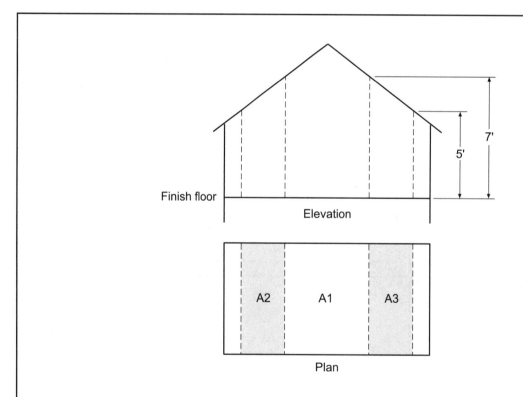

Finish floor

Elevation

7'

5'

A2   A1   A3

Plan

A1 + A2 + A3 ≥ required room floor area per Section R304

A1 ≥ 50% of required room floor area per Section R304

For SI: 1 degree = 0.1745 rad

The minimum required floor area for any habitable room having a sloping ceiling, as would typically be encountered where an attic area is finished for use as a living or sleeping area, must be based on only those portions of the room with a ceiling height of at least 5 feet.

| **Topic:** Minimum Height | **Category:** Building Planning |
| **Reference:** IRC R305.1, R305.1.1 | **Subject:** Ceiling Height |

Code Text: *Habitable space, hallways, bathrooms, toilet rooms, laundry rooms and portions of basements containing these spaces shall have a ceiling height of not less than 7 feet (2134 mm).* See exceptions addressing rooms with sloped ceilings and bathrooms. *Portions of basements that do not contain habitable space, hallways, bathrooms, toilet rooms and laundry rooms shall have a ceiling height of not less than 6 feet 8 inches (2032 mm).* See exception allowing beams, ducts and similar obstructions to project to within 6 feet 4 inches of the finished floor.

Discussion and Commentary: For both safety reasons and usability by the occupants, the minimum ceiling height throughout occupiable areas of a dwelling unit is regulated. Most rooms that are commonly used by the occupants are included, other than closets and storage areas. Where basements are used for habitable purposes, they too must comply with the minimum height requirement of 7 feet.

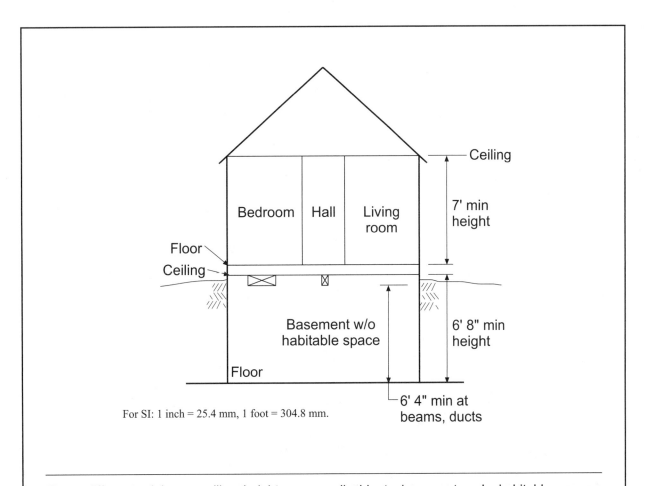

For SI: 1 inch = 25.4 mm, 1 foot = 304.8 mm.

Three different minimum ceiling heights are applicable to basements. In habitable spaces, hallways, bathrooms and laundry rooms, the minimum required height is 7 feet. Other areas of the basement only need a minimum ceiling height of 6 feet 8 inches. Where beams, ducts and other obstructions exist, a headroom clearance of 6 feet 4 inches is mandated.

Topic: Minimum Space Required	**Category:** Building Planning
Reference: IRC R307	**Subject:** Toilet, Bath and Shower Spaces

Code Text: *Fixtures shall be spaced in accordance with Figure R307.1 and in accordance with the requirements of Section P2705.1. Bathtub and shower floors and walls above bathtubs with installed shower heads and in shower compartments shall be finished with a nonabsorbent surface. Such wall surfaces shall extend to a height of not less than 6 feet (1829 mm) above the floor.*

Discussion and Commentary: It is necessary to provide adequate clearances at and around bathroom fixtures to allow for ease of use by the occupants of the dwelling unit. The code addresses clear floor space and clearances for lavatories, water closets, bathtubs, and showers. In addition, the minimum permitted size for a shower is 30 inches by 30 inches with a clearance of at least 24 inches provided in front of the shower opening.

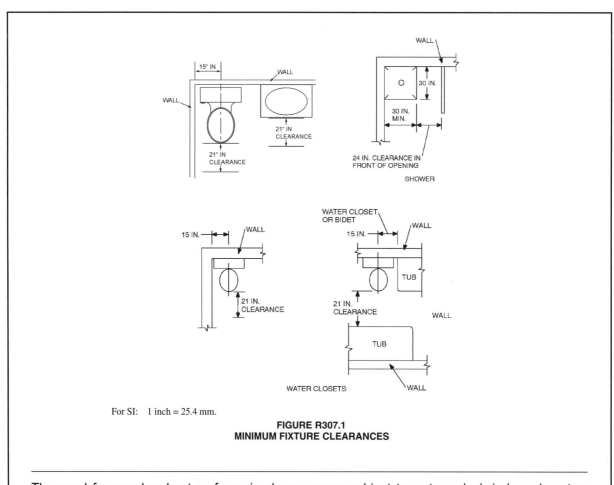

For SI: 1 inch = 25.4 mm.

FIGURE R307.1
MINIMUM FIXTURE CLEARANCES

The need for nonabsorbent surfaces in shower areas subject to water splash is based on two concerns. For sanitation purposes, finish materials must be of a type that can be cleaned easily. Also, continued absorption of moisture will lead to deterioration of the structural components.

Topic: Safety Glazing Identification   **Category:** Building Planning

Reference: IRC R308.1   **Subject:** Glazing

Code Text: *Except as indicated in Section R308.1.1, each pane of glazing installed in hazardous locations as defined in Section R308.4 shall be provided with a manufacturer's designation specifying who applied the designation, designating the type of glass and the safety glazing standard with which it complies, which is visible in the final installation. The designation shall be acid etched, sandblasted, ceramic-fired, laser etched, embossed, or be of a type which once applied cannot be removed without being destroyed. A label shall be permitted in lieu of the manufacturer's designation.* See exceptions for tempered spandrel glass and where certifications of compliance are provided.

Discussion and Commentary: Improper glazing installed in areas subject to human impact can create a serious hazard. Accordingly, it is critical that glazing in such locations be appropriately identified to help ensure that the proper glazing is in place.

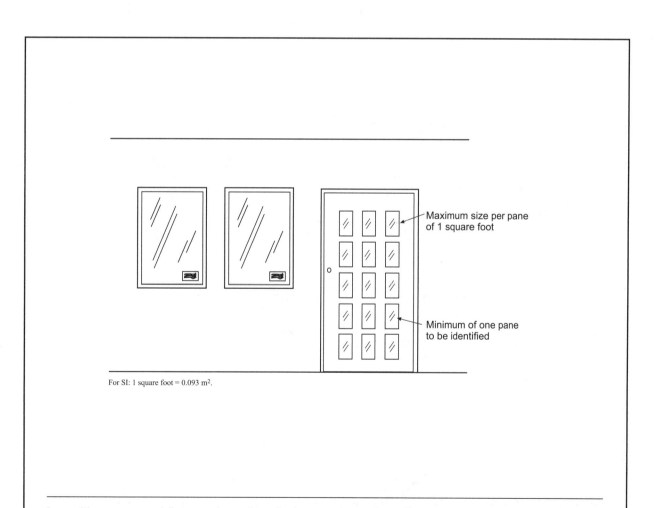

For SI: 1 square foot = 0.093 m².

In multipane assemblies, such as french doors, where the individual panes do not exceed one square foot in exposed area, Section R308.1.1 provides for a reduction in the required information on all but one pane. At least one pane must be fully identified.

Topic: Human Impact Loads

Category: Building Planning

Reference: IRC R308.3

Subject: Glazing

Code Text: *Individual glazed areas, including glass mirrors in hazardous locations such as those indicated as defined in Section R308.4, shall pass the test requirements of . . . CPSC 16 CFR, Part 1201. Glazing shall comply with test criteria for Category II unless otherwise indicated in Table R308.3.1(1).* See exception permitting the use of glazing in limited locations that has met the test criteria of ANSI Z97.1 for Class A glazing unless Class B is permitted by Table R308.3.1(2).

Discussion and Commentary: Glazing located in hazardous locations subject to human impact must pass the test requirements developed by the Consumer Product Safety Commission (CPSC). Two classifications of safety glazing have been established—Category I and Category II. Glazing identified as Category II is required in all locations where safety glazing is mandated except the three locations identified in Table R308.3.1(1) where Class I glazing is permitted.

TABLE R308.3.1(1)
MINIMUM CATEGORY CLASSIFICATION OF GLAZING USING CPSC 16 CFR 1201

EXPOSED SURFACE AREA OF ONE SIDE OF ONE LITE	GLAZING IN STORM OR COMBINATION DOORS (Category Class)	GLAZING IN DOORS (Category Class)	GLAZED PANELS REGULATED BY ITEM 7 OF SECTION R308.4 (Category Class)	GLAZED PANELS REGULATED BY ITEM 6 OF SECTION R308.4 (Category Class)	GLAZING IN DOORS AND ENCLOSURES REGULATED BY ITEM 5 OF SECTION R308.4 (Category Class)	SLIDING GLASS DOORS PATIO TYPE (Category Class)
9 square feet or less	I	I	NR	I	II	II
More than 9 square feet	II	II	II	II	II	II

For SI: 1 square foot = 0.0929 m².
NR means "No Requirement."

TABLE R308.3.1(2)
MINIMUM CATEGORY CLASSIFICATION OF GLAZING USING ANSI Z97.1

EXPOSED SURFACE AREA OF ONE SIDE OF ONE LITE	GLAZED PANELS REGULATED BY ITEM 7 OF SECTION R308.4 (Category Class)	GLAZED PANELS REGULATED BY ITEM 6 OF SECTION R308.4 (Category Class)	DOORS AND ENCLOSURES REGULATED BY ITEM 5 OF SECTION R308.4[a] (Category Class)
9 square feet or less	No requirement	B	A
More than 9 square feet	A	A	A

For SI: 1 square foot = 0.0929 m².
a. Use is permitted only by the exception to Section R308.3.1.

Glazing in doors and glazing in enclosures for hot tubs, whirlpools, saunas, steam rooms, bathtubs and showers must meet the CPSC criteria. In other areas considered to be hazardous locations, Class A or B glazing tested to the criteria of ANSI Z97.1 may be installed. Such glazing must comply with the test criteria for Class A unless indicated in Table R308.3.1(2).

Topic: Glazing in Doors   **Category:** Building Planning
Reference: IRC R308.4.1   **Subject:** Safety Glazing

Code Text: *Glazing in all fixed and operable panels of swinging, sliding and bifold doors shall be considered a hazardous location.* See exceptions for small glazed openings and those openings glazed with decorative glass.

Discussion and Commentary: Any door containing glazing must be glazed with safety glass or other safety glazing material recognized by the code for that intended purpose. Glazing in doors is of particular concern due to the increased likelihood of accidental impact by individuals operating or opening the doors. In addition, a person may push against a glazed portion of the door to gain leverage in pushing it open. Therefore, it is important that only safety glazing materials be used for glazing in doors.

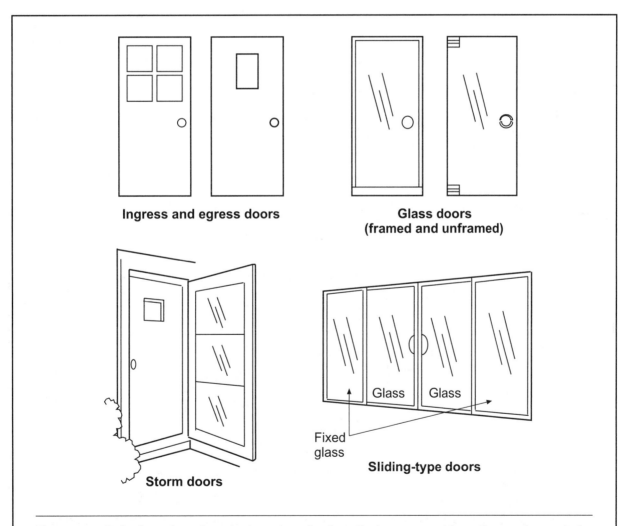

Ingress and egress doors

Glass doors (framed and unframed)

Storm doors

Sliding-type doors

Fixed glass

Glass Glass

There are a limited number of products and applications that are exempt from the requirements for hazardous locations, including small openings in doors through which a 3-inch-diameter sphere will not pass, and specific decorative assemblies, such as leaded, faceted or carved glass.

Code Text: *Glazing in an individual fixed or operable panel adjacent to a door where the nearest vertical edge of the glazing is within a 24-inch (610 mm) arc of either vertical edge of the door in a closed position and where the bottom exposed edge of the glazing is less than 60 inches (1524 mm) above the floor or walking surface shall be considered a hazardous location.* See five exceptions where safety glazing is not mandated.

Discussion and Commentary: When an individual approaches a doorway, areas adjacent to the door pose a risk when glazing is within 60 inches vertically of the walking surface. A person may slip or mistake the glass panel adjacent to a door for a passageway and walk into the glass, or a person may push against the sidelight with one hand for support while opening the door with the other hand. Therefore, safety glazing is required for any glazed opening located within 24 inches horizontally of the vertical edge of the door.

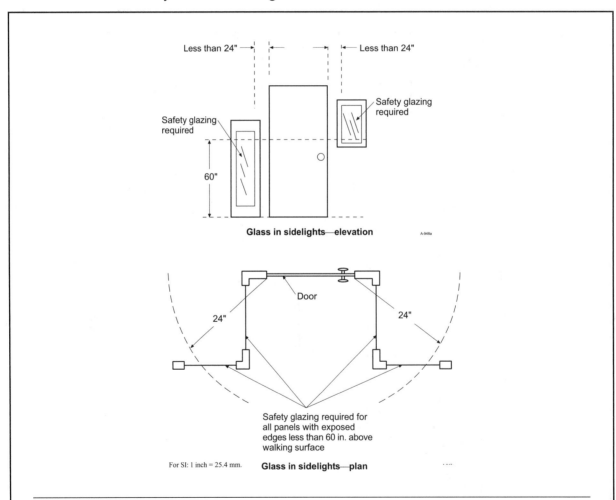

Where there is an intervening wall or similar permanent barrier between the door and the glazing, or where access through the door is to a closet or similar storage area of limited depth, safety glazing is not required, as the potential for contact is greatly reduced. Safety glazing is also not mandated where decorative glazing is installed or for glazing adjacent to the fixed panel of a patio door.

Topic: Glazing in Windows   **Category:** Building Planning
Reference: IRC R308.4.3   **Subject:** Safety Glazing

Code Text: *Glazing in an individual fixed or operable panel that meets all of the following conditions shall be considered as a hazardous location: (1) the exposed area of an individual pane is larger than 9 square feet (0.84 m²), (2) the bottom edge of the glazing is less than 18 inches (457 mm) above the floor, (3) the top edge of the glazing is more than 36 inches (914 mm) above the floor, and (4) one or more walking surface(s) are within 36 inches (914 mm), measured horizontally and in a straight line, of the glazing. See three exceptions.*

Discussion and Commentary: Large pieces of glass create a hazard where located close to a travel path because it is possible to impact glazing where no obstacle or barrier is provided as an alternative impact area. Expansive glazing may also be mistaken for a clear opening in the wall.

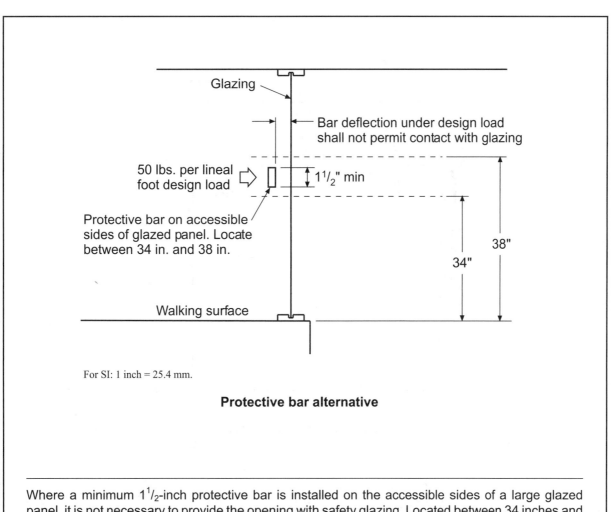

For SI: 1 inch = 25.4 mm.

Protective bar alternative

Where a minimum 1¹⁄₂-inch protective bar is installed on the accessible sides of a large glazed panel, it is not necessary to provide the opening with safety glazing. Located between 34 inches and 38 inches above the floor, the bar must be capable of withstanding a 50 plf horizontal load.

Code Text: *Glazing in walls, enclosures, or fences containing or facing hot tubs, spas, whirlpools, saunas, steam rooms, bathtubs, showers and indoor or outdoor swimming pools where the bottom exposed edge of the glazing is less than 60 inches (1524 mm) measured vertically above any standing or walking surface shall be considered a hazardous location.. See exception for glazing located more than 60 inches horizontally from the water's edge of a hot tub, whirlpool, spa, swimming pool or bathtub.*

Discussion and Commentary: Because the standing surfaces of, and adjacent to bathtubs, showers, hot tubs, swimming pools and similar elements are wet and slippery, glazing adjacent to these elements must be regulated due to the potential for human impact. It is not uncommon for the user to slip while trying to enter or exit. Safety glazing is mandated where any of the glazing within the enclosed area extends to within 60 inches vertically of the standing surface.

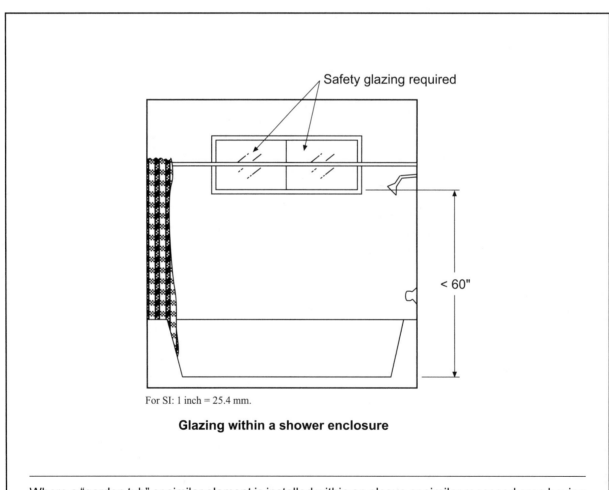

For SI: 1 inch = 25.4 mm.

Glazing within a shower enclosure

Where a "garden tub" or similar element is installed within an alcove or similar recessed area having windows or other glazed openings, the glazing in the walls of the alcove are considered a portion of the enclosure and are thus regulated.

| **Topic:** Skylights and Sloped Glazing | **Category:** Building Planning |
| **Reference:** IRC R308.6.2 | **Subject:** Glazing |

Code Text: *The following types of glazing may be used* (for skylights and sloped glazing)*: 1) laminated glass with a complying polyvinyl butyral interlayer; 2) fully tempered glass; 3) heat-strengthened glass; 4) wired glass; and 5) approved rigid plastics.*

Discussion and Commentary: A skylight or sloped glazing is considered *glass or other transparent or translucent glazing material installed at a slope of 15 degrees or more from vertical. Glazing materials in skylights, including unit skylights, solariums, sunrooms, roofs and sloped walls are included.* The requirements enhance the protection of a building's occupants from the possibility of falling glazing materials. Only certain glazing materials have the necessary characteristics to be permitted in an overhead installation. In addition, the provisions of the IRC address design loads normally attributed to roofs.

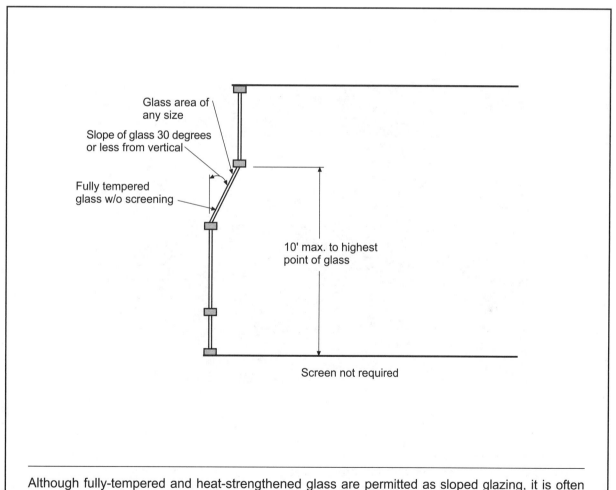

Although fully-tempered and heat-strengthened glass are permitted as sloped glazing, it is often necessary that a protective retaining screen be located below the glass. Fully-tempered glass may spontaneously break into large chunks, whereas heat-strengthened glass can create falling shards.

Topic: Carports	**Category:** Building Planning
Reference: IRC R309.2	**Subject:** Garages and Carports

Code Text: *Carports shall be open on at least two sides. Carport floor surfaces shall be of approved noncombustible material. See exceptions for asphalt surfaces located at ground level. Carports not open on at least two sides shall be considered a garage and shall comply with the provisions of* Section 309 *for garages. The area of floor used for parking of automobiles or other vehicles shall be sloped to facilitate the movement of liquids to a drain or toward the main vehicle entry doorway.*

Discussion and Commentary: Where the structure, either attached to the residence or detached, intended to shelter one or more motor vehicles, is enclosed on at least three sides, the structure is regulated as a garage. Where the structure is open on at least two sides, it is regulated simply as a carport. Carports are considered to have a minor degree of hazard, and only their floor surfaces are regulated.

The limited number of requirements for carports is based primarily on their openness to the exterior. A small chance of a fire incident is expected, and should one occur the cross-ventilation provided by the open sides will not allow smoke to accumulate. Combustible loading within carports is also typically limited.

Topic: Openings Required

Category: Building Planning

Reference: IRC R310.1

Subject: Emergency Escape and Rescue Openings

Code Text: *Basements, habitable attics and every sleeping room shall have at least one operable emergency escape and rescue opening.* See exception for small mechanical equipment areas. *Where basements contain one or more sleeping rooms, emergency egress and rescue openings shall be required in each sleeping room. Where emergency escape and rescue openings are provided they shall have a sill height of not more than 44 inches (1118 mm) measured from the finished floor to the bottom of the clear opening. Emergency escape and rescue openings shall open directly into a public way, or to a yard or court that opens to a public way.*

Discussion and Commentary: Because so many deaths and injuries from fire occur as the result of occupants of residential buildings being asleep at the time of a fire, the code requires that basements and all sleeping rooms have doors or windows that may be used for emergency escape or rescue. The concern is based on the fact that often a fire will have spread before the occupants are aware of the danger; thus, the normal means of escape will most likely be blocked.

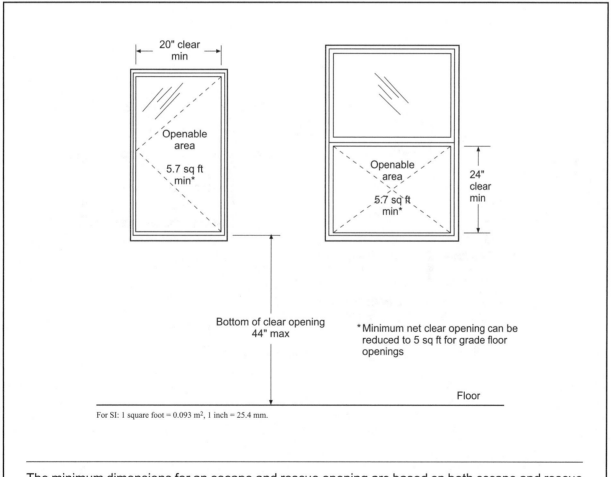

For SI: 1 square foot = 0.093 m², 1 inch = 25.4 mm.

The minimum dimensions for an escape and rescue opening are based on both escape and rescue criteria. Where the opening occurs very near ground level, the clear opening size can be reduced because access from the exterior can be more easily accomplished without the use of a ladder.

Topic: Window Wells   **Category:** Building Planning

Reference: IRC R310.1, R310.2   **Subject:** Emergency Escape and Rescue Openings

Code Text: *Emergency escape and rescue openings with a finished sill height below the adjacent ground elevation shall be provided with a window well. The minimum horizontal area of the window well shall be 9 square feet ($0.9\ m^2$), with a minimum horizontal projection and width of 36 inches (914 mm).* See exception for ladder encroachment into required dimensions. *The area of the window well shall allow the emergency and rescue opening to be fully opened.*

Discussion and Commentary: The window well provisions address those emergency escape and rescue openings that occur below grade. Simply applying the standard opening criteria to these window wells does not provide an adequate clear opening size to transition from the below-grade space to grade level. The increased cross-sectional dimension will be sufficient to provide for the escape of occupants or the entry of firefighters or other rescue personnel. Unless the site is considered a Group I soil, window wells must be designed for proper drainage.

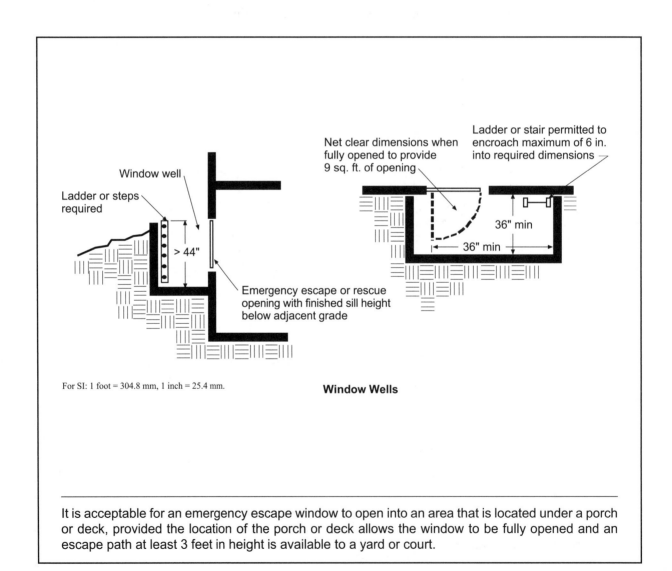

For SI: 1 foot = 304.8 mm, 1 inch = 25.4 mm.

Window Wells

It is acceptable for an emergency escape window to open into an area that is located under a porch or deck, provided the location of the porch or deck allows the window to be fully opened and an escape path at least 3 feet in height is available to a yard or court.

Quiz

Study Session 3
IRC Sections R303 – R310

1. When natural light is used to satisfy the minimum light requirements for habitable rooms, the aggregate glazing area shall be a minimum of _____ of the floor area.

   a. 4 percent

   b. 5 percent

   c. 8 percent

   d. 10 percent

   Reference _____

2. When natural ventilation is used to satisfy the minimum ventilation requirements for habitable rooms, the aggregate open area to the outdoors shall be a minimum of _____ of the floor area.

   a. 4 percent

   b. 5 percent

   c. 8 percent

   d. 10 percent

   Reference _____

3. Garage sprinklers, when required, shall be residential sprinklers or quick-response sprinklers, designed to provide a density of _____ gpm/sf.

   a. 0.05

   b. 0.10

   c. 0.15

   d. 0.25

   Reference _____

4. Where applicable escape and rescue openings are provided and a whole-house mechanical ventilation system is installed, glazing is not required for natural light purposes if artificial light is provided capable of producing an average illumination of _____ footcandle(s) over the area of the room at a height of 30 inches above the floor level.

    a. 1   b. 5

    c. 6   d. 10

Reference _____

5. At least one habitable room with a minimum floor area of _____ square feet shall be provided in every dwelling unit.

    a. 100   b. 120

    c. 150   d. 200

Reference _____

6. The floor area beneath a furred ceiling can be considered to be contributing to the minimum required habitable area for the room where it has a minimum height of _____ above the floor.

    a. 5 feet   b. 6 feet, 4 inches

    c. 6 feet, 8 inches   d. 7 feet

Reference _____

7. Other than in a kitchen, what is the minimum permitted horizontal dimension of any habitable room?

    a. 6 feet   b. 7 feet

    c. 8 feet   d. 9 feet

Reference _____

8. In general, the minimum required ceiling height of all habitable rooms is _____ .

    a. 6 feet, 8 inches   b. 7 feet, 0 inches

    c. 7 feet, 6 inches   d. 8 feet, 0 inches

Reference _____

9. Ceilings in portions of basements without habitable spaces shall have a minimum ceiling height of _____ from the finished floor to beams, ducts, or other obstructions.

   a. 6 feet, 4 inches   b. 6 feet, 6 inches

   c. 6 feet, 8 inches   d. 7 feet, 0 inches

   Reference _____

10. A minimum clearance of _____ shall be provided in front of a water closet.

    a. 21 inches   b. 24 inches

    c. 30 inches   d. 32 inches

    Reference _____

11. The minimum distance between the centerline of a water closet and a side wall shall be _____ .

    a. 15 inches   b. 16 inches

    c. 18 inches   d. 21 inches

    Reference _____

12. Within a shower compartment, the walls shall be finished with a nonabsorbent surface to a minimum height of _____ .

    a. 70 inches   b. 72 inches

    c. 78 inches   d. 84 inches

    Reference _____

13. Complying heating facilities are not required in dwelling units where the winter design temperature of the locale is a minimum of _____ .

    a. 60°F   b. 64°F

    c. 68°F   d. 70°F

    Reference _____

14. Where located less than 5 feet above the walking surface, glazing in a wall enclosing an outdoor hot tub shall be considered to be installed in a hazardous location unless the glazing is a minimum of _____ horizontally from the water's edge.

    a. 3 feet   b. 4 feet

    c. 5 feet   d. 10 feet

Reference _____

15. A skylight is defined by the IRC as glass or other glazing material installed at a minimum slope of _____ .

    a. 15 degrees from the vertical

    b. 30 degrees from the vertical

    c. 15 degrees from the horizontal

    d. 30 degrees from the horizontal

Reference _____

16. Curbs are not required for skylights installed on roofs having a minimum slope of _____ .

    a. 2:12   b. 3:12

    c. 4:12   d. 5:12

Reference _____

17. A screen used to protect an air exhaust opening that terminates outdoors shall have a minimum opening size of _____ inch and a maximum opening size of _____ inch.

    a. $^1/_8$, $^1/_4$ b. $^1/_4$, $^3/_8$

    c. $^1/_4$, $^1/_2$ d. $^1/_2$, $^3/_4$

Reference _____

18. A minimum clearance of _____ inches shall be provided in front of the opening to a shower.

    a. 24   b. 30

    c. 32   d. 36

Reference _____

19. Regular glass used in a louvered window shall be a minimum of _____ inch in nominal thickness.

    a. $^3/_{32}$ b. $^1/_8$

    c. $^5/_{32}$ d. $^3/_{16}$

Reference _____

20. Emergency escape and rescue openings shall have a maximum sill height of _____ above the floor.

    a. 40 inches   b. 42 inches

    c. 44 inches   d. 48 inches

Reference _____

21. Emergency escape and rescue openings, when considered as grade floor openings, shall have a minimum net clear opening of _____ .

    a. 4.0 square feet   b. 4.4 square feet

    c. 5.0 square feet   d. 5.7 square feet

Reference _____

22. For an emergency escape and rescue opening, the minimum clear opening height shall be _____, and the minimum clear opening width shall be _____.

    a. 24 inches, 20 inches   b. 24 inches, 28 inches

    c. 28 inches, 22 inches   d. 28 inches, 24 inches

Reference _____

23. Where a window well is required in conjunction with an escape and rescue opening, the window well shall have a minimum net clear area of _____ square feet with a minimum horizontal dimension of _____ .

    a. 5.7, 20 inches   b. 5.7, 24 inches

    c. 9.0, 30 inches   d. 9.0, 36 inches

Reference _____

24. An asphalt surface is permitted as the floor surface of a carport provided the surface is
_____ .

    a. limited to 400 square feet

    b. located at ground level

    c. sealed with an approved material

    d. sloped a minimum of 1:48

Reference _____

25. A 6-square-foot skylight of laminated glass shall have a minimum _____
polyvinyl butyral interlayer where located 14 feet above the walking surface.

    a. 0.015-inch   b. 0.024-inch

    c. 0.030-inch   d. 0.044-inch

Reference _____

26. Unless located at least 3 feet below the contaminant source, a mechanical outside air
intake opening shall be located a minimum of _____ feet from a plumbing
vent.

    a. 3   b. 5

    c. 10   d. 12

Reference _____

27. The illumination source for interior stairs shall be capable of illuminating the treads and
landings to a minimum level of _____ foot-candle(s).

    a. 1   b. 5

    c. 8   d. 10

Reference _____

28. A minimum ceiling height of _____ is required at the center of the front
clearance area for bathroom fixtures.

    a. 6 feet, 4 inches   b. 6 feet, 6 inches

    c. 6 feet, 8 inches   d. 7 feet, 0 inches

Reference _____

29. Glazing adjacent to a door to a storage closet is not required to be safety glazing provided the closet is a maximum of _____ in depth.

    a. 24 inches   b. 30 inches

    c. 36 inches   d. 42 inches

Reference _____

30. Glazing in a door shall be safety glazing where the opening allows the passage of a minimum _____ sphere.

    a. 3-inch   b. $3\frac{1}{2}$-inch

    c. 4-inch   d. 6-inch

Reference _____

31. Where window wells are provided for escape and rescue openings, window well drainage is not required where the building foundation is supported by Group _____ soils.

    a. I   b. I or II

    c. I, II or III   d. IV

Reference _____

32. Where lighting outlets are installed in interior stairways, a wall switch shall be provided at each floor level to control the light outlet where the stairway has a minimum of _____ risers.

    a. two   b. three

    c. four   d. six

Reference _____

33. A 4-square-foot glazed panel installed in an entry door shall have a minimum glazing category classification of CPSC 16 CFR 1201 Category _____.

    a. I   b. II

    c. III   d. IV

Reference _____

34. Except for those basements with a maximum floor area of _____ square feet used only to house mechanical equipment, all basements shall be provided with at least one operable emergency escape and rescue opening.

    a. 120   b. 150

    c. 200   d. 400

Reference _____

35. An emergency escape window may be installed under a deck or porch, provided a minimum height of _____ inches is maintained to a yard or court.

    a. 24   b. 36

    c. 44   d. 48

Reference _____

Study Session

4

2012 IRC Sections R311 – R323
Building Planning III

OBJECTIVE: To obtain an understanding of the fire and life safety criteria of the code, including stairways, ramps and landings; handrails and guards; automatic sprinkler systems; smoke alarms and carbon monoxide alarms; foam plastic and insulation; protection against decay and termite infestation; accessibility; flood-resistant construction; and storm shelters.

REFERENCE: Sections R311 through R323, 2012 *International Residential Code*

KEY POINTS:
- How many egress doors from a dwelling are required? Under what conditions is an additional egress door required? What is the minimum required size of an egress door?
- What is the minimum size required for a landing? Under what conditions is a landing not required on each side of an exterior door?
- What is the minimum required width of a hallway?
- How wide must a stairway be? Where is this width measured? What type of encroachments into the minimum required width are permitted? How far may they encroach?
- What is the maximum allowable riser height in a stairway? Minimum allowable tread run? Within a flight of stairs, what is the maximum tolerance permitted between the smallest and greatest riser height? Between the smallest and greatest tread run?
- What are the various criteria for tread nosings? When are nosings not required?
- What is the minimum allowable headroom clearance at stairways and landings? How is this height measured?
- How is handrail height measured? What is the minimum height permitted? The maximum height? What sizes and shapes of handrails are acceptable?
- What types of special stairways are addressed? How do they differ in layout and size from traditional stairway design?
- What is the maximum slope permitted for a ramp? When is a handrail required for a ramp? What size landings are required?

KEY POINTS:
(Cont'd)

- When is a guard required? What is the minimum allowable height of the guard above the walking surface? How must intermediate rails be located? How do the provisions differ for guards on open sides of stairs?
- Under what conditions is an automatic residential fire sprinkler system required in a townhouse? What criteria are to be used for the design and installation of the sprinkler system?
- Are one- and two-family dwellings required to be protected by an automatic residential fire sprinkler system? If so, what criteria are to be used for the design and installation of the sprinkler system?
- Where are smoke alarms required to be located? Do they need to be interconnected? How must they be powered? What must be done when alterations or repairs take place?
- Under what conditions are carbon monoxide alarms required? Where must they be located?
- How are foam plastics regulated? What are the restrictions on their use as wall and ceiling finishes? Insulation materials?
- What areas are considered subject to decay damage? Termite damage? How should such conditions be abated?
- How must the premises be identified?
- What type of building constructed under the IRC requires some degree of accessibility?
- How are elevators and platform lifts regulated?
- What are the general requirements for buildings constructed in flood hazard areas?
- What standard is applicable to the design and construction of storm shelters? Under what conditions are storm shelters required?

Topic: Egress Door

Reference: IRC R311.2

Category: Building Planning

Subject: Means of Egress

Code Text: *At least one egress door shall be provided for each dwelling unit. The egress door shall be side-hinged, and shall provide a minimum clear width of 32 inches (813 mm) when measured between the face of the door and the stop, with the door open 90 degrees (1.57 rad). The minimum clear height of the door opening shall not be less than 78 inches (1981 mm) measured from the top of the threshold to the bottom of the stop. Other doors shall not be required to comply with these minimum dimensions. Egress doors shall be readily openable from inside the dwelling unit without the use of a key or special knowledge or effort.*

Discussion and Commentary: Regardless of the size of the dwelling unit, only one exterior egress door is required. Not only must the designated egress door be side-hinged, the door opening must provide 32 inches in clear width and 78 inches in clear height. The other doors within the dwelling unit can be of any size and need not be of the swinging type.

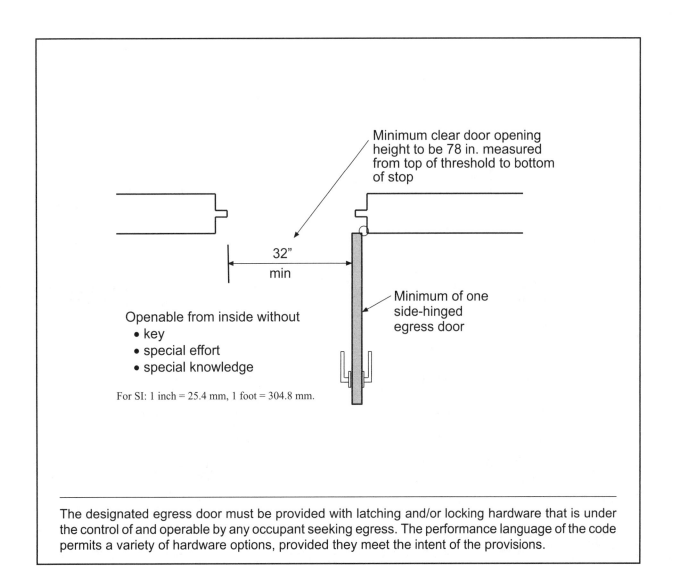

Minimum clear door opening height to be 78 in. measured from top of threshold to bottom of stop

32"
min

Minimum of one side-hinged egress door

Openable from inside without
- key
- special effort
- special knowledge

For SI: 1 inch = 25.4 mm, 1 foot = 304.8 mm.

The designated egress door must be provided with latching and/or locking hardware that is under the control of and operable by any occupant seeking egress. The performance language of the code permits a variety of hardware options, provided they meet the intent of the provisions.

Code Text: *There shall be a floor or landing on each side of each exterior door.* See exception for exterior balconies less than 60 square feet in floor area. *Landings or floors at the required egress door shall not be more than $1^1/_2$ inches (38 mm) lower than the top of the threshold.* See exception for exterior landings at all exterior doorways where the door does not swing over the landing.

Discussion and Commentary: As a general rule, a maximum elevation change of $1^1/_2$ inches is permitted at the exterior side of exterior doors, measured from the top of the threshold to the landing. A commonly utilized exception permits up to a $7^3/_4$-inch height difference, provided the exterior door, other than a screen or storm door, does not swing outward over the exterior landing. The user's typical familiarity with the change in elevation at the dwelling's exterior doors justifies such an allowance.

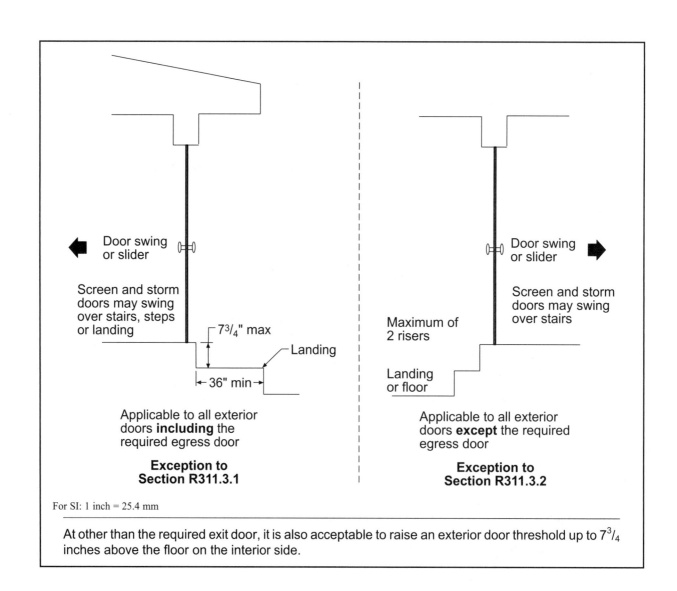

Door swing or slider ←

Screen and storm doors may swing over stairs, steps or landing

$7^3/_4$" max

Landing

36" min

Applicable to all exterior doors **including** the required egress door

Exception to Section R311.3.1

Door swing or slider →

Screen and storm doors may swing over stairs

Maximum of 2 risers

Landing or floor

Applicable to all exterior doors **except** the required egress door

Exception to Section R311.3.2

For SI: 1 inch = 25.4 mm

At other than the required exit door, it is also acceptable to raise an exterior door threshold up to $7^3/_4$ inches above the floor on the interior side.

Topic: Stairway Width and Headroom	**Category:** Building Planning
Reference: IRC R311.7.1 – R311.7.3	**Subject:** Means of Egress

Code Text: *Stairways shall not be less than 36 inches (914 mm) in clear width at all points above the permitted handrail height and below the required headroom height. Handrails shall not project more than 4$\frac{1}{2}$ inches (114 mm) on either side of the stairway and the minimum clear width of the stairway at and below the handrail height, including treads and landings, shall not be less than 31$\frac{1}{2}$ inches (787 mm) where a handrail is installed on one side and 27 inches (698 mm) where handrails are provided on both sides.* See exception for the minimum width of spiral stairways. *The minimum headroom in all parts of the stairway shall not be less than 6 feet 8 inches (2032 mm). A flight of stairs shall not have a vertical rise larger than 12 feet (3658 mm) between floor levels or landings.*

Discussion and Commentary: Although a minimum width of 36 inches is required for stairways, the code is not as concerned about elements such as trim, stringers or other items that may be found below the level of the handrail, as long as they do not exceed the projection of the handrail(s). The width limitations are based on the body's movements as a person walks on a stairway.

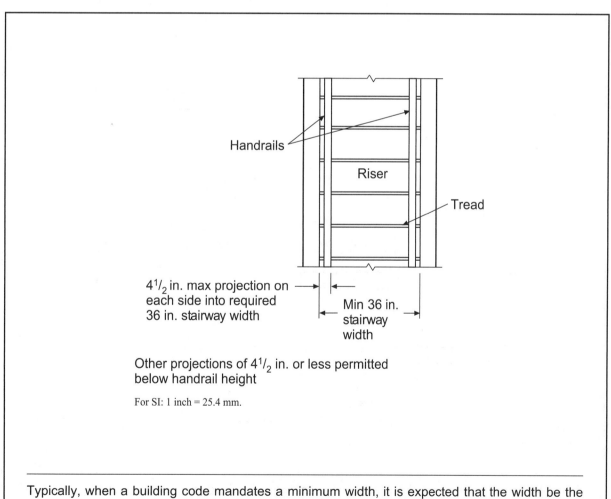

Handrails

Riser

Tread

4$\frac{1}{2}$ in. max projection on each side into required 36 in. stairway width

Min 36 in. stairway width

Other projections of 4$\frac{1}{2}$ in. or less permitted below handrail height

For SI: 1 inch = 25.4 mm.

Typically, when a building code mandates a minimum width, it is expected that the width be the clear, net, usable, unobstructed width. In this case, however, limited projections into the minimum width are acceptable where located at or below the height of the handrail.

Topic: Stair Treads and Risers

Reference: IRC R311.7.5.1, R311.7.5.2

Category: Building Planning

Subject: Means of Egress

Code Text: *The maximum riser height shall be $7^3/_4$ inches (196 mm). The greatest riser height within any flight of stairs shall not exceed the smallest by more than $^3/_8$ inch (9.5 mm). Open risers are permitted provided that the opening between treads does not permit the passage of a 4-inch diameter (102 mm) sphere. See exception for stairs with a maximum rise of 30 inches. The minimum tread depth shall be 10 inches (254 mm). The greatest tread depth within any flight of stairs shall not exceed the smallest by more than $^3/_8$ inch (9.5 mm).*

Discussion and Commentary: Although the rise and run configuration is important in regulating stairway safety, another significant safety factor relative to stairways is the uniformity of risers and treads in any flight (the section of a stairway leading from one landing to the next). It is important that any variation that would interfere with the rhythm of the stair user be avoided. Special allowances are provided for winder treads.

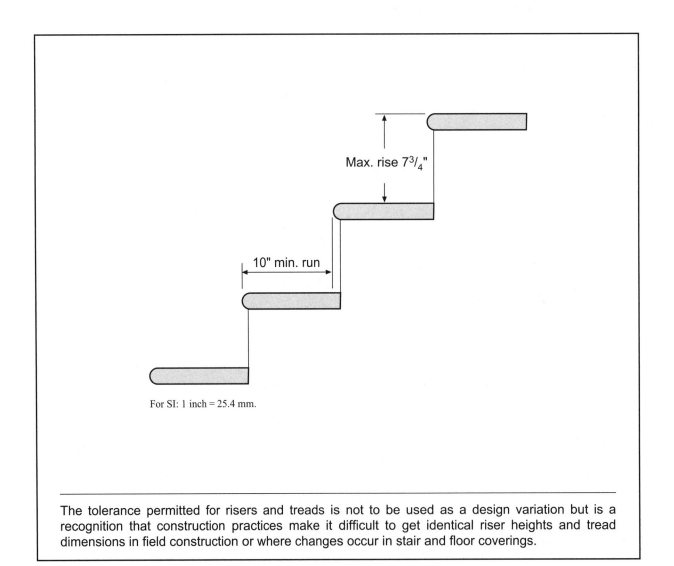

Max. rise $7^3/_4$"

10" min. run

For SI: 1 inch = 25.4 mm.

The tolerance permitted for risers and treads is not to be used as a design variation but is a recognition that construction practices make it difficult to get identical riser heights and tread dimensions in field construction or where changes occur in stair and floor coverings.

Code Text: *Handrails shall be provided on at least one side of each continuous run of treads or flight with four or more risers. Handrail height, measured vertically from the sloped plane adjoining the tread nosing, or finish surface of ramp slope, shall be not less than 34 inches (864 mm) and not more than 38 inches (965 mm). See exceptions. Handrails for stairways shall be continuous for the full length of the flight, from a point directly above the top riser of the flight to a point directly above the lowest riser of the flight. Handrail ends shall be returned or shall terminate in newel posts or safety terminals. See exceptions.*

Discussion and Commentary: Any stairway consisting of four or more risers must be provided with a handrail on at least one side. The rail is a proven safety feature for users of stairways, particularly those who fail to pay proper attention to stair use. Two types of handrail shapes are described, differing because of their perimeter measurement.

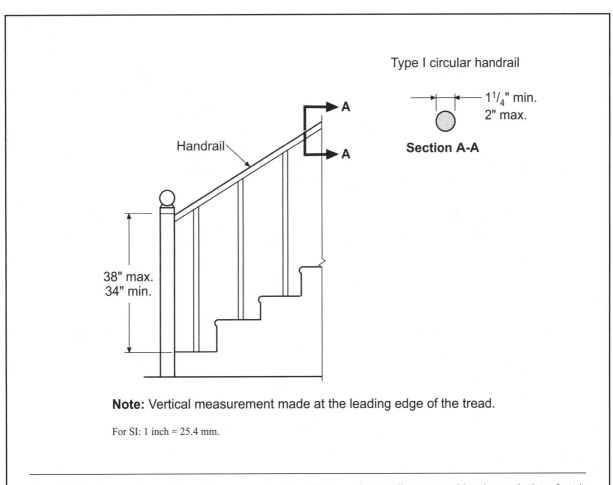

Type I circular handrail

1$\frac{1}{4}$" min.
2" max.

Section A-A

Handrail

A
A

38" max.
34" min.

Note: Vertical measurement made at the leading edge of the tread.

For SI: 1 inch = 25.4 mm.

An effective handrail must be of a size and shape that can be easily grasped by the majority of stair users. Whereas the minimum and maximum cross sections of 1$\frac{1}{4}$ inches and 2 inches are specific for circular rails, other shapes with equivalent grasping surfaces may be acceptable.

Code Text: *Type II. Handrails with a perimeter greater than 6¼ inches (160 mm) shall have a graspable finger recess area on both sides of the profile. The finger recess shall begin within a distance of ¾ inch (19 mm) measured vertically from the tallest portion of the profile and achieve a depth of at least 5/16 inch (8 mm) within 7/8 inch (22 mm) below the widest portion of the profile. This required depth shall continue for at least 3/8 inch (10 mm) to a level that is not less than 1¾ inches (45 mm) below the tallest portion of the profile. The minimum width of the handrail above the recess shall be 1¼ inches (32 mm) to a maximum of 2¾ inches (70 mm).*

Discussion and Commentary: The key features of the graspability of Type II handrails are graspable finger recesses on both sides of the handrail. These recesses allow users to firmly grip a properly proportioned grasping surface on the top of the handrail, ensuring that the user can tightly retain a grip on the handrail for all forces that are associated with attempts to arrest a fall.

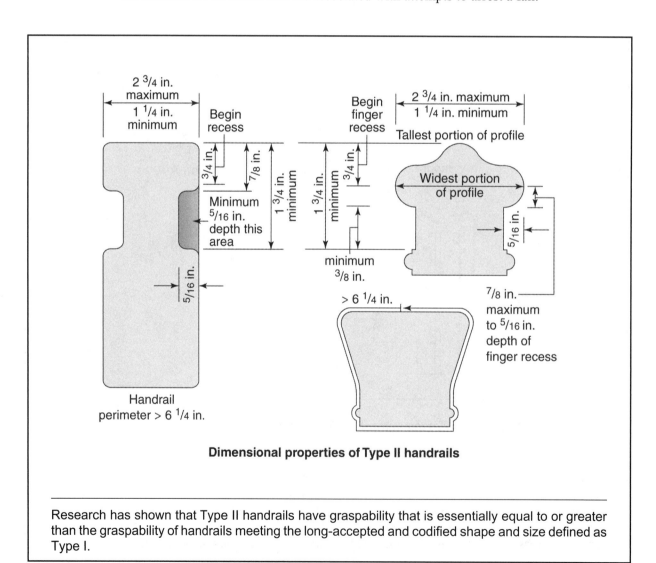

Dimensional properties of Type II handrails

Research has shown that Type II handrails have graspability that is essentially equal to or greater than the graspability of handrails meeting the long-accepted and codified shape and size defined as Type I.

Topic: Special Stairways	**Category:** Building Planning
Reference: IRC R311.7.10	**Subject:** Means of Egress

Code Text: *Spiral stairways are permitted, provided the minimum width at and below the handrail shall be 26 inches (660 mm) with each tread having a 7 $\frac{1}{2}$-inch (190 mm) minimum tread depth at 12 inches (305 mm) from the narrower edge. All treads shall be identical, and the rise shall be not more than 9 $\frac{1}{2}$ inches (241 mm). Stairways serving bulkhead enclosures, not part of the required building egress, providing access from the outside grade level to the basement shall be exempt from the requirements of Sections R311.3* (landings at doors) *and R311.7* (stairways) *where the maximum height from the basement finished floor level to grade adjacent to the stairway does not exceed 8 feet (2438 mm).*

Discussion and Commentary: These special types of stairways are allowed for use in dwelling units, provided they are constructed in compliance with the specific limitations mandated by the code.

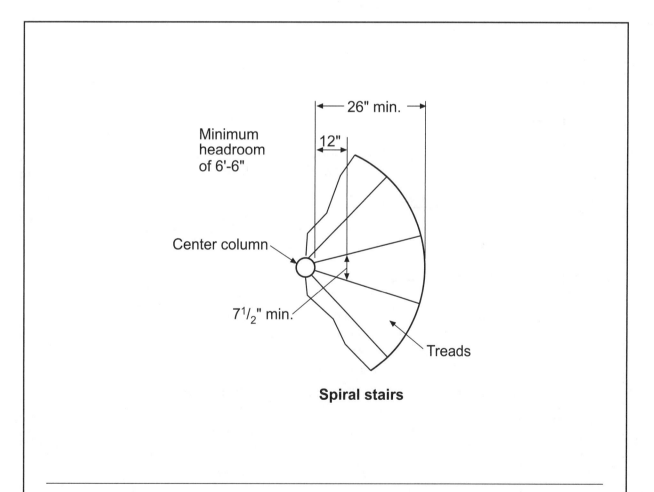

Spiral stairs

Spiral stairs are not as easy to use as a traditionally designed stairway. Although the rise and run may differ from the fundamental criteria, the travel path must be consistent in size and shape to reduce the hazard level.

Code Text: *Ramps shall have a maximum slope of one unit vertical in twelve units horizontal (8.3-percent slope). See exception where technically infeasible due to site constraints. A minimum 3-foot-by-3-foot (914 mm by 914 mm) landing shall be provided: at the top and bottom of ramps, where doors open onto ramps, and where ramps change direction. Handrails shall be provided on at least one side of all ramps exceeding a slope of one unit vertical in 12 units horizontal (8.33-percent slope).*

Discussion and Commentary: A ramp is defined as a walking surface that has a running slope steeper than 1 unit vertical in 20 units horizontal (5-percent slope). For general use purposes as well as exiting, the maximum permitted slope is 1:12. Ramps with slopes of 1:12 to 1:20 do not require a handrail, as the rise or descent is gradual enough to provide a safe travel path. Because ramps are generally limited to a 1:12 slope, the only ramps that will require a handrail are those permitted to be steeper than 1:12 by the exception for site constraints.

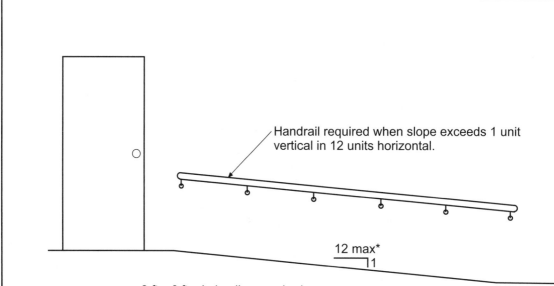

Handrail required when slope exceeds 1 unit vertical in 12 units horizontal.

12 max*
1

3 ft x 3 ft min landing required:
- At the top and bottom of ramp
- Where doors open onto ramp
- Where ramp changes direction

*1:8 maximum permitted where technically infeasible due to site constraints

For SI: 1 foot = 304.8 mm.

The ramp provisions in the IRC differ significantly from those in the IBC because of the building code's emphasis on accessibility. The IRC requirements are simply to allow safe movement through the dwelling unit and do not intend to address use by persons with physical disabilities.

Topic: Location and Height	**Category:** Building Planning
Reference: IRC R312.1.1, R312.1.2	**Subject:** Guards

Code Text: *Guards shall be located along open-sided walking surfaces, including stairs, ramps and landings, that are located more than 30 inches (762 mm) measured vertically to the floor or grade below at any point within 36 inches (914 mm) horizontally to the edge of the open side. Required guards shall be not less than 36 inches (914 mm) high measured vertically above the adjacent walking surface, adjacent fixed seating or the line connecting the leading edges of the treads.* See exceptions for guards at the open sides of stairs.

Discussion and Commentary: Some form of protection is necessary at elevated floor areas with a vertical drop of more than 30 inches, as a fall from that height can potentially result in injury. Where required, the guard must be of an adequate height to prevent someone from falling over the edge of the protected area. Where fixed seating is installed adjacent to an elevation change of more than 30 inches, the guard must be extended at least 36 inches above the level of the seating.

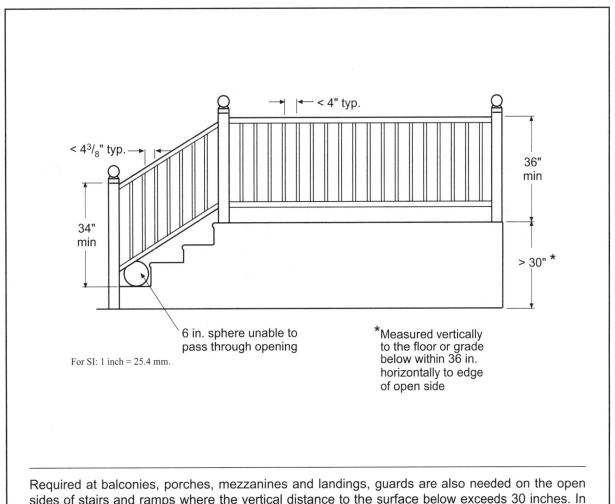

< 4" typ.

< 4³/₈" typ.

36"
min

34"
min

> 30" *

6 in. sphere unable to
pass through opening

For SI: 1 inch = 25.4 mm.

*Measured vertically
to the floor or grade
below within 36 in.
horizontally to edge
of open side

Required at balconies, porches, mezzanines and landings, guards are also needed on the open sides of stairs and ramps where the vertical distance to the surface below exceeds 30 inches. In such stairway situations, it is acceptable to use a complying handrail as the guard.

Topic: Opening Limitations

Category: Building Planning

Reference: IRC R312.1.3

Subject: Guards

Code Text: *Required guards shall not have openings from the walking surface to the required guard height which allow passage of a sphere 4 inches (102 mm) in diameter.* See exceptions for 1) triangular openings formed by riser, tread, and bottom rail, and 2) guards on open sides of stairs.

Discussion and Commentary: Guards must be constructed so that they not only prevent people from falling over them but also prevent children from crawling through them. The criteria spacing was chosen after many years of research and discussion. The chance of even a very small child being able to get through such a narrow opening is very low.

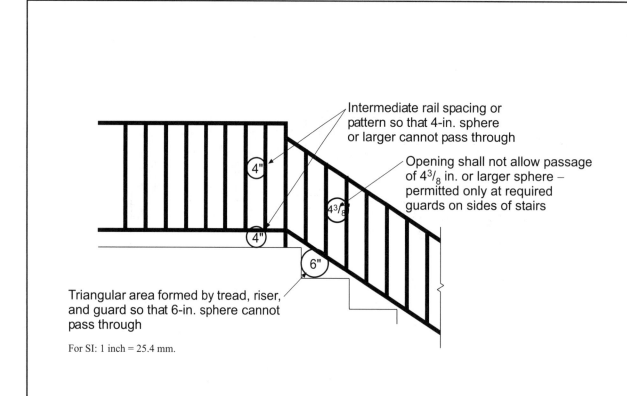

Intermediate rail spacing or pattern so that 4-in. sphere or larger cannot pass through

Opening shall not allow passage of 4³/₈ in. or larger sphere – permitted only at required guards on sides of stairs

Triangular area formed by tread, riser, and guard so that 6-in. sphere cannot pass through

For SI: 1 inch = 25.4 mm.

In addition to limiting the opening size along the open sides of stairways, the code addresses the hazard created by open risers. Section R311.7.5.1 allows the use of open risers but only where the opening between the treads does not permit the passage of a 4-inch-diameter sphere.

Topic: Window Sill Height

Category: Building Planning

Reference: IRC R312.2

Subject: Window Fall Protection

Code Text: *In dwelling units, where the opening of an operable window is located more than 72 inches (1829 mm) above the finished grade or surface below, the lowest part of the clear opening of the window shall be a minimum of 24 inches (610 mm) above the finished floor of the room in which the window is located. Operable sections of windows shall not permit openings that allow passage of a 4-inch (102 mm) diameter sphere where such openings are located within 24 inches (610 mm) of the finished floor.* See exceptions for windows that 1) do not open far enough to allow passage of a 4-inch sphere, or 2) are protected by complying window guards, fall prevention devices or opening limiting devices.

Discussion and Commentary: Historical data have shown that each year a considerable number of children fall from windows. It has been estimated that a sizable percentage of those falls occurred through windows with a low sill height. The restrictions on window sill height are intended to raise the height of the opening at the sill above the center of gravity of a small child, thus reducing the number of falls.

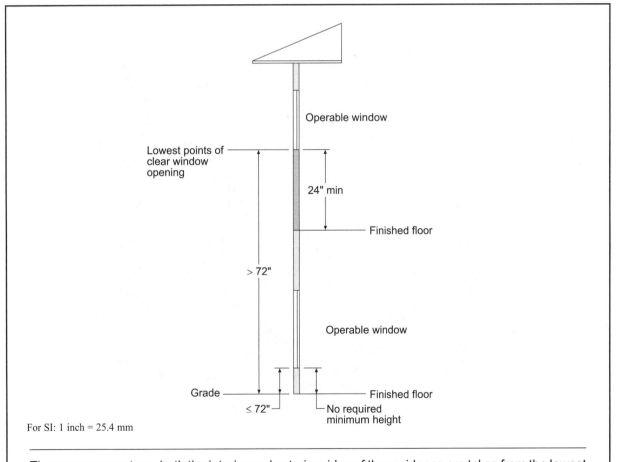

For SI: 1 inch = 25.4 mm

The measurements on both the interior and exterior sides of the residence are taken from the lowest part of the clear window opening, providing for consistent application of the provision. Where the lower window panel is inoperable, the measurement is to be taken from the lowest point of the lowest operable panel.

Code Text: *An automatic residential fire sprinkler system shall be installed in townhouses.* See exception for existing townhouses undergoing additions or alterations. *Automatic residential fire sprinkler systems for townhouses shall be designed and installed in accordance with Section P2904. An automatic residential fire sprinkler system shall be installed in one- and two-family dwellings.* See exception for existing one- and two-family dwellings undergoing additions or alterations. *Automatic residential fire sprinkler systems shall be designed and installed in accordance with Section P2904 or NFPA 13D.*

Discussion and Commentary: An automatic sprinkler system designed for a dwelling unit aids in the detection and control of fires. When installed, the system is expected to prevent total fire involvement (flashover) in the room of fire origin. A properly installed and maintained automatic sprinkler system will improve the likelihood of occupants escaping or being evacuated during a fire incident.

IRC Section P2904 provides a simple, prescriptive approach to the design of dwelling fire sprinkler systems. The provisions are consistent with those of NFPA 13D but have been simplified. The option of using either Section P2904 or NFPA 13D is available for the design and installation of residential sprinklers mandated by the IRC.

Code Text: *Smoke alarms shall be installed in the following locations: 1) in each sleeping room, 2) outside of each separate sleeping area in the immediate vicinity of the bedrooms, and 3) on each additional story of the dwelling, including basements but not including crawl spaces and unhabitable attics. When more than one smoke alarm is required to be installed within an individual dwelling unit, the alarm devices shall be interconnected in such a manner that the actuation of one alarm will activate all of the alarms in the individual unit.* See exception for dwelling units undergoing alterations or repairs. *Physical interconnection of smoke alarms shall not be required where listed wireless alarms are installed and all alarms sound upon activation of one alarm.* See exception for existing alarms.

Discussion and Commentary: A majority of deaths from fire in residential structures occur because of the delay in detecting and responding to a fire. Therefore, smoke alarms are mandated throughout the typically occupied areas of a dwelling unit to not only detect the products of combustion but also to alert the occupants of the fire.

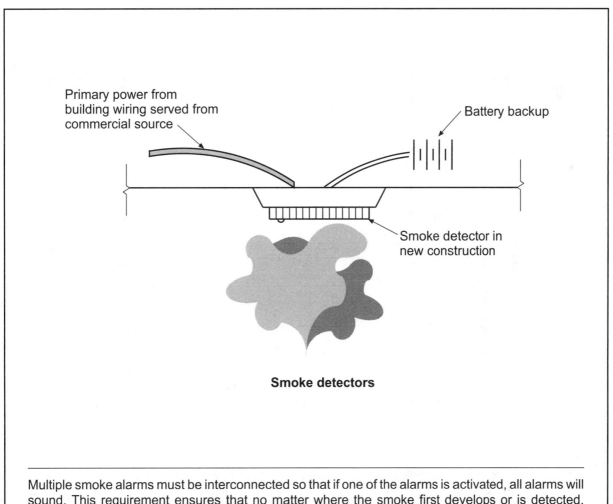

Smoke detectors

Multiple smoke alarms must be interconnected so that if one of the alarms is activated, all alarms will sound. This requirement ensures that no matter where the smoke first develops or is detected, occupants throughout the dwelling unit are made aware of the situation.

Code Text: *For new construction, an approved carbon monoxide alarm shall be installed outside of each separate sleeping area in the immediate vicinity of the bedrooms in dwelling units within which fuel-fired appliances are installed and in dwelling units that have attached garages. Single station carbon monoxide alarms shall be listed as complying with UL 2034 and shall be installed in accordance with* the IRC *and the manufacturer's installation instructions.*

Discussion and Commentary: Carbon monoxide accumulates in the body over time relative to its concentration in the air. Accordingly, carbon monoxide detectors sound an alarm based on the concentration of carbon monoxide and the amount of time that certain levels are detected, simulating an accumulation of the toxic gas in the body. High levels of carbon monoxide will trigger an alarm within a short period of time, while lower levels must be present over a longer period of time for the alarm to sound.

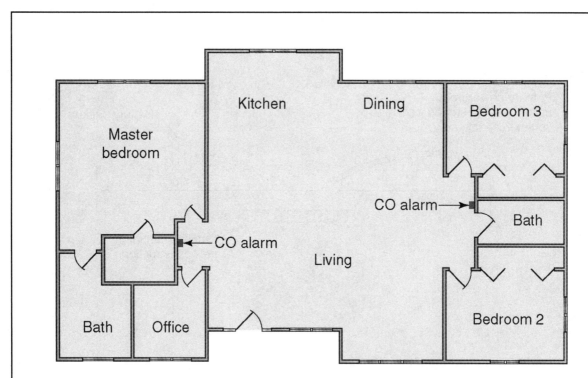

Carbon monoxide (CO) alarm installed in the immediate vicinity of each sleeping area

Because the source of unsafe levels of carbon monoxide in the home is typically from faulty operation of a fuel-fired furnace or water heater, or from the exhaust of an automobile, the mandate for carbon monoxide alarms applies only to homes containing fuel-fired appliances or those having an attached garage.

Topic: Thermal Barrier
Reference: IRC R316.4

Category: Building Planning
Subject: Foam Plastic

Code Text: *Unless otherwise allowed in Section R316.5 or Section R316.6, foam plastic shall be separated from the interior of a building by an approved thermal barrier of minimum $^1/_2$-inch (12.7 mm) gypsum wallboard or a material that is tested in accordance with and meets the acceptance criteria of both the Temperature Transmission Fire Test and the Integrity Fire Test of NFPA 275.*

Discussion and Commentary: Foam plastic must not be exposed to the building's interior during the early stages of a fire; rather, it must be isolated from the fire for a short period of time. The adequacy of the barrier is based on its ability to remain in place for 15 minutes. The NFPA 275 fire test methods identify specific sample construction, fire exposures and acceptance criteria to qualify a material or product for use as a thermal barrier.

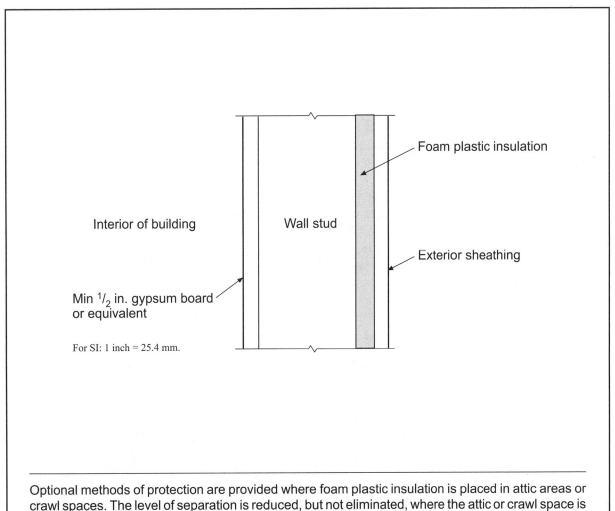

Interior of building

Wall stud

Foam plastic insulation

Exterior sheathing

Min $^1/_2$ in. gypsum board or equivalent

For SI: 1 inch = 25.4 mm.

Optional methods of protection are provided where foam plastic insulation is placed in attic areas or crawl spaces. The level of separation is reduced, but not eliminated, where the attic or crawl space is accessed only for the service of utilities.

Topic: Location Required

Reference: IRC R317.1

Category: Building Planning

Subject: Protection Against Decay

Code Text: *Protection of wood and wood-based products from decay shall be provided in the following locations by the use of naturally durable wood or wood that is preservative treated in accordance with AWPA U1 for the species, product, preservative and end use. Preservatives shall be listed in Section 4 of AWPA U1. See seven specific locations where such lumber is required.*

Discussion and Commentary: For those portions of a wood-framed structure that are subject to damage by decay, lumber must be pressure preservatively treated or of a species of wood having a natural resistance to decay. Such naturally durable wood includes the heartwood of decay-resistant redwood, cedars, black locust and black walnut. Crawl spaces and unexcavated areas under a building usually contain moisture-laden air. Foundation walls and floor slabs-on-grade absorb moisture from the ground and, by capillary action, move it to framing members to which they are in contact. Other locations also have similar conditions where damage to the wood is quite possible unless adequate clearance is maintained or the appropriate type of material is used.

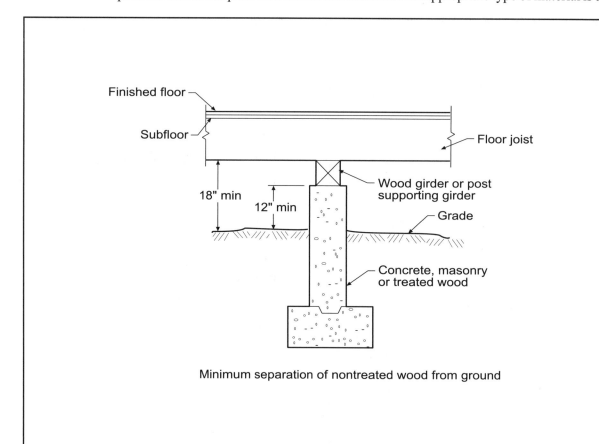

Minimum separation of nontreated wood from ground

Wood members designed to be in contact with the ground must be suitable for ground contact use. This provision applies to all wood members that support permanent structures designed for human occupancy. Untreated wood is permitted where it is used entirely below groundwater level or continuously submerged in fresh water.

Topic: Subterranean Control Methods

Reference: IRC R318.1

Category: Building Planning

Subject: Protection Against Termites

Code Text: *In areas subject to damage from termites as indicated by Table R301.2(1), methods of protection against termites shall be one of the following methods or a combination of these methods: 1) chemical termiticide treatment, 2) termite baiting system, 3) pressure-preservative-treated wood, 4) naturally durable termite-resistant wood, 5) physical barriers and 6) cold-formed steel framing.*

Discussion and Commentary: Figure R301.2(6) depicts the geographical areas where termite damage to structures is probable. In those areas, the structure must be protected from termite damage in an appropriate manner. Although there are a number of approved methods, the most common is soil poisoning. Alternatives include the use of pressure preservatively treated wood and termite shields over perimeter walls. Often, a combination of these methods is necessary to establish the desired level of protection.

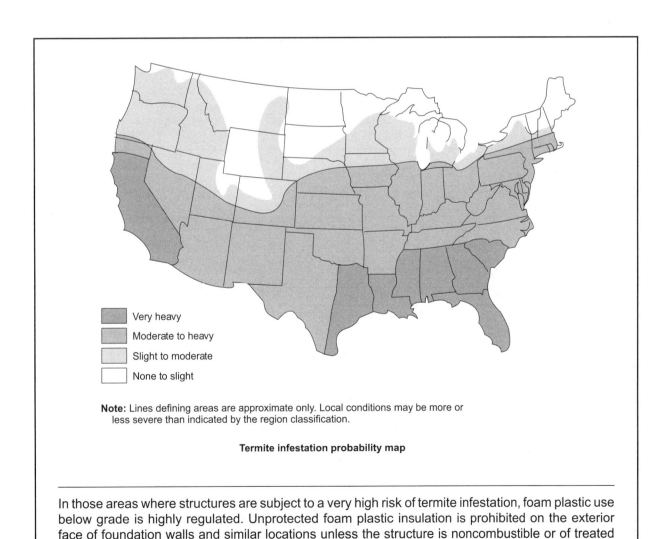

Very heavy
Moderate to heavy
Slight to moderate
None to slight

Note: Lines defining areas are approximate only. Local conditions may be more or less severe than indicated by the region classification.

Termite infestation probability map

In those areas where structures are subject to a very high risk of termite infestation, foam plastic use below grade is highly regulated. Unprotected foam plastic insulation is prohibited on the exterior face of foundation walls and similar locations unless the structure is noncombustible or of treated wood.

Code Text: *Buildings shall have approved address numbers, building numbers or approved building identification placed in a position that is plainly legible and visible from the street or road fronting the property. Numbers shall be a minimum of 4 inches (102 mm) high with a minimum stroke width of $^1/_2$ inch (12.7 mm).*

Discussion and Commentary: Buildings should have plainly visible address numbers posted directly on the building or in such a place on the property that the building may be identified by emergency services such as fire, medical or police personnel. The primary concern is that emergency forces should be able to locate the building without going through a lengthy search procedure.

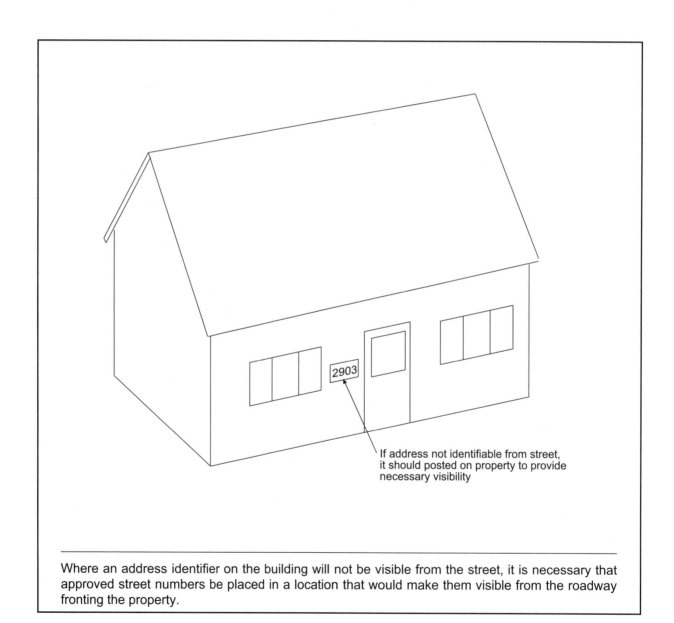

2903

If address not identifiable from street, it should posted on property to provide necessary visibility

Where an address identifier on the building will not be visible from the street, it is necessary that approved street numbers be placed in a location that would make them visible from the roadway fronting the property.

Topic: Scope

Category: Building Planning

Reference: IRC R320, IBC 1107.6.3

Subject: Accessibility

Code Text: *Where there are four or more dwelling units or sleeping units in a single structure, the provisions of Chapter 11 of the* International Building Code *for Group R-3 shall apply. In Group R-3 occupancies where there are four or more dwelling units or sleeping units intended to be occupied as a residence in a single structure, every dwelling unit and sleeping unit intended to be occupied as a residence shall be a Type B unit.* See exception for reduction in number of Type B units.

Discussion and Commentary: Although most buildings constructed under the provisions of the *International Residential Code* are exempt from the accessibility provisions, structures containing four or more townhouses will be regulated. The required Type B units are described in ICC A117.1 and are consistent with the design and construction requirements of the federal Fair Housing Act.

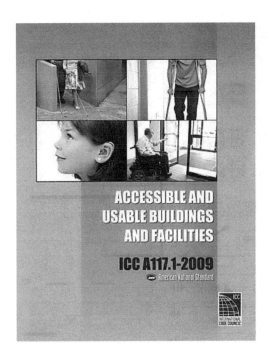

The technical requirements for Type B dwelling units are found in Section 1004 of ICC A117.1. The issues addressed by the accessibility standard include the primary entrance, accessible route, walking surfaces, doors, ramps, elevators and lifts, operable parts, bathrooms and kitchens.

Topic: General Provisions	Category: Building Planning
Reference: IRC R323.1	Subject: Storm Shelters

Code Text: Section R323 *applies to the construction of storm shelters when constructed as separated detached buildings or when constructed as safe rooms within buildings for the purpose of provided safe refuge from storms that produce high winds, such as tornados and hurricanes. In addition to other applicable requirements in* the IRC, *storm shelters shall be constructed in accordance with ICC/NSSA-500.*

Discussion and Commentary: The IRC does not mandate that storm shelters be provided in areas where high winds are probable, but when provided such shelters must be constructed in accordance with ICC/NSSA-500 *Standard on the Design and Construction of Storm Shelters*. Storm shelters are intended to protect occupants from serious injury from flying debris in the event of a tornado, hurricane or other high-wind event while maintaining a minimum interior environment.

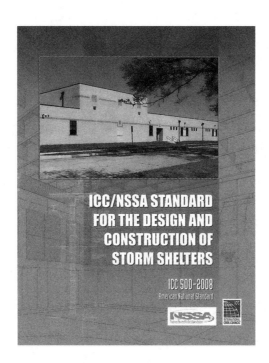

Shelters conforming to the ICC-500 standard are designed to withstand impact from windborne projectiles such as construction materials and natural material debris that are common to high-wind events. Shelters may be prefabricated or site-built, below or above ground level, detached from the dwelling or located with the residence.

Quiz

Study Session 4
IRC Sections R311 – R323

1. The minimum width of a hallway shall be _____ .

   a. 32 inches   b. 36 inches

   c. 42 inches   d. 44 inches

   Reference _____

2. Where the stairway has a straight run, a stairway landing shall have a minimum depth of _____ measured in the direction of travel.

   a. 30 inches   b. 32 inches

   c. 36 inches   d. 42 inches

   Reference _____

3. Unless technically infeasible due to site constraints, the maximum slope of a ramp shall be one unit vertical in _____ units horizontal.

   a. five   b. eight

   c. ten   d. twelve

   Reference _____

4. Handrails are not required on ramps having a maximum slope of _____ .

   a. 1:5   b. 1:8

   c. 1:12   d. 1:15

   Reference _____

5. Where a door opens onto a ramp, a minimum _____ landing shall be provided.

    a. 32-inch-by-32-inch   b. 36-inch-by-36-inch

    c. 36-inch-by-60-inch   d. 60-inch-by-60-inch

Reference _____

6. Stairways shall be a minimum _____ wide at all points above the permitted handrail height and below the required headroom height.

    a. 30 inches   b. 34 inches

    c. 36 inches   d. 38 inches

Reference _____

7. Where a handrail is provided on one side of a stairway, what is the minimum required clear width at and below the handrail height?

    a. 27 inches   b. 29 inches

    c. $31^1/_2$ inches   d. 36 inches

Reference _____

8. Stairways shall have a maximum riser height of _____ and a minimum tread run of _____ .

    a. $8^1/_4$ inches, 9 inches   b. 8 inches, 9 inches

    c. $7^3/_4$ inches, 10 inches   d. $7^1/_2$ inches, 10 inches

Reference _____

9. The maximum variation between the greatest riser height and the smallest riser height within any flight of stairs shall be _____ .

    a. $^1/_4$ inch   b. $^3/_8$ inch

    c. $^1/_2$ inch   d. $^5/_8$ inch

Reference _____

10. Unless a minimum tread depth of _____ is provided, a minimum $^3/_4$-inch nosing is required at the leading edge of all treads.

    a. $9^1/_2$ inches

    b. 10 inches

    c. 11 inches

    d. 12 inches

Reference _____

11. Unless the stair has a total rise of no more than 30 inches, the opening between treads at open risers shall be such that a minimum _____ sphere shall not pass through.

    a. 3-inch

    b. 4-inch

    c. 6-inch

    d. no limitation is mandated

Reference _____

12. In the identification of flood hazard areas, areas that have been determined to be subject to maximum wave heights of _____ are not required to be designated coastal high-hazard areas feet unless subject to high-velocity wave action or wave-induced erosion.

    a. 3

    b. 4

    c. 5

    d. 6

Reference _____

13. What is the maximum riser height permitted for spiral stairways?

    a. $7^3/_4$ inches

    b. 8 inches

    c. $8^1/_4$ inches

    d. $9^1/_2$ inches

Reference _____

14. Stairway handrail height shall be, measured vertically from the nosing of the treads, a minimum of _____ and a maximum of _____.

    a. 30 inches, 34 inches

    b. 30 inches, 38 inches

    c. 34 inches, 38 inches

    d. 34 inches, 42 inches

Reference _____

15. Where a stairway handrail is located adjacent to a wall, a minimum clearance of _____ shall be provided between the wall and the handrail.

    a. $1^1/_4$ inches   b. $1^1/_2$ inches

    c. 2 inches   d. $3^1/_2$ inches

Reference _____

16. A Type I handrail having a circular cross section shall have a minimum outside diameter of _____ and a maximum outside diameter of _____ .

    a. $1^1/_4$ inches, 2 inches   b. $1^1/_4$ inches, $2^5/_8$ inches

    c. $1^1/_2$ inches, 2 inches   d. $1^1/_2$ inches, $2^5/_8$ inches

Reference _____

17. A porch, balcony or similar raised floor surface more than 30 inches above the floor or grade below shall be provided with a guard having a minimum height of _____ .

    a. 32 inches   b. 34 inches

    c. 36 inches   d. 42 inches

Reference _____

18. Required guards on open sides of raised floor areas shall be provided with intermediate rails or ornamental closures such that a minimum _____ sphere cannot pass through.

    a. 3-inch   b. 4-inch

    c. 6-inch   d. 9-inch

Reference _____

19. Which one of the following areas in a dwelling unit does not specifically require the installation of a smoke alarm?

    a. basement   b. kitchen

    c. habitable attic   d. sleeping room

Reference _____

20. In general, foam plastic shall be separated from the interior of a dwelling by an approved thermal barrier of minimum _____ .

    a. $^3/_8$-inch gypsum board   b. $^1/_2$-inch gypsum board

    c. $^5/_8$-inch gypsum board   d. $^5/_8$-inch Type X gypsum board

Reference _____

21. When carbon monoxide alarms are required in new construction, they shall be installed _____ .

    a. within every bedroom

    b. on all floor levels of multistory units

    c. outside of each separate sleeping area

    d. in the garage

Reference _____

22. An automatic residential fire sprinkler system shall be provided in all buildings containing _____ or more townhouses.

    a. 3 (all townhouses require a sprinkler system)

    b. 8

    c. 16

    d. sprinkler systems are not required in townhouses

Reference _____

23. In new construction, carbon monoxide alarms are required in dwelling units _____ .

    a. where fuel-fired appliances are installed

    b. that have attached garages

    c. with habitable space below grade

    d. either a or b

Reference _____

24. Unless they are made of pressure preservatively treated or naturally durable wood, wood sill plates that rest on masonry or concrete exterior foundation walls shall be located a minimum of _____ from exposed ground.

    a. 6 inches   b. 8 inches

    c. 12 inches   d. 18 inches

    Reference _____

25. In the establishment of a design flood elevation, the depth of peak elevation of flooding for a _____ flood is used.

    a. 10-year   b. 20-year

    c. 50-year   d. 100-year

    Reference _____

26. Where stair risers are not vertical, they shall be sloped under the tread above at a maximum angle of _____ from the vertical.

    a. 10°   b. 15°

    c. 22°   d. 30°

    Reference _____

27. Where foam plastic is spray applied to a sill plate without a thermal barrier, it shall have a maximum density of _____ pcf.

    a. 1.0   b. 1.5

    c. 2.0   d. 3.0

    Reference _____

28. The maximum width above the recess of a Type II handrail shall be _____ inches.

    a. $1^1/_2$ b. 2

    c. $2^5/_8$ d. $2^3/_4$

    Reference _____

29. Handrails shall be provided on at least one side of a stairway having a minimum of _____ risers.

    a. 2   b. 3

    c. 4   d. 5

Reference _____

30. What is the minimum required clear opening for an egress door from a dwelling unit?

    a. 32 inches by 78 inches   b. 32 inches by 80 inches

    c. 36 inches by 78 inches   d. 36 inches by 80 inches

Reference _____

31. Where the door does not swing over the landing, the exterior landing at a required egress door shall be located a maximum of _____ inches below the top of the threshold.

    a. 7   b. $7^1/_2$

    c. $7^3/_4$ d. 8

Reference _____

32. Where nosings are required on a stairway with solid risers, the nosings shall extend a minimum of _____ inch(es) and a maximum of _____ inch(es) beyond the risers.

    a. $^1/_2$, 1   b. $^3/_4$, $1^1/_4$

    c. 1, $1^1/_2$ d. $1^1/_4$, 2

Reference _____

33. Address numbers used for building identification shall have a minimum height of _____ inches.

    a. 4   b. 6

    c. 8   d. 9

Reference _____

34. Openings for required guards on the sides of stair treads shall be provided with intermediate rails or ornamental closures such that a minimum _____ sphere cannot pass through.

    a. $3\frac{1}{2}$

    b. 4

    c. $4\frac{3}{8}$

    d. 6

Reference _____

35. Where a safe room is constructed as a storm shelter in order to provide safe refuge from high winds, the room shall be constructed in accordance with _____ .

    a. ICC/NSSA-500

    b. CPSC 16 CFR, Part 1201

    c. DOC PS 1-07

    d. FEMA TB-2—93

Reference _____

5

2012 IRC Chapter 4
Foundations

OBJECTIVE: To gain an understanding of the requirements relating to foundation systems, including soil evaluation; footing size, depth and slope; foundation design and reinforcing; frost-protected shallow foundations; wood foundations; insulating form foundation walls; foundation drainage; waterproofing and dampproofing; and under-floor spaces.

REFERENCE: Chapter 4, 2012 *International Residential Code*

KEY POINTS:
- How must lots be graded? What is the minimum required slope of the grade away from the foundation?
- When are soils tests required? When are presumptive load-bearing values to be used?
- What criteria determines the minimum size of footings? What is the minimum required footing thickness? Minimum projection distance? Minimum depth?
- What additional conditions must be met for footings located in Seismic Design Categories D_0, D_1 and D_2? Where must any required reinforcing be located?
- What is the maximum slope permitted for the top surface of footings? The bottom surface?
- In what manner must the sill plate be connected to the foundation wall? What is the minimum size and spacing for anchor bolts? How must the connection be made in Seismic Design Categories D_0, D_1 and D_2?
- How shall buildings be located in respect to an adjacent ascending slope? Descending slope?
- How are foundations and floor slabs for buildings located on expansive soils to be regulated?
- Where a building is constructed on or adjacent to a slope, what is the minimum required height of a foundation above a street gutter or other point of drainage discharge?
- What are the provisions for the installation of wood foundation systems?
- How does a frost-protected shallow foundation work? What are the primary conditions for compliance?
- What criteria are necessary to address plain concrete and masonry foundation walls? Reinforced concrete and masonry foundation walls?

- What special conditions are required for foundation walls of buildings constructed in Seismic Design Categories D_0, D_1 and D_2? Where must reinforcement be located?
- When are pier foundations acceptable? What are the minimum requirements?
- What is an insulating concrete form (ICF) foundation wall system? What are the different types of ICF systems? How are they regulated?
- How must retaining walls be designed?
- When are drains required around foundation walls? What drainage methods are acceptable?
- How are soils classified? What are the characteristics of the various soil groups?
- When must a foundation wall be dampproofed? Waterproofed? What methods are used for dampproofing and waterproofing?
- What is the minimum required size for under-floor ventilation openings? Where shall such openings be located? How may the required area of openings be reduced?
- Under what conditions is an unvented crawl space permitted?
- How shall access to an under-floor space be provided?
- When is a drainage system needed for an under-floor space?

Topic: Drainage

Category: Foundations

Reference: IRC R401.3

Subject: General Requirements

Code Text: *Surface drainage shall be diverted to a storm sewer conveyance or other approved point of collection that does not create a hazard. Lots shall be graded so as to drain surface water away from foundation walls. The grade shall fall a minimum of 6 inches (152 mm) within the first 10 feet (3048 mm).* See exception for use of drains or swales.

Discussion and Commentary: Proper site drainage is an important element in preventing wet basements, damp crawl spaces, eroded banks and the possible failure of a foundation system. Drainage patterns should result in adequate slopes to approved drainage devices that are capable of carrying concentrated runoff. It is often necessary to use gutters and downspouts to direct roof water to the appropriate drainage points.

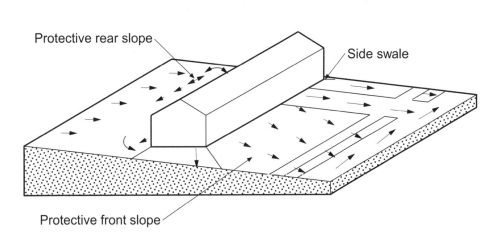

Grading method for lot where slope is from rear to front lot. Drainage swales are located at rear and sides of dwelling.

Drains or swales are effective measures where the site does not allow for the necessary fall away from the structure. Minimum slope gradients for swales are often based on ground frost and moisture conditions. The permeability of soil and type of drainage devices are also important.

Topic: Soil Tests

Reference: IRC R401.4, R401.4.1

Category: Foundations

Subject: General Requirements

Code Text: *Where quantifiable data created by accepted soil science methodologies indicate expansive, compressible, shifting or other questionable soil characteristics are likely to be present, the building official shall determine whether to require a soil test to determine the soil's characteristics at a particular location. This test shall be made by an approved agency using an approved method. In lieu of a complete geotechnical evaluation, the load-bearing values in Table R401.4.1 shall be assumed.*

Discussion and Commentary: Expansive soils, soil instability and increased lateral pressure due to a high water table or to surcharge loads from adjacent structures are special conditions that must be considered in the design of a foundation system. Where such conditions occur, they are beyond the scope of the prescriptive provisions of the code for foundation systems. In some cases, a soil test may be required to evaluate the soil's characteristics.

TABLE R401.4.1
PRESUMPTIVE LOAD–BEARING VALUES OF
FOUNDATION MATERIALS[a]

CLASS OF MATERIAL	LOAD-BEARING PRESSURE (pounds per square foot)
Crystalline bedrock	12,000
Sedimentary and foliated rock	4,000
Sandy gravel and/or gravel (GW and GP)	3,000
Sand, silty sand, clayey sand, silty gravel and clayey gravel (SW, SP, SM, SC, GM and GC)	2,000
Clay, sandy clay, silty clay, clayey silt, silt and sandy silt (CL, ML, MH and CH)	1,500[b]

For SI: 1 pound per square foot = 0.0479 kPa.

a. When soil tests are required by Section R401.4, the allowable bearing capacities of the soil shall be part of the recommendations.

b. Where the building official determines that in-place soils with an allowable bearing capacity of less than 1,500 psf are likely to be present at the site, the allowable bearing capacity shall be determined by a soils investigation.

Where the bearing capacity of the soil has not been determined by geotechnical analysis such as borings, field load tests, laboratory tests and/or engineering analysis, it is a common practice to use presumptive bearing values for the design of the foundation system.

Topic: Wood and Concrete

Category: Foundations

Reference: IRC R402.1.2, R402.2

Subject: Materials

Code Text: *All lumber and plywood* used in wood foundation systems *shall be pressure-preservative treated and dried after treatment in accordance with AWPA U1, and shall bear the label of an accredited agency. Concrete shall have a minimum specified compressive strength of f'$_c$, as shown in Table R402.2. Concrete subject to moderate or severe weathering as indicated in Table R301.2(1) shall be air entrained as specified in Table R402.2.*

Discussion and Commentary: Freezing and thawing cycles can be the most destructive weathering factors for concrete. For nonair entrained concrete, deterioration is caused by the water freezing in the cement matrix, in the aggregate, or both. Studies have documented that concrete provided with proper air entrainment is highly resistant to this deterioration.

TABLE R402.2
MINIMUM SPECIFIED COMPRESSIVE STRENGTH OF CONCRETE

TYPE OR LOCATION OF CONCRETE CONSTRUCTION	MINIMUM SPECIFIED COMPRESSIVE STRENGTH[a] (f'$_c$)		
	Weathering Potential[b]		
	Negligible	Moderate	Severe
Basement walls, foundations and other concrete not exposed to the weather	2,500	2,500	2,500[c]
Basement slabs and interior slabs on grade, except garage floor slabs	2,500	2,500	2,500[c]
Basement walls, foundation walls, exterior walls and other vertical concrete work exposed to the weather	2,500	3,000[d]	3,000[d]
Porches, carport slabs and steps exposed to the weather, and garage floor slabs	2,500	3,000[d,e,f]	3,500[d,e,f]

For SI: 1 pound per square inch = 6.895 kPa.

a. Strength at 28 days psi.

b. See Table R301.2(1) for weathering potential.

c. Concrete in these locations that may be subject to freezing and thawing during construction shall be air-entrained concrete in accordance with Footnote d.

d. Concrete shall be air-entrained. Total air content (percent by volume of concrete) shall be not less than 5 percent or more than 7 percent.

e. See Section R402.2 for maximum cementitious materials content.

f. For garage floors with a steel troweled finish, reduction of the total air content (percent by volume of concrete) to not less than 3 percent is permitted if the specified compressive strength of the concrete is increased to not less than 4,000 psi.

Severe

Moderate

Negligible

a. Alaska and Hawaii are classified as severe and negligible, respectively.

b. Lines defining areas are approximate only. Local conditions may be more or less severe than indicated by region classification. A severe classification is where weather conditions result in significant snowfall combined with extended periods during which there is little or no natural thawing causing deicing salts to be used extensively.

Weathering probability map for concrete

Performance of wood foundation systems is dependent on the use of properly treated materials. Verification of the proper materials is provided by requiring identification showing the approval of an accredited inspection agency.

Code Text: *Minimum sizes for concrete and masonry footings shall be as set forth in Table R403.1 and Figure R403.3.1(1). The footing width, W, shall be based on the load-bearing value of the soil in accordance with Table R401.4.1. Spread footings shall be at least 6 inches (152 mm) in thickness, T. Footing projections, P, shall be at least 2 inches (51 mm) and shall not exceed the thickness of the footing.*

Discussion and Commentary: To avoid overstressing the footing, minimum size requirements are established based on construction type, number of stories and soil load-bearing value. The projection limitation is critical for 6-inch-thick footings supported by poor soil conditions. Excessive projections could result in the footing being cracked in the same plane as the foundation wall, which could occur if the allowable stress in the concrete is exceeded.

TABLE R403.1
MINIMUM WIDTH OF CONCRETE,
PRECAST OR MASONRY FOOTINGS
(inches)[a]

	LOAD-BEARING VALUE OF SOIL (psf)			
	1,500	2,000	3,000	≥ 4,000
Conventional light-frame construction				
1-story	12	12	12	12
2-story	15	12	12	12
3-story	23	17	12	12
4-inch brick veneer over light frame or 8-inch hollow concrete masonry				
1-story	12	12	12	12
2-story	21	16	12	12
3-story	32	24	16	12
8-inch solid or fully grouted masonry				
1-story	16	12	12	12
2-story	29	21	14	12
3-story	42	32	21	16

For SI: 1 inch = 25.4 mm, 1 pound per square foot = 0.0479 kPa.
a. Where minimum footing width is 12 inches, use of a single wythe of solid or fully grouted 12-inch nominal concrete masonry units is permitted.

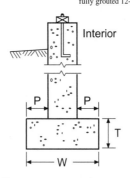

Basement or crawl space
with concrete wall and
spread footing

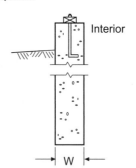

Basement or crawl space
with foundation wall
bearing directly on soil

In buildings constructed in Seismic Design Categories D_0, D_1 or D_2, footing reinforcement is mandated. Interconnection of the stem wall and its supporting footing is necessary to resist the tendency to slip during an earthquake. Various methods address different foundation systems.

Topic: Minimum Depth   **Category:** Foundations
Reference: IRC R403.1.4, R403.1.4.1   **Subject:** Footings

Code Text: *All exterior footings shall be placed at least 12 inches (305 mm) below the undisturbed ground surface. Except where otherwise protected from frost, foundation walls, piers and other permanent supports of buildings and structures shall be protected from frost by one or more of the following methods: 1) extending below the frost line specified in Table R301.2(1); 2) constructing in accordance with Section R403.3 (frost protected shallow foundations); 3) constructing in accordance with ASCE 32* (Design and Construction of Frost Protected Shallow Foundations)*; or 4) erected on solid rock.* See exceptions for small freestanding accessory structures and decks not supported by a dwelling.

Discussion and Commentary: The volume changes (frost heave) that take place during freezing and thawing cycles produce excessive stresses in the foundations and create extensive damage to the supported walls. Thus, foundations must be extended below the depth of frost penetration.

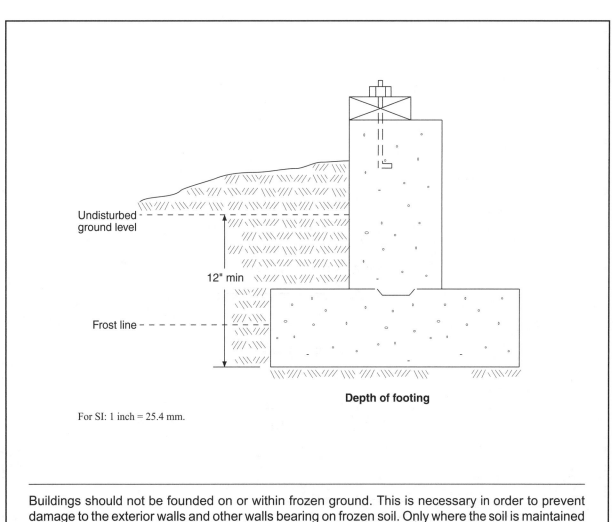

Depth of footing

For SI: 1 inch = 25.4 mm.

Buildings should not be founded on or within frozen ground. This is necessary in order to prevent damage to the exterior walls and other walls bearing on frozen soil. Only where the soil is maintained in a permanently frozen condition can such construction be permitted.

Code Text: *The top surface of footings shall be level. The bottom surface of footings shall not have a slope exceeding one unit vertical in 10 units horizontal (10-percent slope). Footings shall be stepped where it is necessary to change the elevation of the top surface of the footings or where the slope of the bottom surface of the footings will exceed one unit vertical in ten units horizontal (10-percent slope).*

Discussion and Commentary: Although the code places no restriction on a stepped foundation, there is a recommended overlap of the top of the foundation wall beyond the step in the foundation. It should be larger than the vertical step in the foundation wall at that point. This recommendation is based on possible crack propagation at an angle of 45 degrees.

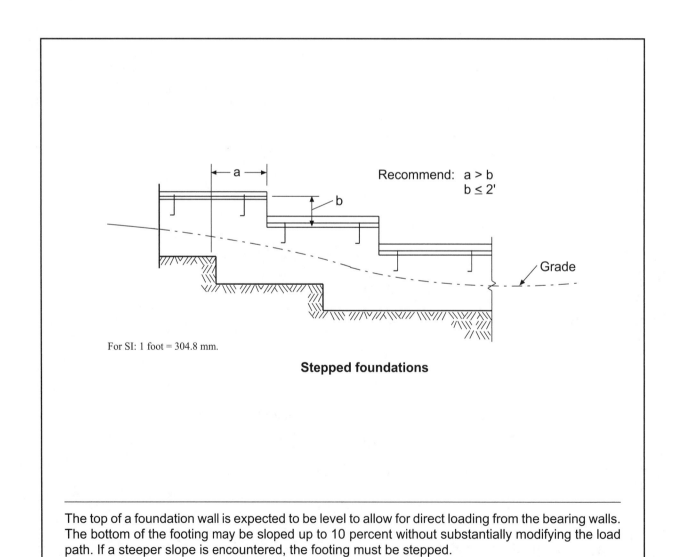

Recommend: a > b
b ≤ 2'

Grade

For SI: 1 foot = 304.8 mm.

Stepped foundations

The top of a foundation wall is expected to be level to allow for direct loading from the bearing walls. The bottom of the footing may be sloped up to 10 percent without substantially modifying the load path. If a steeper slope is encountered, the footing must be stepped.

Topic: Foundation Anchorage

Category: Foundations

Reference: IRC R403.1.6

Subject: Footings

Code Text: *Sill plates and walls supported directly on continuous foundations shall be anchored to the foundation in accordance with Section R403.1.6. Wood sole plates at exterior walls on monolithic slabs, wood sole plates of braced wall panels at building interiors on monolithic slabs and all wood sill plates shall be anchored to the foundation with anchor bolts spaced a maximum of 6 feet (1829 mm) on center. Bolts shall be at least $^1/_2$ inch (13 mm) in diameter and shall extend a minimum of 7 inches (178 mm) into concrete or grouted cells of concrete masonry units. There shall be a minimum of two bolts per plate section with one bolt located not more than 12 inches (305 mm) or less than seven bolt diameters from each end of the plate section. See exceptions for foundation anchor straps and short walls connecting offset braced wall panels.*

Discussion and Commentary: To prevent walls and floors from shifting under lateral loads, anchorage to the supporting foundation is needed. The minimum required connection is supplied by anchor bolts. Foundation straps are also acceptable if installed in accordance with the manufacturer's instructions and spaced to provide equivalent hold-down strength.

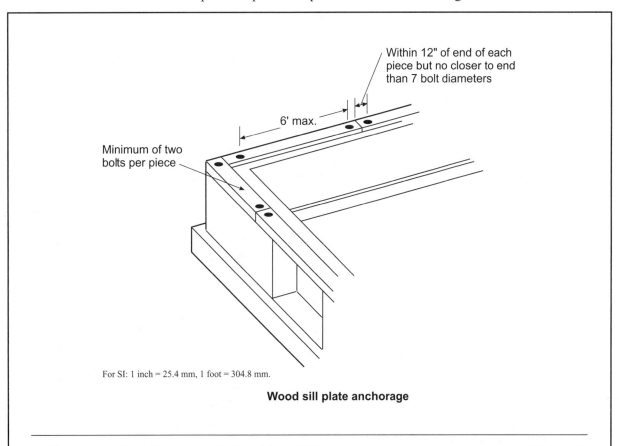

Within 12" of end of each piece but no closer to end than 7 bolt diameters

6' max.

Minimum of two bolts per piece

For SI: 1 inch = 25.4 mm, 1 foot = 304.8 mm.

Wood sill plate anchorage

In higher seismic risk areas, additional foundation anchorage requirements are established in Section R403.1.6.1. For example, large square plate washers are mandated to compensate for the practice of oversized bolt holes. These washers enable the nuts to be tightened enough to achieve an increased clamping action between the foundation wall and the sill plate. See Section R602.11.1 for anchorage details.

| **Topic:** Foundation Elevation | **Category:** Foundation |
| **Reference:** IRC R403.1.7 | **Subject:** Footings |

Code Text: *The placement of buildings and structures on or adjacent to slopes steeper than 1 unit vertical in 3 units horizontal (33.3-percent slope) shall conform to Sections R403.1.7.1 through R403.1.7.4. On graded sites, the top of any exterior foundation shall extend above the elevation of the street gutter at point of discharge or the inlet of an approved drainage device a minimum of 12 inches (305 mm) plus 2 percent. Alternate elevations are permitted subject to the approval of the building official, provided it can be demonstrated that required drainage to the point of discharge and away from the structure is provided at all locations on the site.*

Discussion and Commentary: Where natural drainage away from a building is not available, the site must be graded so that water will not drain toward, or accumulate at, the exterior foundation wall. A prescriptive elevation is set forth that will ensure positive drainage to a street gutter or other drainage point; however, any other method that moves water away from the building can be accepted by the building official.

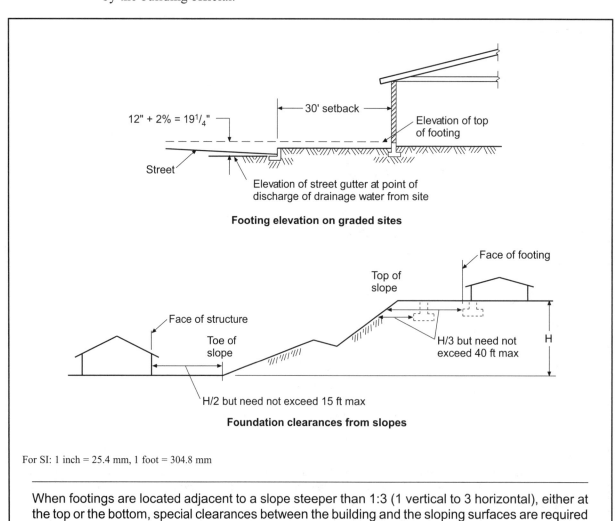

Footing elevation on graded sites

Foundation clearances from slopes

For SI: 1 inch = 25.4 mm, 1 foot = 304.8 mm

When footings are located adjacent to a slope steeper than 1:3 (1 vertical to 3 horizontal), either at the top or the bottom, special clearances between the building and the sloping surfaces are required to protect against slope drainage, erosion and shallow failures.

Code Text: *For buildings where the monthly mean temperature of the building is maintained at a minimum of 64°F (18°C), footings are not required to extend below the frost line when protected from frost by insulation in accordance with Figure R403.3(1) and Table R403.3(1).* Not permitted for unheated areas such as porches, garages and crawl spaces.

Discussion and Commentary: As a fundamental rule, footings must be placed below the frost line. However, the use of a frost protected foundation is an acceptable alternative, allowing placement of the footing/foundation above the frost line. This foundation system uses insulation to reduce the heat loss at the slab edge. By holding heat from the dwelling in the ground under the foundation, the insulation eliminates the potential for freezing, and thus the consequences of freeze/thaw conditions are avoided.

TABLE R403.3(1)
MINIMUM FOOTING DEPTH AND INSULATION REQUIREMENTS FOR FROST-PROTECTED FOOTINGS IN HEATED BUILDINGS[a]

AIR FREEZING INDEX (°F-days)[b]	MINIMUM FOOTING DEPTH, D (inches)	VERTICAL INSULATION R-VALUE[c, d]	HORIZONTAL INSULATION R-VALUE[c, e]		HORIZONTAL INSULATION DIMENSIONS PER FIGURE R403.3(1) (inches)		
			Along walls	At corners	A	B	C
1,500 or less	12	4.5	Not required	Not required	Not required	Not required	Not required
2,000	14	5.6	Not required	Not required	Not required	Not required	Not required
2,500	16	6.7	1.7	4.9	12	24	40
3,000	16	7.8	6.5	8.6	12	24	40
3,500	16	9.0	8.0	11.2	24	30	60
4,000	16	10.1	10.5	13.1	24	36	60

a. Insulation requirements are for protection against frost damage in heated buildings. Greater values may be required to meet energy conservation standards.
b. See Figure R403.3(2) or Table R403.3(2) for Air Freezing Index values.
c. Insulation materials shall provide the stated minimum R-values under long-term exposure to moist, below-ground conditions in freezing climates. The following R-values shall be used to determine insulation thicknesses required for this application: Type II expanded polystyrene—2.4R per inch; Type IV extruded polystyrene—4.5R per inch; Type VI extruded polystyrene—4.5R per inch; Type IX expanded polystyrene—3.2R per inch; Type X extruded polystyrene—4.5R per inch.
d. Vertical insulation shall be expanded polystyrene insulation or extruded polystyrene insulation.
e. Horizontal insulation shall be extruded polystyrene insulation.

Insulation detail

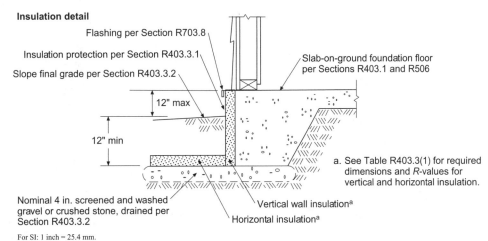

Flashing per Section R703.8

Insulation protection per Section R403.3.1

Slope final grade per Section R403.3.2

Slab-on-ground foundation floor per Sections R403.1 and R506

12" max

12" min

Nominal 4 in. screened and washed gravel or crushed stone, drained per Section R403.3.2

a. See Table R403.3(1) for required dimensions and R-values for vertical and horizontal insulation.

Vertical wall insulation[a]

Horizontal insulation[a]

For SI: 1 inch = 25.4 mm.

It is often necessary to extend the insulation away from the foundation horizontally. By doing so, it is probable that the insulating material will be damaged by landscape work or other activities unless an approved method of protection is provided.

Topic: Masonry Walls

Category: Foundations

Reference: IRC R404.1.1.1

Subject: Foundation Walls

Code Text: *Concrete masonry and clay masonry foundation walls shall be constructed as set forth in Table R404.1.1(1) (Plain Masonry), R404.1.1(2) (8-inch Reinforced Masonry), R404.1.1(3) (10-inch Reinforced Masonry) or R404.1.1(4) (12-inch Reinforced Masonry) and shall also comply with the applicable provisions of Sections R606, R607 and R608. In buildings assigned to Seismic Design Categories D_0, D_1 or D_2, concrete masonry and clay masonry foundation walls shall also comply with Section R404.1.4.1.*

Discussion and Commentary: Foundation walls are typically designed and constructed to carry the vertical loads from the structure above, resist wind and any lateral forces transmitted to the foundation, and sustain earth pressures exerted against the wall, including any forces that may be imposed by frost action.

TABLE R404.1.1(2)
8-INCH MASONRY FOUNDATION WALLS WITH REINFORCING
WHERE d > 5 INCHES[a]

WALL HEIGHT	HEIGHT OF UNBALANCED BACKFILL[e]	MINIMUM VERTICAL REINFORCEMENT[b,c]		
		Soil classes and lateral soil load[d] (psf per foot below grade)		
		GW, GP, SW and SP soils 30	GM, GC, SM, SM-SC and ML soils 45	SC, ML-CL and inorganic CL soils 60
6 feet 8 inches	4 feet (or less)	#4 at 48″ o.c.	#4 at 48″ o.c.	#4 at 48″ o.c.
	5 feet	#4 at 48″ o.c.	#4 at 48″ o.c.	#4 at 48″ o.c.
	6 feet 8 inches	#4 at 48″ o.c.	#5 at 48″ o.c.	#6 at 48″ o.c.
7 feet 4 inches	4 feet (or less)	#4 at 48″ o.c.	#4 at 48″ o.c.	#4 at 48″ o.c.
	5 feet	#4 at 48″ o.c.	#4 at 48″ o.c.	#4 at 48″ o.c.
	6 feet	#4 at 48″ o.c.	#5 at 48″ o.c.	#5 at 48″ o.c.
	7 feet 4 inches	#5 at 48″ o.c.	#6 at 48″ o.c.	#6 at 40″ o.c.
8 feet	4 feet (or less)	#4 at 48″ o.c.	#4 at 48″ o.c.	#4 at 48″ o.c.
	5 feet	#4 at 48″ o.c.	#4 at 48″ o.c.	#4 at 48″ o.c.
	6 feet	#4 at 48″ o.c.	#5 at 48″ o.c.	#5 at 48″ o.c.
	7 feet	#5 at 48″ o.c.	#6 at 48″ o.c.	#6 at 40″ o.c.
	8 feet	#5 at 48″ o.c.	#6 at 48″ o.c.	#6 at 32″ o.c.
8 feet 8 inches	4 feet (or less)	#4 at 48″ o.c.	#4 at 48″ o.c.	#4 at 48″ o.c.
	5 feet	#4 at 48″ o.c.	#4 at 48″ o.c.	#5 at 48″ o.c.
	6 feet	#4 at 48″ o.c.	#5 at 48″ o.c.	#6 at 48″ o.c.
	7 feet	#5 at 48″ o.c.	#6 at 48″ o.c.	#6 at 40″ o.c.
	8 feet 8 inches	#6 at 48″ o.c.	#6 at 32″ o.c.	#6 at 24″ o.c.
9 feet 4 inches	4 feet (or less)	#4 at 48″ o.c.	#4 at 48″ o.c.	#4 at 48″ o.c.
	5 feet	#4 at 48″ o.c.	#4 at 48″ o.c.	#5 at 48″ o.c.
	6 feet	#4 at 48″ o.c.	#5 at 48″ o.c.	#6 at 48″ o.c.
	7 feet	#5 at 48″ o.c.	#6 at 48″ o.c.	#6 at 40″ o.c.
	8 feet	#6 at 48″ o.c.	#6 at 40″ o.c.	#6 at 24″ o.c.
	9 feet 4 inches	#6 at 40″ o.c.	#6 at 24″ o.c.	#6 at 16″ o.c.
10 feet	4 feet (or less)	#4 at 48″ o.c.	#4 at 48″ o.c.	#4 at 48″ o.c.
	5 feet	#4 at 48″ o.c.	#4 at 48″ o.c.	#5 at 48″ o.c.
	6 feet	#4 at 48″ o.c.	#5 at 48″ o.c.	#6 at 48″ o.c.
	7 feet	#5 at 48″ o.c.	#6 at 48″ o.c.	#6 at 32″ o.c.
	8 feet	#6 at 48″ o.c.	#6 at 32″ o.c.	#6 at 24″ o.c.
	9 feet	#6 at 40″ o.c.	#6 at 24″ o.c.	#6 at 16″ o.c.
	10 feet	#6 at 32″ o.c.	#6 at 16″ o.c.	#6 at 16″ o.c.

For SI: 1 inch = 25.4 mm, 1 foot = 304.8 mm, 1 pound per square foot per foot = 0.157 kPa/mm.

a. Mortar shall be Type M or S and masonry shall be laid in running bond.
b. Alternative reinforcing bar sizes and spacings having an equivalent cross-sectional area of reinforcement per lineal foot of wall shall be permitted provided the spacing of the reinforcement does not exceed 72 inches.
c. Vertical reinforcement shall be Grade 60 minimum. The distance from the face of the soil side of the wall to the center of vertical reinforcement shall be at least 5 inches.
d. Soil classes are in accordance with the Unified Soil Classification System and design lateral soil loads are for moist conditions without hydrostatic pressure. Refer to Table R405.1.
e. Unbalanced backfill height is the difference in height between the exterior finish ground level and the lower of the top of the concrete footing that supports the foundation wall or the interior finish ground level. Where an interior concrete slab-on-grade is provided and is in contact with the interior surface of the foundation wall, measurement of the unbalanced backfill height from the exterior finish ground level to the top of the interior concrete slab is permitted.

In lieu of an engineered foundation wall system, the IRC provides prescriptive tables for both plain (unreinforced) and reinforced masonry foundation walls. If reinforced walls are under consideration, three different wall widths (8-inch, 10-inch and 12-inch) are addressed. The provisions of ACI 530/ASCE 5/TMS 402 or NCMA TR68-A may be used to design masonry walls as well.

Code Text: *Concrete foundation walls that support light-frame walls shall be designed and constructed in accordance with the provisions of* Section R404.1.2, *ACI 318, ACI 332 or PCA 100. Concrete foundation walls shall be laterally supported at the top and bottom. Horizontal reinforcement shall be provided in accordance with Table R404.1.2(1). Vertical reinforcement shall be provided in accordance with Table R404.1.2(2) through R404.1.2(8), as applicable.*

Discussion and Commentary: Under the prescriptive provisions of the IRC, the minimum vertical reinforcement required in a concrete foundation wall is based upon multiple factors. The necessary information required for determining the proper reinforcement includes the 1) maximum wall height, 2) maximum unbalanced fill height, 3) type of soil, and 4) minimum wall thickness.

TABLE R404.1.2(8)
MINIMUM VERTICAL REINFORCEMENT FOR 6-, 8-, 10-INCH AND 12-INCH NOMINAL FLAT BASEMENT WALLS[b, c, d, e, f, h, i, k, n]

MAXIMUM WALL HEIGHT (feet)	MAXIMUM UNBALANCED BACKFILL HEIGHT[g] (feet)	GW, GP, SW, SP 30				GM, GC, SM, SM-SC and ML 45				SC, ML-CL and inorganic CL 60			
		6	8	10	12	6	8	10	12	6	8	10	12
5	4	NR	NR	NR	NR	NR	NR	NR	NR	NR	NR	NR	NR
	5	NR	NR	NR	NR	NR	NR	NR	NR	NR	NR	NR	NR
6	4	NR	NR	NR	NR	NR	NR	NR	NR	NR	NR	NR	NR
	5	NR	NR	NR	NR	NR	NR^i	NR	NR	4 @ 35	NR^i	NR	NR
	6	NR	NR	NR	NR	5 @ 48	NR	NR	NR	5 @ 36	NR	NR	NR
7	4	NR	NR	NR	NR	NR	NR	NR	NR	NR	NR	NR	NR
	5	NR	NR	NR	NR	NR	NR	NR	NR	5 @ 47	NR	NR	NR
	6	NR	NR	NR	NR	5 @ 42	NR	NR	NR	6 @ 43	5 @ 48	NR^i	NR
	7	5 @ 46	NR	NR	NR	6 @ 42	5 @ 46	NR^i	NR	6 @ 34	6 @ 48	NR	NR
8	4	NR	NR	NR	NR	NR	NR	NR	NR	NR	NR	NR	NR
	5	NR	NR	NR	NR	4 @ 38	NR^i	NR	NR	5 @ 43	NR	NR	NR
	6	4 @ 37	NR^i	NR	NR	5 @ 37	NR	NR	NR	6 @ 37	5 @ 43	NR^i	NR
	7	5 @ 40	NR	NR	NR	6 @ 37	5 @ 41	NR^i	NR	6 @ 34	6 @ 43	NR	NR
	8	6 @ 43	5 @ 47	NR^i	NR	6 @ 34	6 @ 43	NR	NR	6 @ 27	6 @ 32	6 @ 44	NR
9	4	NR	NR	NR	NR	NR	NR	NR	NR	NR	NR	NR	NR
	5	NR	NR	NR	NR	4 @ 35	NR^i	NR	NR	5 @ 40	NR	NR	NR
	6	4 @ 34	NR^i	NR	NR	6 @ 48	NR	NR	NR	6 @ 36	6 @ 39	NR^i	NR
	7	5 @ 36	NR	NR	NR	6 @ 34	5 @ 37	NR	NR	6 @ 33	6 @ 38	5 @ 37	NR^i
	8	6 @ 38	5 @ 41	NR^i	NR	6 @ 33	6 @ 38	5 @ 37	NR^i	6 @ 24	6 @ 29	6 @ 39	4 @ 48^m
	9	6 @ 34	6 @ 46	NR	NR	6 @ 26	6 @ 30	6 @ 41		6 @ 19	6 @ 23	6 @ 30	6 @ 39
10	4	NR	NR	NR	NR	NR	NR	NR	NR	NR	NR	NR	NR
	5	NR	NR	NR	NR	4 @ 33	NR^i	NR	NR	5 @ 38	NR	NR	NR
	6	5 @ 48	NR^i	NR	NR	6 @ 45	NR	NR	NR	6 @ 34	5 @ 37	NR	NR
	7	6 @ 47	NR	NR	NR	6 @ 34	6 @ 48	NR	NR	6 @ 30	6 @ 35	6 @ 48	NR^i
	8	6 @ 34	5 @ 38	NR	NR	6 @ 30	6 @ 34	6 @ 47	NR^i	6 @ 22	6 @ 26	6 @ 35	6 @ 45^m
	9	6 @ 34	6 @ 41	4 @ 48	NR^i	6 @ 23	6 @ 27	6 @ 35	4 @ 48^m	DR	6 @ 22	6 @ 27	6 @ 34
	10	6 @ 28	6 @ 33	6 @ 45	NR	DR^j	6 @ 23	6 @ 29	6 @ 38	DR	6 @ 22	6 @ 22	6 @ 28

MINIMUM VERTICAL REINFORCEMENT—BAR SIZE AND SPACING (inches); Soil classes[a] and design lateral soil (psf per foot of depth); Minimum nominal wall thickness (inches)

For SI: 1 foot = 304.8 mm; 1 inch = 25.4 mm; 1 pound per square foot per foot = 0.1571 kПа^2/m, 1 pound per square inch = 6.895 kPa.

a. Soil classes are in accordance with the Unified Soil Classification System. Refer to Table R405.1.
b. Table values are based on reinforcing bars with a minimum yield strength of 60,000 psi.
c. Vertical reinforcement with a yield strength of less than 60,000 psi and/or bars of a different size than specified in the table are permitted in accordance with Section R404.1.2.3.7.6 and Table R404.1.2(9).
d. NR indicates no vertical wall reinforcement is required, except for 6-inch nominal walls formed with stay-in-place forming systems in which case vertical reinforcement shall be #4 @ 48 inches on center.
e. Allowable deflection criterion is $L/240$, where L is the unsupported height of the basement wall in inches.
f. Interpolation is not permitted.
g. Where walls will retain 4 feet or more of unbalanced backfill, they shall be laterally supported at the top and bottom before backfilling.
h. Vertical reinforcement shall be located to provide a cover of 1.25 inches measured from the inside face of the wall. The center of the steel shall not vary from the specified location by more than the greater of 10 percent of the wall thickness or $^3/_8$-inch.
i. Concrete cover for reinforcement measured from the inside face of the wall shall not be less than $^3/_4$-inch. Concrete cover for reinforcement measured from the outside face of the wall shall not be less than $1^1/_2$ inches for No. 5 bars and smaller, and not less than 2 inches for larger bars.
j. DR means design is required in accordance with the applicable building code, or where there is no code in accordance with ACI 318.
k. Concrete shall have a specified compressive strength, f'_c, of not less than 2,500 psi at 28 days, unless a higher strength is required by footnote l or m.
l. The minimum thickness is permitted to be reduced 2 inches, provided the minimum specified compressive strength of concrete, f'_c, is 4,000 psi.
m. A plain concrete wall with a minimum nominal thickness of 12 inches is permitted, provided minimum specified compressive strength of concrete, f'_c, is 3,500 psi.
n. See Table R611.3 for tolerance from nominal thickness permitted for flat walls.

The notes to Table R404.1.2(8) provide additional information necessary to the proper application of the provisions. Of considerable importance, the minimum concrete cover required for the reinforcement is established.

Code Text: *Stay-in-place concrete forms shall comply with* Section R404.1.2.3.6.1. *The surface burning characteristics of foam plastic used in insulating concrete forms shall comply with Section R302. Stay-in-place forms constructed of rigid foam plastic shall be protected on the interior of the building as required by Section R316. Where gypsum board is used to protect the foam plastic, it shall be installed with a mechanical fastening system. Exterior surfaces* shall be *protected from sunlight and physical damage by the application of an approved exterior wall covering complying with* the IRC.

Discussion and Commentary: An insulating concrete form (ICF) system uses stay-in-place forms of rigid foam plastic insulation, a hybrid of cement and foam insulation, a hybrid of cement and wood chips, or other insulating material for constructing cast-in-place concrete walls. Three types of ICF systems are prescriptively addressed in the IRC; waffle-grid, flat-wall and screen grid.

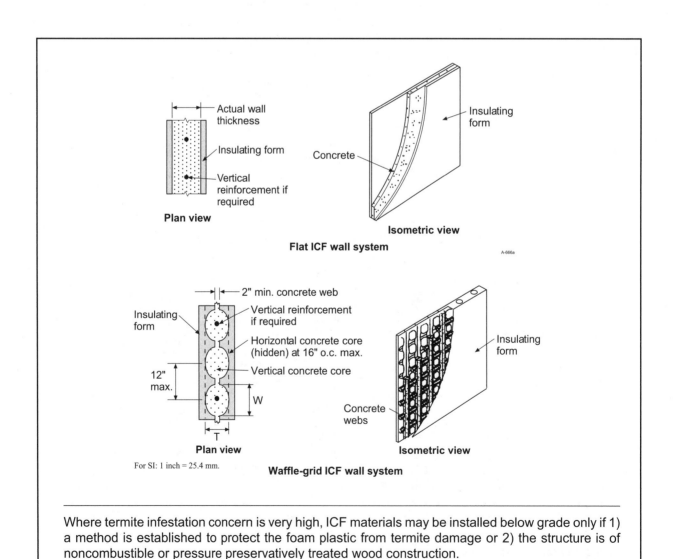

Where termite infestation concern is very high, ICF materials may be installed below grade only if 1) a method is established to protect the foam plastic from termite damage or 2) the structure is of noncombustible or pressure preservatively treated wood construction.

Code Text: *Retaining walls that are not laterally supported at the top and that retain in excess of 24 inches (610 mm) of unbalanced fill shall be designed to ensure stability against overturning, sliding, excessive foundation pressure and water uplift. Retaining walls shall be designed for a safety factor of 1.5 against lateral sliding and overturning.*

Discussion and Commentary: Retaining walls are a typical feature at many residential sites. As such, the IRC provides specific criteria for which a retaining wall must be designed. Soil moisture greatly influences the lateral pressure of the soil against the retaining wall. The placement of drains near the base of the retaining wall lessens the time period that the backfill is saturated.

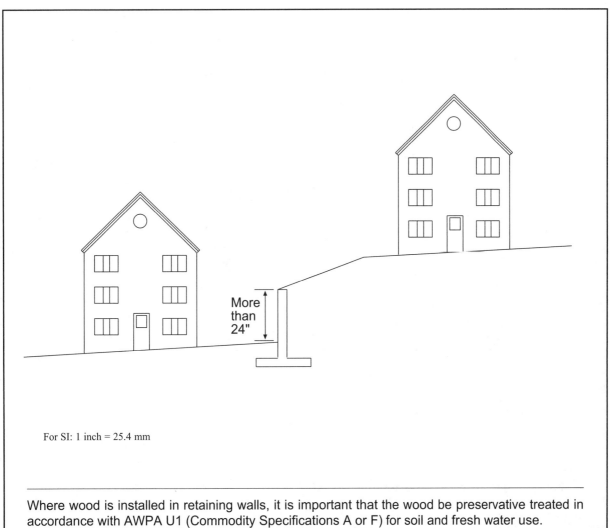

More than 24"

For SI: 1 inch = 25.4 mm

Where wood is installed in retaining walls, it is important that the wood be preservative treated in accordance with AWPA U1 (Commodity Specifications A or F) for soil and fresh water use.

Topic: Concrete or Masonry Foundations   **Category:** Foundations
Reference: IRC R405.1   **Subject:** Foundation Drainage

Code Text: *Drains shall be provided around all concrete or masonry foundations that retain earth and enclose habitable or usable spaces located below grade. See exception for well-drained ground or sand-gravel mixture soils. Drainage tiles, gravel or crushed stone drains, perforated pipe or other approved systems or materials shall be installed at or below the area to be protected and shall discharge by gravity or mechanical means into an approved drainage system.*

Discussion and Commentary: To allow free groundwater that may be present adjacent to the foundation wall to be removed, drains are usually placed around houses. Such drains are intended to prevent leakage into interior spaces below grade. Drainage tiles are extremely important in areas having moderate to heavy rainfall and soils with a low percolation rate.

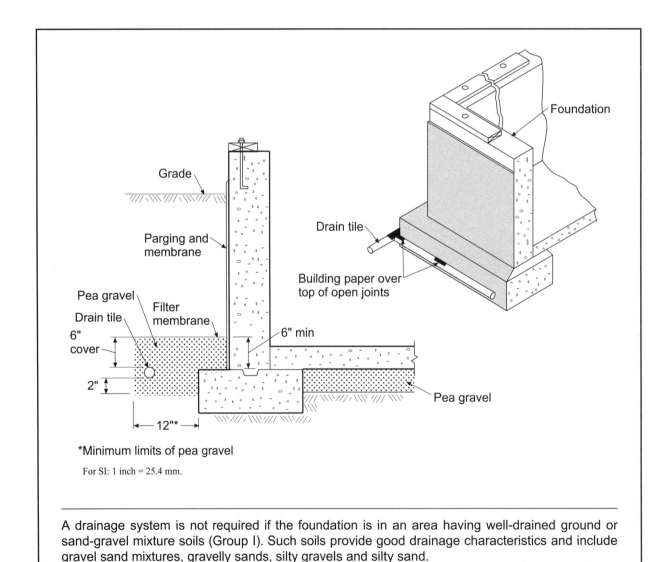

*Minimum limits of pea gravel

For SI: 1 inch = 25.4 mm.

A drainage system is not required if the foundation is in an area having well-drained ground or sand-gravel mixture soils (Group I). Such soils provide good drainage characteristics and include gravel sand mixtures, gravelly sands, silty gravels and silty sand.

Code Text: *Except where required by Section R406.2 to be waterproofed, foundation walls that retain earth and enclose interior spaces and floors below grade shall be dampproofed from the top of the footing to the finished grade. In areas where a high water table or other severe soil-water conditions are known to exist, exterior foundation walls that retain earth and enclose interior spaces and floors below grade shall be waterproofed from the top of the footing to the finished grade.*

Discussion and Commentary: Dampproofing installations generally consist of the application of one or more coatings of impervious compounds that are intended to prevent the passage of water vapor through walls under slight pressure. Waterproofing installations consist of the application of a combination of sealing materials and impervious coatings used to prevent the passage of moisture in either a vapor or liquid form under conditions of significant hydrostatic pressure.

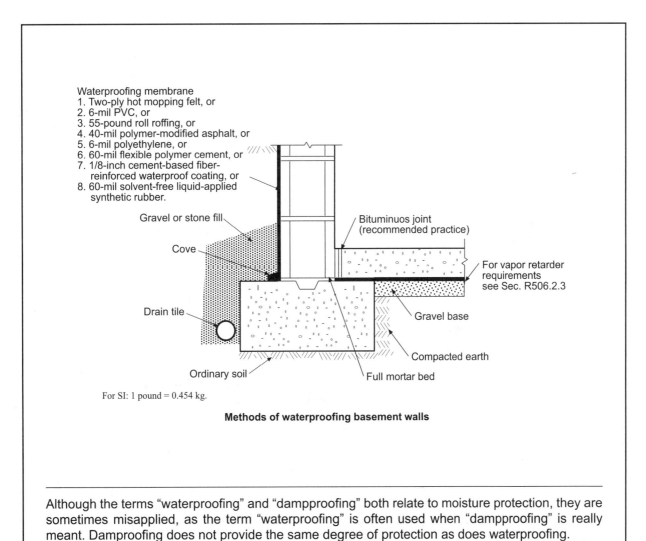

For SI: 1 pound = 0.454 kg.

Methods of waterproofing basement walls

Although the terms "waterproofing" and "dampproofing" both relate to moisture protection, they are sometimes misapplied, as the term "waterproofing" is often used when "dampproofing" is really meant. Dampproofing does not provide the same degree of protection as does waterproofing.

Code Text: *The under-floor space between the bottom of the floor joists and the earth under any building (except space occupied by a basement) shall have ventilation openings through foundation walls or exterior walls. The minimum net area of ventilation openings shall not be less than 1 square foot ($0.0929m^2$) for each 150 square feet ($14 m^2$) of under-floor space area, unless the ground surface is covered by a Class 1 vapor retarder material* (where openings are required only at a ratio of 1 to 1,500). Unvented crawl spaces are also permitted under the provisions of Section R408.3.

Discussion and Commentary: To control condensation in crawl space areas and thus reduce the chance of dry rot, natural ventilation of such spaces by reasonably distributed openings through exterior foundation walls is required. Susceptibility of a structure to condensation is a function of the geographical location and the climatic conditions.

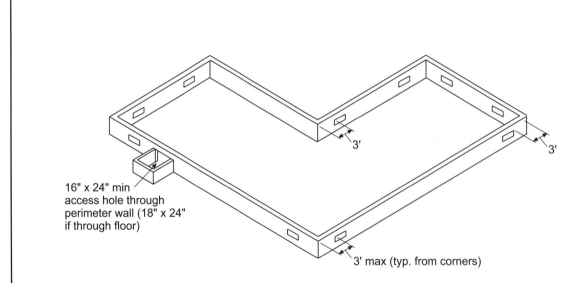

16" x 24" min access hole through perimeter wall (18" x 24" if through floor)

3'

3'

3' max (typ. from corners)

Covered openings through foundation wall to crawl space

Minimum total net clear area of openings equal to the crawl space area divided by 150*

For SI: 1 inch = 25.4 mm, 1 foot = 304.8 mm.

*Minimum net area of openings may be reduced to 1 sq ft for each 1,500 sq ft of area where Class 1 vapor retarder material covers ground surface

The installation of a covering material over the ventilation opening keeps rodents and vermin from entering the crawl space. The code lists a variety of materials including corrosion-resistant wire mesh having a least dimension of $^1/_8$ inch.

Quiz

Study Session 5
IRC Chapter 4

1. Unless grading is prohibited by physical barriers, lots shall be graded away from the foundation with a minimum fall of _____ within the first _____.

   a. 6 inches, 5 feet

   b. 6 inches, 10 feet

   c. 12 inches, 5 feet

   d. 12 inches, 10 feet

   Reference _____

2. In the absence of a complete geotechnical evaluation to determine the soil's characteristics, clayey sand material shall be assumed to have a presumptive load-bearing value of _____ .

   a. 1,500 psf

   b. 2,000 psf

   c. 3,000 psf

   d. 4,000 psf

   Reference _____

3. Concrete used in a basement slab shall have a minimum compressive strength of _____ where a severe weathering potential exists.

   a. 2,000 psi

   b. 2,500 psi

   c. 3,000 psi

   d. 3,500 psi

   Reference _____

4. Air entrainment for concrete subjected to weathering, when required for locations other than garage floors with a steel troweled finish, shall have a total air content of _____ minimum and _____ maximum.

    a. 4 percent, 7 percent   b. 4 percent, 8 percent

    c. 5 percent, 7 percent   d. 5 percent, 8 percent

Reference _____

5. A two-story dwelling of conventional light-frame construction is to be constructed in an area where the load-bearing value of the soil is 2,000 psf. What is the minimum required width of the concrete footing?

    a. 8 inches   b. 11 inches

    c. 12 inches   d. 15 inches

Reference _____

6. The minimum required thickness for concrete spread footings is _____ .

    a. 4 inches   b. 6 inches

    c. 7 inches   d. 8 inches

Reference _____

7. What are the minimum and maximum required footing projections for a concrete footing having a thickness of 8 inches?

    a. 2 inches, 4 inches   b. 2 inches, 8 inches

    c. 4 inches, 8 inches   d. 8 inches, 12 inches

Reference _____

8. Where a permanent wood foundation basement wall system is used, the wall shall be supported by a minimum _____ footing plate resting on gravel or crushed stone fill a minimum of _____ in width.

    a. 2-inch-by-6-inch, 12 inches   b. 2-inch-by-8-inch, 16 inches

    c. 2-inch-by-12-inch, 16 inches   d. 2-inch-by-12-inch, 24 inches

Reference _____

9. For concrete foundation systems constructed in Seismic Design Category D_2, foundations with stem walls shall be provided with a minimum of _____ bar(s) at the top of the wall and _____ bar(s) near the bottom of the footing.

    a. one #4, one #4   b. one #5, one #5

    c. one #5, two #4   d. two #4, two #5

Reference _____

10. All exterior footings shall be placed a minimum of _____ below the undisturbed ground surface.

    a. 6 inches   b. 8 inches

    c. 9 inches   d. 12 inches

Reference _____

11. The maximum permitted slope for the bottom surface of footings shall be _____ .

    a. 1:8   b. 1:10

    c. 1:12   d. 1:20

Reference _____

12. For a two-story dwelling assigned to Seismic Design Category B, anchor bolts used to attach wood sill plates to foundation walls shall be spaced a maximum of _____ on center.

    a. 4 feet   b. 5 feet

    c. 6 feet   d. 7 feet

Reference _____

13. Anchor bolts used to attach a wood sole plate to a concrete foundation shall be a minimum of _____ in diameter and extend a minimum of _____ into the concrete.

    a. $\frac{1}{2}$ inch, 7 inches   b. $\frac{1}{2}$ inch, 15 inches

    c. $\frac{5}{8}$ inch, 7 inches   d. $\frac{5}{8}$ inch, 15 inches

Reference _____

14. Soil described as a "sandy clay" has a Unified Soil Classification System Symbol of _____ and is considered to be in Soil Group _____ .

    a. SC, I   b. CL, II

    c. SM, III   d. CH, IV

Reference _____

15. In order to use frost protected shallow foundations, the monthly mean temperature of the building must be maintained at a minimum of _____ .

    a. 60°F   b. 64°F

    c. 68°F   d. 70°F

Reference _____

16. A dwelling located in an area with an air freezing index of 3,500 is constructed with a frost-protected shallow foundation. What is the minimum required horizontal insulation R-value at the corners of the foundation?

    a. 8.0   b. 8.6

    c. 11.2   d. 13.1

Reference _____

17. A 9-foot-high plain masonry foundation wall is subjected to 7 feet of unbalanced backfill. If the soil class is GC, what is the minimum required nominal wall thickness?

    a. 8 inches   b. 10 inches

    c. 12 inches   d. 12 inches solid grouted

Reference _____

18. An 8-foot-high, flat concrete foundation wall of nominal 10-inch thickness is subjected to 6 feet of unbalanced backfill. If the soil class is GP, what minimum vertical reinforcement is required?

    a. #4 at 38 inches on center   b. #5 at 47 inches on center

    c. #6 at 43 inches on center   d. no vertical reinforcement required

Reference _____

19. Where masonry veneer is used, concrete and masonry foundations shall extend a minimum of _____ above adjacent finished grade.

    a. 4 inches   b. 6 inches

    c. 8 inches   d. 12 inches

Reference _____

20. Where the foundation wall supports a minimum of _____ of unbalanced backfill, backfill shall not be placed against the wall until the wall has sufficient strength and has been anchored to the floor above or has been sufficiently braced.

    a. 2 feet   b. 4 feet

    c. 6 feet   d. 8 feet

Reference _____

21. Where wood studs having an F_b value of 1320 are used in a wood foundation wall, the minimum stud size shall be _____ and the maximum stud spacing shall be _____ on center.

    a. 2 inches by 4 inches, 24 inches

    b. 2 inches by 4 inches, 16 inches

    c. 2 inches by 6 inches, 24 inches

    d. 2 inches by 6 inches, 16 inches

Reference _____

22. What is the maximum height of backfill permitted against a wood foundation wall not designed by AF&PA Report # 7?

    a. 3 feet   b. 4 feet

    c. 6 feet   d. 7 feet

Reference _____

23. Required foundation drains of gravel or crushed stone shall extend a minimum of _____ beyond the outside edge of the footing and _____ above the top of the footing.

    a. 6 inches, 4 inches   b. 6 inches, 6 inches

    c. 12 inches, 4 inches   d. 12 inches, 6 inches

Reference _____

24. As a general rule, the under-floor space between the bottom of floor joists and the earth shall be provided with ventilation openings sized for a minimum net area of 1 square foot for each _____ square feet of under-floor area.

    a. 100   b. 120

    c. 150   d. 200

Reference _____

25. Access to an under-floor space through a perimeter wall shall be provided by a minimum _____ access opening.

    a. 16-inch-by-20-inch   b. 16-inch-by-24-inch

    c. 18-inch-by-24-inch   d. 18-inch-by-30-inch

Reference _____

26. Frost protection is not required for a foundation that supports a freestanding accessory structure of light-framed construction that has a maximum floor area of _____ square feet and a maximum eave height of _____ feet.

    a. 200, 10   b. 400, 10

    c. 400, 12   d. 600, 10

Reference _____

27. Gravel used as a footing material for a wood foundation shall be washed, well graded and have a maximum stone size of _____ inch.

    a. $^1/_4$ b. $^1/_2$

    c. $^3/_4$ d. 1

Reference _____

28. Horizontal insulation used in a frost protected shallow foundation system shall be protected against damage if it is located less than _____ inches below the ground surface.

    a. 12   b. 15

    c. 18   d. 24

Reference _____

29. Where waterproofing of a masonry foundation wall is necessary due to the presence of a high water table, a minimum membrane thickness of _____ is required if the waterproofing material is polymer-modified asphalt.

    a. 6-mil   b. 30-mil

    c. 40-mil   d. 60-mil

Reference _____

30. Ventilating openings providing under-floor ventilation shall be located so that at least one such opening is installed a maximum of _____ feet from each corner of the building.

    a. 2   b. 3

    c. 5   d. 6

Reference _____

31. A minimum of _____ anchor bolt(s) located _____ of the plate section is required to anchor maximum 24-inch-long walls connecting offset braced wall panels.

    a. 1, within the middle third

    b. 1, within the middle one-half

    c. 2, in the outer one-fourths

    d. 2, within 6 inches of each end

Reference _____

32. Unless specifically approved, what is the maximum slump permitted for concrete placed in removable foundation wall forms?

    a. 4 inches   b. $4^1/_2$ inches

    c. 5 inches   d. 6 inches

Reference _____

33. Where one coat of complying surface-bonding cement is used as the required dampproofing over the parging of a masonry foundation wall, the cement coating must be a minimum of _____ inch thick.

    a. $^1/_8$ b. $^3/_{16}$

    c. $^1/_4$ d. $^3/_8$

Reference _____

34. Steel columns used as a part of a foundation system shall be a minimum of _____ inches in diameter standard pipe size or approved equivalent.

    a. 3   b. $3^1/_2$

    c. 4   d. 5

Reference _____

35. The continuous vapor barrier required in an unvented crawl space shall extend a minimum of _____ inches up the stem wall.

    a. 3   b. 4

    c. 6   d. 8

Reference _____

Study Session

6

2012 IRC Chapter 5
Floors

OBJECTIVE: To develop an understanding of the provisions regulating floor construction, including the sizing and installation of floor joists and girders; drilling and notching of floor framing members; engineered wood products such as trusses, glued-laminated members and I-joists; floor sheathing spans and installation; steel floor framing; concrete slab-on-ground floors; and decks attached to exterior walls.

REFERENCE: Chapter 5, 2012 *International Residential Code*

KEY POINTS:
- What type of membrane must be applied to the underside of floor framing members for those floor assemblies not required to be fire-resistance-rated?
- How are the maximum allowable spans of floor joists determined? What design live loads are to be used for joists supporting sleeping rooms? For areas other than sleeping rooms? For attics accessed by a permanent stairway?
- What is the general limitation for cantilever spans of floor joists? What spans are permitted for cantilevered floor joists supporting a bearing wall and a roof? Supporting an exterior balcony?
- How shall bearing partitions be supported by floor joists? What method is acceptable where piping or vents are installed at the point of support?
- How must bearing partitions perpendicular to floor joists be located?
- What conditions regulate the sizing of girders for exterior bearing walls? For interior bearing walls?
- How much bearing is required for the ends of floor joists supported by wood or metal? Supported by concrete or masonry? What other methods are acceptable?
- What methods of connection are permitted where joists frame from opposite sides of a bearing wall or girder?
- How are floor joists to be restrained to resist twisting and lateral movement?
- What are the limitations for notches in floor joists? Bored holes?
- When are single headers and single trimmers permitted at floor openings? When are framing anchors or joist hangers mandated?

- How are wood floor trusses regulated? What type of information must be provided on the truss design drawings?

- In what type of floor system is draftstopping required? At what intervals? What materials are permitted for draftstopping?

- How is wood structural panel sheathing identified? How does the panel span rating reflect the maximum allowable floor span? How is the allowable span for particleboard determined?

- What special concerns are addressed for pressure preservatively treated wood basement floors and floors on ground?

- What are the limitations for using the prescriptive provisions for steel floor framing?

- What materials are permitted as steel load-bearing members? How must they be identified? What are the fundamental fastening requirements for steel-to-steel floor connections? Floor to foundation or bearing wall connections? How are the maximum allowable spans determined?

- What is the minimum permitted thickness for concrete slab-on-ground floors? How must the site be prepared?

- How is any fill below a concrete slab-on-ground floor regulated? When is a base course needed? When is a vapor barrier required?

- When provided, where is reinforcement for slab-on-ground floors required to be located?

- What ledger connection details are required for decks attached to exterior walls?

Code Text: *Floor assemblies, not required elsewhere in this code to be fire-resistance rated, shall be provided with a $^1/_2$-inch gypsum wallboard membrane, $^5/_8$-inch wood structural panel membrane, or equivalent on the underside of the floor framing member.* See exceptions for floor assemblies: 1) located directly over a sprinklered space, 2) located directly over a crawl space not intended for storage or fuel-fired appliances, 3) of limited size where fireblocked, and 4) constructed with minimum 2x10 dimensional lumber, structural composite lumber or equivalent.

Discussion and Commentary: The application of gypsum wallboard or other approved material on the underside of floor framing members provides some protection to the floor system against the effects of fire and delay collapse of the floor. This protection addresses concerns for firefighter safety and incidents of injury or death to firefighters due to floor collapse while fighting residential fires. The emphasis is placed on light-frame floor construction consisting of I-joists, manufactured floor trusses, cold-formed steel framing, and other materials and manufactured products considered most susceptible to failure during a fire.

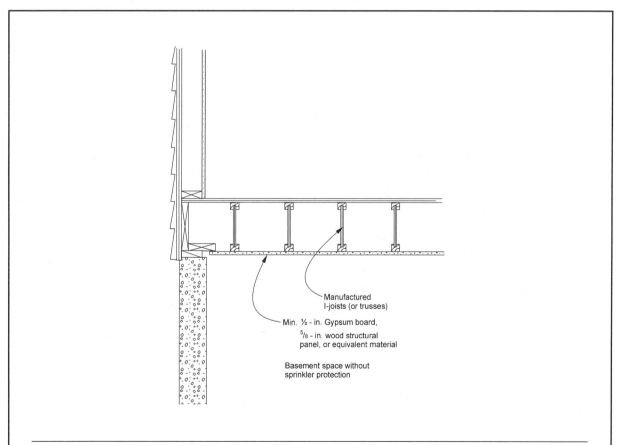

Manufactured
I-joists (or trusses)

Min. ½ - in. Gypsum board,
$^5/_8$ - in. wood structural
panel, or equivalent material

Basement space without
sprinkler protection

Floors framed with solid-sawn lumber and structural composite lumber of substantial size are exempt because they perform fairly well in retaining adequate strength under fire conditions. Similarly, if sprinklers are installed to protect the space below the floor assembly, membrane protection is not required. Exceptions also address crawl spaces without storage and small areas.

Topic: Framing at Braced Wall Lines

Category: Floors

Reference: IRC R502.2.1

Subject: Wood Floor Framing

Code Text: *A load path for lateral forces shall be provided between floor framing and braced wall panels located above or below a floor, as specified in Section R602.10.8 (Connections).*

Discussion and Commentary: The lateral forces induced by wind and earthquake are transferred to the building foundation through the roof and floor systems and the braced wall lines to which these framing members are attached. The braced wall lines in turn transfer such loads to the foundation by the sill plate and its connection to the foundation system. The proper and successful transfer of all loads from the building envelope system to the foundation is accomplished by proper creation of a complete load path.

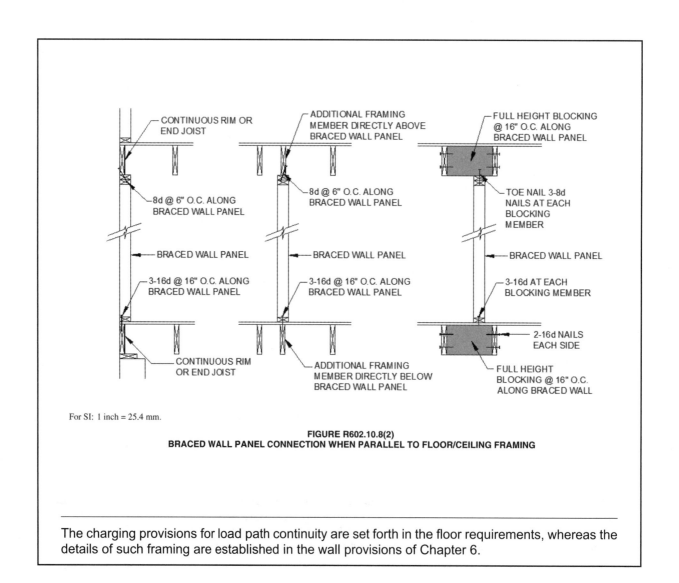

For SI: 1 inch = 25.4 mm.

FIGURE R602.10.8(2)
BRACED WALL PANEL CONNECTION WHEN PARALLEL TO FLOOR/CEILING FRAMING

The charging provisions for load path continuity are set forth in the floor requirements, whereas the details of such framing are established in the wall provisions of Chapter 6.

Topic: Allowable Joist Spans **Category:** Floors
Reference: IRC R502.3.1, R502.3.2 **Subject:** Wood Floor Framing

Code Text: *Table R502.3.1(1) shall be used to determine the maximum allowable span of floor joists that support sleeping areas and attics that are accessed by means of a fixed stairway in accordance with Section R311.7* (Stairways) *provided that the design live load does not exceed 30 psf (1.44 kPa) and the design dead load does not exceed 20 psf (0.96 kPa). Table R502.3.1(2) shall be used to determine the maximum allowable span of floor joists that support all other areas of the building, other than sleeping rooms and attics, provided that the design live load does not exceed 40 psf (1.92 kPa) and the design dead load does not exceed 20 psf (0.96 kPa).*

Discussion and Commentary: The joist span tables are based on common lumber species and grades, joist spacing and design loads. Although a separate span table is available for sleeping rooms based on 30 psf, it is not uncommon to use the 40 psf table for the entire dwelling unit. Both tables assume a maximum design dead load of 20 psf.

TABLE R502.3.1(2)
FLOOR JOIST SPANS FOR COMMON LUMBER SPECIES
(Residential living areas, live load = 40 psf, L/D = 360)[b]

JOIST SPACING (inches)	SPECIES AND GRADE		DEAD LOAD = 10 psf				DEAD LOAD = 20 psf			
			2×6	2×8	2×10	2×12	2×6	2×8	2×10	2×12
			Maximum floor joist spans							
			(ft - in.)	(ft - in.)	(ft - in.)	(ft - in.)	(ft - in.)	(ft - in.)	(ft - in.)	(ft - in.)
12	Douglas fir-larch	SS	11- 4	15- 0	19- 1	23- 3	11- 4	15- 0	19- 1	23- 3
	Douglas fir-larch	#1	10-11	14- 5	18- 5	22- 0	10-11	14- 2	17- 4	20- 1
	Douglas fir-larch	#2	10- 9	14- 2	17- 9	20- 7	10- 6	13- 3	16- 3	18-10
	Douglas fir-larch	#3	8- 8	11- 0	13- 5	15- 7	7-11	10- 0	12- 3	14- 3
	Hem-fir	SS	10- 9	14- 2	18- 0	21-11	10- 9	14- 2	18- 0	21-11
	Hem-fir	#1	10- 6	13-10	17- 8	21- 6	10- 6	13-10	16-11	19- 7
	Hem-fir	#2	10- 0	13- 2	16-10	20- 4	10- 0	13- 1	16- 0	18- 6
	Hem-fir	#3	8- 8	11- 0	13- 5	15- 7	7-11	10- 0	12- 3	14- 3
	Southern pine	SS	11- 2	14- 8	18- 9	22-10	11- 2	14- 8	18- 9	22-10
	Southern pine	#1	10-11	14- 5	18- 5	22- 5	10-11	14- 5	18- 5	22- 5
	Southern pine	#2	10- 9	14- 2	18- 0	21- 9	10- 9	14- 2	16-11	19-10
	Southern pine	#3	9- 4	11-11	14- 0	16- 8	8- 6	10-10	12-10	15- 3
	Spruce-pine-fir	SS	10- 6	13-10	17- 8	21- 6	10- 6	13-10	17- 8	21- 6
	Spruce-pine-fir	#1	10- 3	13- 6	17- 3	20- 7	10- 3	13- 3	16- 3	18-10
	Spruce-pine-fir	#2	10- 3	13- 6	17- 3	20- 7	10- 3	13- 3	16- 3	18-10
	Spruce-pine-fir	#3	8- 8	11- 0	13- 5	15- 7	7-11	10- 0	12- 3	14- 3
16	Douglas fir-larch	SS	10- 4	13- 7	17- 4	21- 1	10- 4	13- 7	17- 4	21- 0
	Douglas fir-larch	#1	9-11	13- 1	16- 5	19- 1	9- 8	12- 4	15- 0	17- 5
	Douglas fir-larch	#2	9- 9	12- 7	15- 5	17-10	9- 1	11- 6	14- 1	16- 3
	Douglas fir-larch	#3	7- 6	9- 6	11- 8	13- 6	6-10	8- 8	10- 7	12- 4
	Hem-fir	SS	9- 9	12-10	16- 5	19-11	9- 9	12-10	16- 5	19-11
	Hem-fir	#1	9- 6	12- 7	16- 0	18- 7	9- 6	12- 0	14- 8	17- 0
	Hem-fir	#2	9- 1	12- 0	15- 2	17- 7	8-11	11- 4	13-10	16- 1
	Hem-fir	#3	7- 6	9- 6	11- 8	13- 6	6-10	8- 8	10- 7	12- 4
	Southern pine	SS	10- 2	13- 4	17- 0	20- 9	10- 2	13- 4	17- 0	20- 9
	Southern pine	#1	9-11	13- 1	16- 9	20- 4	9-11	13- 1	16- 4	19- 6
	Southern pine	#2	9- 9	12-10	16- 1	18-10	9- 6	12- 4	14- 8	17- 2
	Southern pine	#3	8- 1	10- 3	12- 2	14- 6	7- 4	9- 5	11- 1	13- 2
	Spruce-pine-fir	SS	9- 6	12- 7	16- 0	19- 6	9- 6	12- 7	16- 0	19- 6
	Spruce-pine-fir	#1	9- 4	12- 3	15- 5	17-10	9- 1	11- 6	14- 1	16- 3
	Spruce-pine-fir	#2	9- 4	12- 3	15- 5	17-10	9- 1	11- 6	14- 1	16- 3
	Spruce-pine-fir	#3	7- 6	9- 6	11- 8	13- 6	6-10	8- 8	10- 7	12- 4

(continued)

The span tables account for a uniform load condition. They also permit isolated concentrated loads such as nonbearing partitions. The use of additional joists and other adequate supports may be necessary where larger concentrated loads are encountered.

Code Text: *Floor cantilever spans shall not exceed the nominal depth of the wood floor joist. Floor cantilevers constructed in accordance with Table R502.3.3(1) shall be permitted when supporting a light-frame bearing wall and roof only. Floor cantilevers supporting an exterior balcony are permitted to be constructed in accordance with Table R502.3.3(2).*

Discussion and Commentary: Under the prescriptive limitations of the IRC, there are three conditions where the maximum allowable span of cantilevered floor joists can be determined. First, regardless of loading conditions, a maximum span equal to the depth of the floor joists is permitted. This requirement allows a very short cantilevered element. The second condition considers the support of both a single-story bearing wall and roof. The third situation is based on the cantilevered support of an exterior balcony with no roof load.

TABLE R502.3.3(2)
CANTILEVER SPANS FOR FLOOR JOISTS SUPPORTING EXTERIOR BALCONY[a, b, e, f]

| Member Size | Spacing | Maximum Cantilever Span (Uplift Force at Backspan Support in lb)[c, d] | | |
| | | Ground Snow Load | | |
		≤ 30 psf	50 psf	70 psf
2 × 8	12″	42″ (139)	39″ (156)	34″ (165)
2 × 8	16″	36″ (151)	34″ (171)	29″ (180)
2 × 10	12″	61″ (164)	57″ (189)	49″ (201)
2 × 10	16″	53″ (180)	49″ (208)	42″ (220)
2 × 10	24″	43″ (212)	40″ (241)	34″ (255)
2 × 12	16″	72″ (228)	67″ (260)	57″ (268)
2 × 12	24″	58″ (279)	54″ (319)	47″ (330)

For SI:　1 inch = 25.4 mm, 1 pound per square foot = 0.0479 kPa.

a. Spans are based on No. 2 Grade lumber of Douglas fir-larch, hem-fir, southern pine, and spruce-pine-fir for repetitive (3 or more) members.

b. Ratio of backspan to cantilever span shall be at least 2:1.

c. Connections capable of resisting the indicated uplift force shall be provided at the backspan support.

d. Uplift force is for a backspan to cantilever span ratio of 2:1. Tabulated uplift values are permitted to be reduced by multiplying by a factor equal to 2 divided by the actual backspan ratio provided (2/backspan ratio).

e. A full-depth rim joist shall be provided at the cantilevered end of the joists. Solid blocking shall be provided at the cantilevered support.

f. Linear interpolation shall be permitted for ground snow loads other than shown.

The ratio of back span to cantilever span must be a minimum of 3:1 where the cantilevered floor joists support a bearing wall and a roof. Where only supporting an exterior balcony, the minimum ratio must be 2:1. It is important that connections be capable of resisting the indicated uplift force.

Code Text: *Joists under parallel bearing partitions shall be of adequate size to support the load. Double joists, sized to adequately support the load, that are separated to permit the installation of piping or vents shall be full depth solid blocked with lumber not less than 2 inches (51 mm) in nominal thickness spaced not more than 4 feet (1219 mm) on center. Bearing partitions perpendicular to joists shall not be offset from supporting girders, walls or partitions more than the joist depth unless such joists are of sufficient size to carry the additional load.*

Discussion and Commentary: It is often necessary to double or triple floor joists under bearing partitions in order to provide support for the additional load from above. The multiple joists create a beam-type member that is capable of properly transmitting the loads to the bearing points at the joist ends. Therefore, a beam of equivalent size may be substituted for the multiple joists.

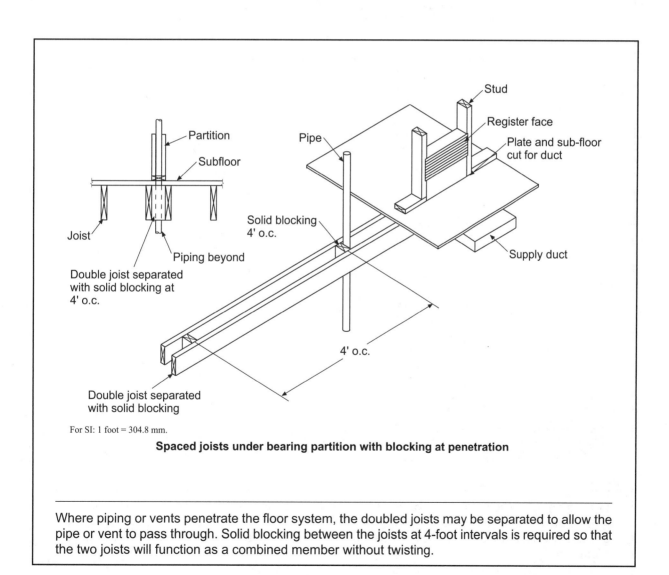

For SI: 1 foot = 304.8 mm.

Spaced joists under bearing partition with blocking at penetration

Where piping or vents penetrate the floor system, the doubled joists may be separated to allow the pipe or vent to pass through. Solid blocking between the joists at 4-foot intervals is required so that the two joists will function as a combined member without twisting.

Code Text: *The allowable spans of girders fabricated of dimension lumber shall not exceed the values set forth in Tables R502.5(1) and R502.5(2).*

Discussion and Commentary: Table R502.5(1) provides the maximum span of girders in exterior walls based on several criteria. The loads supported by the header, building width, built-up member size and ground snow load are needed to use the prescriptive table. The built-up members listed vary as to depth and number of members. The more commonly encountered condition of girders for interior bearing walls is addressed in Table R502.5(2). In using this table, only the loading condition, building span, and built-up member size are needed to apply the table.

TABLE R502.5(2)
GIRDER SPANS[a] AND HEADER SPANS[a] FOR INTERIOR BEARING WALLS
(Maximum spans for Douglas fir-larch, hem-fir, southern pine and spruce-pine-fir[b] and required number of jack studs)

| HEADERS AND GIRDERS SUPPORTING | SIZE | BUILDING WIDTH[c] (feet) | | | | | |
| | | 20 | | 28 | | 36 | |
		Span	NJ[d]	Span	NJ[d]	Span	NJ[d]
One floor only	2-2×4	3-1	1	2-8	1	2-5	1
	2-2×6	4-6	1	3-11	1	3-6	1
	2-2×8	5-9	1	5-0	2	4-5	2
	2-2×10	7-0	2	6-1	2	5-5	2
	2-2×12	8-1	2	7-0	2	6-3	2
	3-2×8	7-2	1	6-3	1	5-7	2
	3-2×10	8-9	1	7-7	2	6-9	2
	3-2×12	10-2	2	8-10	2	7-10	2
	4-2×8	9-0	1	7-8	1	6-9	1
	4-2×10	10-1	1	8-9	1	7-10	2
	4-2×12	11-9	1	10-2	2	9-1	2
Two floors	2-2×4	2-2	1	1-10	1	1-7	1
	2-2×6	3-2	2	2-9	2	2-5	2
	2-2×8	4-1	2	3-6	2	3-2	2
	2-2×10	4-11	2	4-3	2	3-10	3
	2-2×12	5-9	2	5-0	3	4-5	3
	3-2×8	5-1	2	4-5	2	3-11	2
	3-2×10	6-2	2	5-4	2	4-10	2
	3-2×12	7-2	2	6-3	2	5-7	3
	4-2×8	6-1	1	5-3	2	4-8	2
	4-2×10	7-2	2	6-2	2	5-6	2
	4-2×12	8-4	2	7-2	2	6-5	2

For SI: 1 inch = 25.4 mm, 1 foot = 304.8 mm.

a. Spans are given in feet and inches.
b. Tabulated values assume #2 grade lumber.
c. Building width is measured perpendicular to the ridge. For widths between those shown, spans are permitted to be interpolated.
d. NJ - Number of jack studs required to support each end. Where the number of required jack studs equals one, the header is permitted to be supported by an approved framing anchor attached to the full-height wall stud and to the header.

In addition to providing the maximum allowable girder spans, the table sets forth the minimum required number of jack studs. Typically, two jack studs are required to support the ends of the girder, providing at least 3 inches of bearing. Only under limited conditions is a single jack stud permitted.

Topic: Bearing

Category: Floors

Reference: IRC R502.6

Subject: Wood Floor Framing

Code Text: *The ends of each joist, beam or girder shall have not less than 1.5 inches (38 mm) of bearing on wood or metal and not less than 3 inches (76 mm) on masonry or concrete except where supported on a 1-inch-by-4-inch (25.4 mm by 102 mm) ribbon strip and nailed to the adjacent stud or by the use of approved joist hangers.*

Discussion and Commentary: Joists, beams and girders are typically provided with bearing surfaces. The required bearing length for masonry and concrete is twice that needed for wood or metal. As an alternative, bearing may be accomplished through the use of a ribbon strip, provided each member is directly connected to an adjoining stud. Approved joist hangers are also acceptable means of support in lieu of direct bearing. Direct bearing is required on masonry or concrete unless a 2x4 sill plate with a minimum bearing area of 48 square inches is provided.

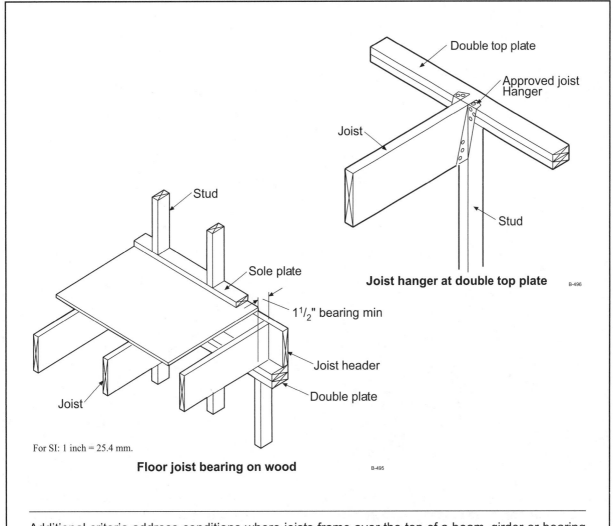

Joist hanger at double top plate B-496

For SI: 1 inch = 25.4 mm.

Floor joist bearing on wood B-495

Additional criteria address conditions where joists frame over the top of a beam, girder or bearing wall. A minimum 3-inch (76 mm) lap is required unless a splice having equivalent strength is installed. Framing anchors or ledgers are mandated when framing into girder sides.

Topic: Lateral Restraint at Supports

Category: Floors

Reference: IRC R502.7

Subject: Wood Floor Framing

Code Text: *Joists shall be supported laterally at the ends by full-depth solid blocking not less than 2 inches (51 mm) nominal in thickness; or by attachment to a header, band, or rim joist, or to an adjoining stud; or shall be otherwise provided with lateral support to prevent rotation. See exception for Seismic Design Categories D_0, D_1 and D_2, where lateral restraint must also be provided at each intermediate support. In addition, engineered wood products are to be supported laterally as required by the manufacturer's recommendations.*

Discussion and Commentary: The potential for rotation of floor joists is reduced by providing complying blocking at the end points. This is typically accomplished at exterior walls by nailing the rim joist directly to the joist ends. Full-depth blocking may be used where the joist ends bear at interior bearing walls. In addition, any other approved methods that can be shown to prevent the rotation of the joists are acceptable.

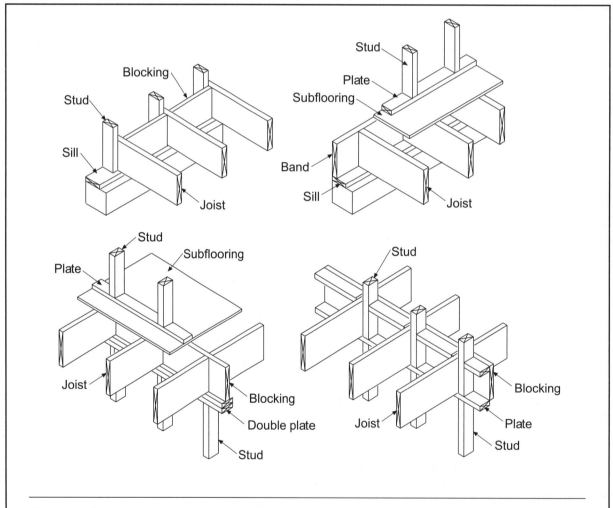

The potential for very deep members (2 inches by 14 inches or greater) to buckle laterally is mitigated by requiring solid blocking or diagonal bridging located at maximum intervals of 8 feet. As an alternative, a continuous 1-inch by 3-inch strip can be nailed to the bottoms of the joists at similar intervals.

Code Text: *Notches in solid lumber joists, rafters and beams shall not exceed one-sixth of the depth of the member, shall not be longer than one-third of the depth of the member and shall not be located in the middle one-third of the span. Notches at the ends of the member shall not exceed one-fourth the depth of the member. The tension side of members 4 inches (102 mm) or greater in nominal thickness shall not be notched except at the ends of the members.*

Discussion and Commentary: Notches potentially reduce the structural strength of framing members to unacceptable levels. Therefore, prescriptive limits on the size and location of any notches are set forth. The specified limitations are only applicable to solid sawn lumber members. Where engineered wood products such as trusses and I-joists are notched, the effects of the penetrations must be considered in the member design.

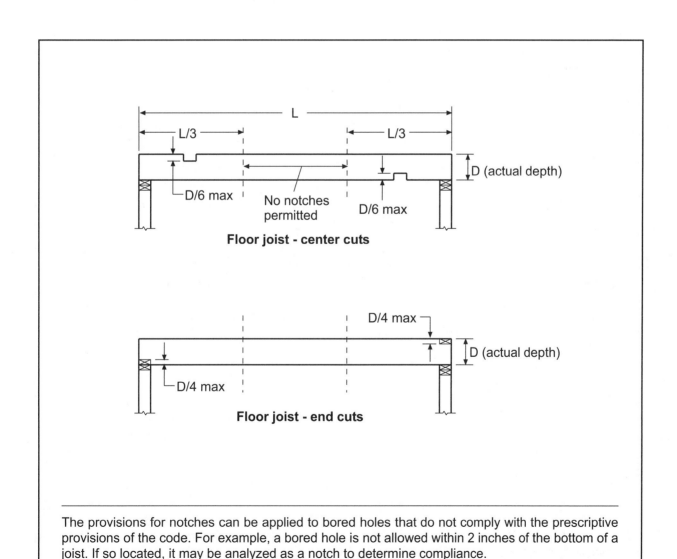

Floor joist - center cuts

Floor joist - end cuts

The provisions for notches can be applied to bored holes that do not comply with the prescriptive provisions of the code. For example, a bored hole is not allowed within 2 inches of the bottom of a joist. If so located, it may be analyzed as a notch to determine compliance.

Code Text: *The diameter of holes bored or cut into members shall not exceed one-third the depth of the member. Holes shall not be closer than 2 inches (51 mm) to the top or bottom of the member, or to any other hole located in the member. Where the member is also notched, the hole shall not be closer than 2 inches (51 mm) to the notch.*

Discussion and Commentary: Where holes are drilled or bored in joists and other wood framing members, they tend to have the same effect as notches. The bored hole reduces the structural strength of the member. Holes cannot be oversized and must be located a minimum distance from the edges of the members. By placing any bored hole within the center portion of the structural member, the effect on the bending stresses is much less of a concern. Holes bored in structural composite lumber, structural glue-laminated members or I-joists are prohibited unless permitted by the manufacturer's recommendations or specifically considered in the member's design.

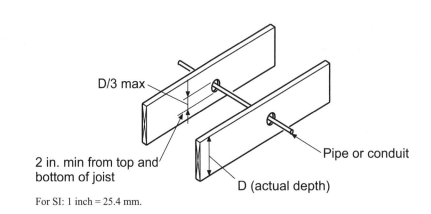

For SI: 1 inch = 25.4 mm.

Although the code does not limit the number of bored holes within a member, it does restrict how far a bored hole can be located from a notch. Common sense should dictate that as few holes as necessary should be bored, with the holes well-distributed along the member span.

Code Text: *Openings in floor framing shall be framed with a header and trimmer joists. When the header joist span does not exceed 4 feet (1219 mm), the header joist may be a single member the same size as the floor joist. Single trimmer joists may be used to carry a single header joist that is located within 3 feet (914 mm) of the trimmer joist bearing. When the header joist span exceeds 4 feet (1219 mm), the trimmer joists and the header joist shall be doubled and of sufficient cross section to support the floor joists framing into the header. Approved hangers shall be used for the header joist to trimmer joist connections when the header joist span exceeds 6 feet (1829 mm). Tail joists over 12 feet (3658 mm) long shall be supported at the header by framing anchors or on ledger strips not less than 2 inches by 2 inches (51 mm by 51 mm).*

Discussion and Commentary: Where larger floor openings are necessary to provide passage of such things as stairways or chases, the adequate transfer of floor loads to their supporting members must be provided.

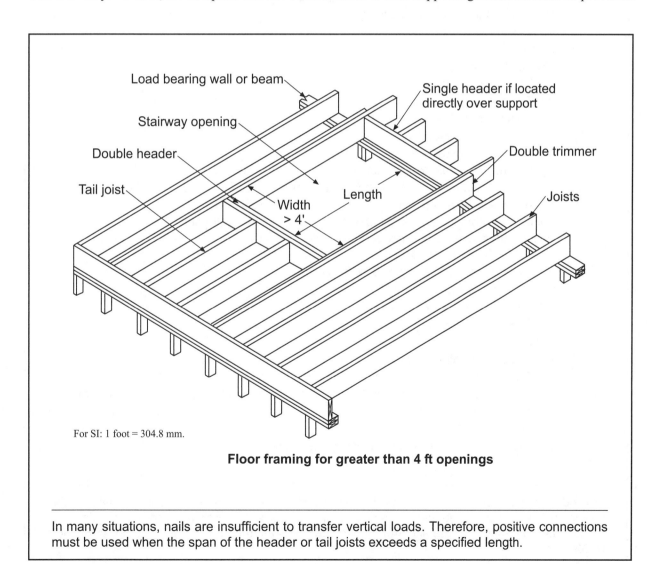

For SI: 1 foot = 304.8 mm.

Floor framing for greater than 4 ft openings

In many situations, nails are insufficient to transfer vertical loads. Therefore, positive connections must be used when the span of the header or tail joists exceeds a specified length.

Code Text: *Wood trusses shall be designed in accordance with approved engineering practice. The design and manufacture of metal plate connected wood trusses shall comply with ANSI/TPI 1. Trusses shall be braced to prevent rotation and provide lateral stability in accordance with the requirements specified in the construction documents for the building and on the individual truss design drawings. Truss members and components shall not be cut, notched, spliced or otherwise altered in any way without the approval of a registered design professional.*

Discussion and Commentary: There are no prescriptive provisions for wood trusses in the IRC. All trusses must be designed, and any modification or alteration must be approved by an architect or engineer.

Minimum Truss Design Drawing Information

1. Slope or depth, span, and spacing.
2. Location of all joints.
3. Required bearing widths.
4. Design loads as applicable.
    4.1 Top chord live load.
    4.2 Top chord dead load.
    4.3 Bottom chord live load.
    4.4 Bottom chord dead load.
    4.5 Concentrated loads and their points of application.
    4.6 Controlling wind and earthquake loads.
5. Adjustments to lumber and joint connector design values for conditions of use.
6. Each reaction force and direction.
7. Joint connector type and description (e.g., size, thickness or gage); and the dimensioned location of each joint connector except where symmetrically located relative to the joint interface.
8. Lumber size, species and grade for each member.
9. Connection requirements for:
    9.1 Truss-to-truss girder.
    9.2 Truss ply-to ply.
    9.3 Field splices.
10. Calculated deflection ratio and/or maximum description for live and total load.
11. Maximum axial compression forces in the truss members to enable the building designer to design the size, connections and anchorage of the permanent continuous lateral bracing. Forces shall be shown on the truss drawing or on supplemental documents.
12. Required permanent truss member bracing location.

Truss design drawings must be prepared by a registered design professional and submitted to the building official and approved prior to installation of the trusses. The drawings shall be provided with the shipment of trusses to the site and must contain at least the information specified in Section R502.11.4.

Code Text: *Draftstopping shall be provided in accordance with Section R302.12. In combustible construction where there is usable space both above and below the concealed space of a floor/ceiling assembly, draftstops shall be installed so that the area of the concealed space does not exceed 1,000 square feet (92.9 m²). Draftstopping shall divide the concealed space into approximately equal areas. Where the assembly is enclosed by a floor membrane above and a ceiling membrane below, draftstopping shall be provided in floor/ceiling assemblies under the following circumstances: 1) ceiling is suspended under the floor framing, or 2) floor framing is constructed of truss-type open-web or perforated members.*

Discussion and Commentary: Draftstopping, like fireblocking, is only mandated in combustible construction. Although the construction of draftstops is not as substantial as that for fireblocking, its integrity is just as critical and must be maintained.

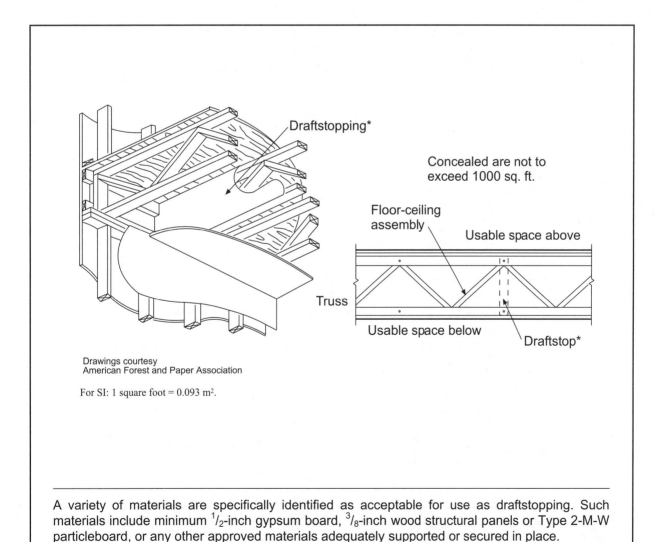

Draftstopping*

Concealed are not to exceed 1000 sq. ft.

Floor-ceiling assembly

Usable space above

Truss

Usable space below

Draftstop*

Drawings courtesy
American Forest and Paper Association

For SI: 1 square foot = 0.093 m².

A variety of materials are specifically identified as acceptable for use as draftstopping. Such materials include minimum ¹/₂-inch gypsum board, ³/₈-inch wood structural panels or Type 2-M-W particleboard, or any other approved materials adequately supported or secured in place.

Code Text: *Where used as subflooring or combination subfloor underlayment, wood structural panels shall be of one of the grades specified in Table R503.2.1.1(1). The maximum allowable span for wood structural panels used as subfloor or combination subfloor underlayment shall be as set forth in Table R503.2.1.1(1) or APA E30.*

Discussion and Commentary: The maximum spans for wood structural panel floor sheathing are limited by the stress and deflection imposed by the design live loads. The edges of the panels between floor supports are prevented from moving relative to each other by use of tongue and groove panel edges, the addition of wood blocking or by the use of an approved underlayment or structural finished floor system.

TABLE R503.2.1.1(1)
ALLOWABLE SPANS AND LOADS FOR WOOD STRUCTURAL PANELS FOR ROOF AND SUBFLOOR SHEATHING AND COMBINATION SUBFLOOR UNDERLAYMENT[a, b, c]

SPAN RATING	MINIMUM NOMINAL PANEL THICKNESS (inch)	ALLOWABLE LIVE LOAD (psf)[h, l]		MAXIMUM SPAN (inches)		LOAD (pounds per square foot, at maximum span)		MAXIMUM SPAN (inches)
		SPAN @ 16[2] o.c.	SPAN @ 24[2] o.c.	With edge support[d]	Without edge support	Total load	Live load	
Sheathing[e]				Roof[f]				Subfloor[l]
12/0	$5/16$	—	—	12	12	40	30	0
16/0	$5/16$	30	—	16	16	40	30	0
20/0	$5/16$	50	—	20	20	40	30	0
24/0	$3/8$	100	30	24	20[g]	40	30	0
24/16	$7/16$	100	40	24	24	50	40	16
32/16	$15/32, 1/2$	180	70	32	28	40	30	16[h]
40/20	$19/32, 5/8$	305	130	40	32	40	30	20[h,i]
48/24	$23/32, 3/48$	—	175	48	36	45	35	24
60/32	$7/8$	—	305	60	48	45	35	32
Underlayment, C-C plugged, single floor[e]				Roof[f]				Combination subfloor underlayment[k]
16 o.c.	$19/32, 5/8$	100	40	24	24	50	40	16[i]
20 o.c.	$19/32, 5/8$	150	60	32	32	40	30	20[i,j]
24 o.c.	$23/32, 3/4$	240	100	48	36	35	25	24
32 o.c.	$7/8$	—	185	48	40	50	40	32
48 o.c.	$13/32, 1 1/8$	—	290	60	48	50	40	48

For SI: 1 inch = 25.4 mm, 1 pound per square foot = 0.0479 kPa.

a. The allowable total loads were determined using a dead load of 10 psf. If the dead load exceeds 10 psf, then the live load shall be reduced accordingly.
b. Panels continuous over two or more spans with long dimension perpendicular to supports. Spans shall be limited to values shown because of possible effect of concentrated loads.
c. Applies to panels 24 inches or wider.
d. Lumber blocking, panel edge clips (one midway between each support, except two equally spaced between supports when span is 48 inches), tongue-and-groove panel edges, or other approved type of edge support.
e. Includes Structural 1 panels in these grades.
f. Uniform load deflection limitation: $1/180$ of span under live load plus dead load, $1/240$ of span under live load only.
g. Maximum span 24 inches for $15/32$-and $1/2$-inch panels.
h. Maximum span 24 inches where $3/4$-inch wood finish flooring is installed at right angles to joists.
i. Maximum span 24 inches where 1.5 inches of lightweight concrete or approved cellular concrete is placed over the subfloor.
j. Unsupported edges shall have tongue-and-groove joints or shall be supported with blocking unless minimum nominal $1/4$-inch thick underlayment with end and edge joints offset at least 2 inches or 1.5 inches of lightweight concrete or approved cellular concrete is placed over the subfloor, or $3/4$-inch wood finish flooring is installed at right angles to the supports. Allowable uniform live load at maximum span, based on deflection of $1/360$ of span, is 100 psf.
k. Unsupported edges shall have tongue-and-groove joints or shall be supported by blocking unless nominal $1/4$-inch-thick underlayment with end and edge joints offset at least 2 inches or $3/4$-inch wood finish flooring is installed at right angles to the supports. Allowable uniform live load at maximum span, based on deflection of $1/360$ of span, is 100 psf, except panels with a span rating of 48 on center are limited to 65 psf total uniform load at maximum span.
l. Allowable live load values at spans of 16" o.c. and 24" o.c taken from reference standard APA E30, APA Engineered Wood Construction Guide. Refer to reference standard for allowable spans not listed in the table.

Wood structural panels are the class of panels manufactured with fully waterproof adhesive and include plywood, oriented strand board (OSB), and composite panels made up of a combination of wood veneers and reconstructed wood layers.

Code Text: *Load-bearing floor framing members shall comply with Figure R505.2(1) and with the dimensional and minimum thickness requirements specified in Tables R505.2(1) and R505.2(2). Screws for steel-to-steel connections shall be installed with a minimum edge distance and center-to-center spacing of $^1/_2$ inch (12.7 mm), shall be self-drilling tapping, and shall conform to ASTM C 1513.*

Discussion and Commentary: The provisions of Section R505 apply to the construction of floor systems using cold-formed steel framing. Cold-formed steel structural members have profiles with rounded corners and slender flat elements with large width-to-thickness ratios. Cold-formed steel structural members are produced from carbon or low alloy steel sheet, strip, plate or bars not more than one inch thick. Forming of the steel shapes is done at or near room temperature using bending brakes, press brakes or roll forming machines.

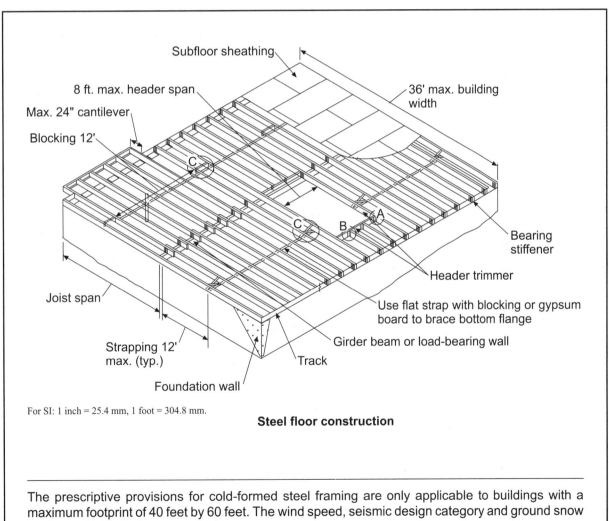

For SI: 1 inch = 25.4 mm, 1 foot = 304.8 mm.

Steel floor construction

The prescriptive provisions for cold-formed steel framing are only applicable to buildings with a maximum footprint of 40 feet by 60 feet. The wind speed, seismic design category and ground snow load are also limited.

Topic: Concrete Slab-on-Ground Floors	**Category:** Floors
Reference: IRC R506.1, R506.2	**Subject:** Concrete Floors on Ground

Code Text: *Concrete slab-on-ground floors shall be a minimum 3.5 inches (89 mm) thick (for expansive soils, see Section R403.1.8). The specified compressive strength of concrete shall be as set forth in Section R402.2. The area within the foundation walls shall have all vegetation, top soil and foreign material removed. Fill shall be compacted to assure uniform support of the slab, and except where approved, the fill depths shall not exceed 24 inches (610 mm) for clean sand or gravel and 8 inches (203 mm) for earth. A 4-inch-thick (102 mm) base course consisting of clean graded sand, gravel, crushed stone or crushed blast-furnace slag passing a 2-inch (51 mm) sieve shall be placed on the prepared subgrade when the slab is below grade. See exception for base course where slab is installed on well-drained or sand-gravel mixture soils.*

Discussion and Commentary: Removal of construction debris and foreign materials, such as lumber formwork, stakes, tree stumps and other vegetation, limits the attraction of termites, insects and vermin. Top soil and vegetation should also be removed in that it is generally loosely compacted to such an extent as to allow soil settlement.

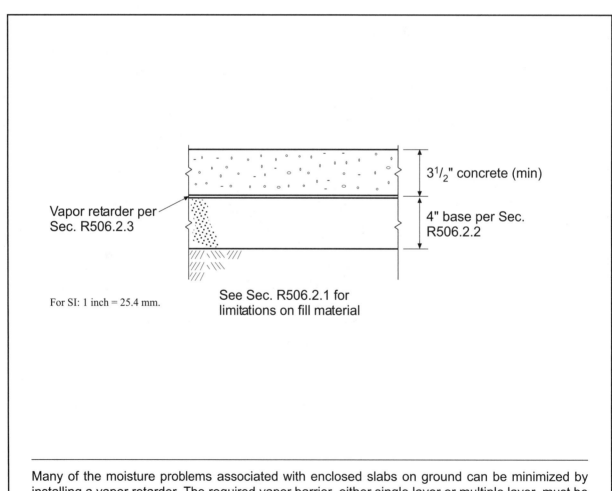

Many of the moisture problems associated with enclosed slabs on ground can be minimized by installing a vapor retarder. The required vapor barrier, either single layer or multiple layer, must be properly installed with lapped joints, and the barrier cannot be punctured during construction.

Topic: Reinforcement Support	**Category:** Floors
Reference: IRC R506.2.4	**Subject:** Concrete Floors on Ground

Code Text: *Where provided in slabs on ground, reinforcement shall be supported to remain in place from the center to upper one third of the slab for the duration of the concrete placement.*

Discussion and Commentary: One of the major problems of slab-on-ground construction is the placement of reinforcement, whether it consists of welded wire mesh or reinforcing bars, prior to and during the placement of concrete. Although the IRC does not mandate such reinforcement, the proper installation is addressed in those cases where it is installed.

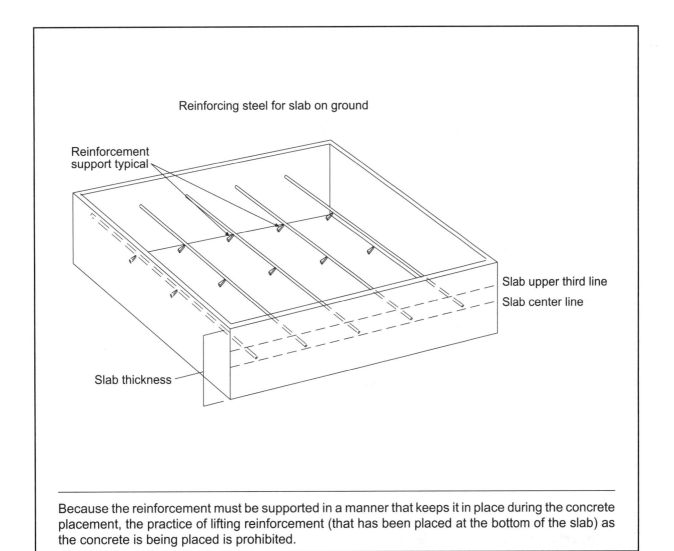

Reinforcing steel for slab on ground

Reinforcement support typical

Slab upper third line

Slab center line

Slab thickness

Because the reinforcement must be supported in a manner that keeps it in place during the concrete placement, the practice of lifting reinforcement (that has been placed at the bottom of the slab) as the concrete is being placed is prohibited.

Code Text: *Where supported by attachment to an exterior wall, decks shall be positively anchored to the primary structure and designed for both vertical and lateral loads. Such attachment shall not be accomplished by the use of toenails or nails subject to withdrawal. For decks supporting a total design load of 50 pounds per square foot (2394 Pa) [40 pounds per square foot (1915 Pa) live load plus 10 pounds per square foot (479 Pa) dead load], the connection ... shall be constructed with $^1/_2$-inch (12.7 mm) lag screws or bolts with washers in accordance with Table R507.2. Deck ledger connections not conforming to Table R507.2 shall be designed in accordance with accepted engineering practice. Girders supporting deck joists shall not be supported on deck ledgers or band joists. Deck ledgers shall not be supported on stone or masonry veneer.*

Discussion and Commentary: Where an exterior wall supports a deck, the deck framing must be positively attached to the building structure. The connection method must address both vertical and lateral loads, and the connection must be available for inspection.

TABLE R507.2
**FASTENER SPACING FOR A SOUTHERN PINE OR HEM-FIR DECK LEDGER AND
A 2-INCH-NOMINAL SOLID-SAWN SPRUCE-PINE-FIR BAND JOIST[c, f, g]**
(Deck live load = 40 psf, deck dead load = 10 psf)

JOIST SPAN	6′ and less	6′1″ to 8′	8′1″ to 10′	10′1″ to 12′	12′1″ to 14′	14′1″ to 16′	16′1″ to 18′
Connection details	On-center spacing of fasteners[d, e]						
$^1/_2$ inch diameter lag screw with $^{15}/_{32}$ inch maximum sheathing[a]	30	23	18	15	13	11	10
$^1/_2$ inch diameter bolt with $^{15}/_{32}$ inch maximum sheathing	36	36	34	29	24	21	19
$^1/_2$ inch diameter bolt with $^{15}/_{32}$ inch maximum sheathing and $^1/_2$ inch stacked washers[b, h]	36	36	29	24	21	18	16

For SI: 1 inch = 25.4 mm, 1 foot = 304.8 mm. 1 pound per square foot = 0.0479 kPa.

a. The tip of the lag screw shall fully extend beyond the inside face of the band joist.

b. The maximum gap between the face of the ledger board and face of the wall sheathing shall be $^1/_2$ inch.

c. Ledgers shall be flashed to prevent water from contacting the house band joist.

d. Lag screws and bolts shall be staggered in accordance with Section R507.2.1.

e. Deck ledger shall be minimum 2 × 8 pressure-preservative-treated No. 2 grade lumber, or other approved materials as established by standard engineering practice.

f. When solid-sawn pressure-preservative-treated deck ledgers are attached to a minimum 1-inch-thick engineered wood product (structural composite lumber, laminated veneer lumber or wood structural panel band joist), the ledger attachment shall be designed in accordance with accepted engineering practice.

g. A minimum 1 × 9$^1/_2$ Douglas Fir laminated veneer lumber rimboard shall be permitted in lieu of the 2-inch nominal band joist.

h. Wood structural panel sheathing, gypsum board sheathing or foam sheathing not exceeding 1 inch in thickness shall be permitted. The maximum distance between the face of the ledger board and the face of the band joist shall be 1 inch.

Wood/plastic composite deck boards, stair treads, handrails and guardrail systems are permitted for exterior decks provided they are properly labeled and installed in accordance with the manufacturer's instructions. The label shall indicate compliance with ASTM D 7032.

Quiz

Study Session 6
IRC Chapter 5

1. A minimum design live load of _____ shall be used for the determination of the maximum allowable floor joist spans in attics that are accessed by a fixed stairway.

    a. 10 psf

    b. 20 psf

    c. 30 psf

    d. 40 psf

    Reference _____

2. Where 2-inch by 10-inch floor joists of #2 hem-fir are spaced at 16 inches on center, what is the maximum allowable span where such joists support a sleeping room and a dead load of 10 psf?

    a. 16 feet, 0 inches

    b. 16 feet, 10 inches

    c. 17 feet, 8 inches

    d. 19 feet, 8 inches

    Reference _____

3. Where 2-inch by 10-inch floor joists of #1 spruce-pine-fir are spaced at 16 inches on center, what is the maximum allowable span where such joists support a living room and a dead load of 20 psf?

    a. 14 feet, 1 inch

    b. 15 feet, 5 inches

    c. 16 feet, 0 inches

    d. 16 feet, 9 inches

    Reference _____

4. Where the ground snow load is 30 psf, what is the maximum allowable span of a built-up spruce-pine-fir girder consisting of three 2 by 12s when it is located at an exterior bearing wall and supports a roof, a ceiling and one center-bearing floor in a building having a width of 28 feet?

    a. 5 feet, 2 inches   b. 7 feet, 8 inches

    c. 8 feet, 0 inches   d. 8 feet, 11 inches

Reference _____

5. Where a built-up girder of three Southern pine 2 by 10s is used for an exterior bearing wall, what is the minimum number of jack studs required to support a girder that carries a roof, ceiling and one clear span floor in a building having a width of 20 feet?

    a. 1   b. 2

    c. 3   d. 4

Reference _____

6. What is the maximum allowable girder span for an interior bearing wall using a built-up girder consisting of three hem fir 2 by 10s and supporting one floor in a building having a width of 36 feet?

    a. 5 feet, 5 inches   b. 6 feet, 9 inches

    c. 7 feet, 7 inches   d. 8 feet, 10 inches

Reference _____

7. Doubled joists under parallel bearing partitions that are separated to permit the installation of piping or vents shall be blocked at maximum intervals of _____ on center.

    a. 16 inches   b. 32 inches

    c. 48 inches   d. 60 inches

Reference _____

8. The ends of floor joists shall bear a minimum of _____ on wood or metal.

    a. 1 inch   b. $1^1/_2$ inches

    c. $2^1/_2$ inches   d. 3 inches

Reference _____

9. Where floor joists frame from opposite sides across the top of a wood girder and are lapped, the minimum lap shall be _____ .

    a. 3 inches   b. 4 inches

    c. 6 inches   d. 8 inches

Reference _____

10. Bridging shall be provided to support floor joists laterally where the minimum 2x joist depth exceeds _____ nominal size.

    a. 6 inches   b. 8 inches

    c. 10 inches   d. 12 inches

Reference _____

11. What is the maximum permitted length of a notch in a floor joist?

    a. 2 inches

    b. twice the notch depth

    c. $^1/_3$ the depth of the joist

    d. $^1/_6$ the depth of the joist

Reference _____

12. A hole bored through a floor joist shall have a maximum diameter of _____ .

    a. 2 inches

    b. $^1/_6$ the depth of the joist

    c. $^1/_4$ the depth of the joist

    d. $^1/_3$ the depth of the joist

Reference _____

13. Tail joists exceeding _____ in length shall be supported by framing anchors or on complying ledger strips.

    a. 4 feet   b. 6 feet

    c. 8 feet   d. 12 feet

Reference _____

14. Where a ceiling is suspended below the floor framing, draftstops shall be installed so that the maximum area of any concealed space is _____ square feet.

    a. 100   b. 400

    c. 1,000   d. 1,500

Reference _____

15. Where wood structural panels are used for required draftstops in concealed floor/ceiling assemblies, what is the minimum thickness mandated?

    a. $^5/_{16}$ inch   b. $^3/_8$ inch

    c. $^{15}/_{32}$ inch   d. $^{23}/_{32}$ inch

Reference _____

16. Wood structural panels $^{15}/_{32}$-inch thick are to be used as subfloor sheathing and are to be covered with $^3/_4$-inch wood finish flooring installed at right angles to the joists. What is the maximum allowable span of the wood structural panels if the span rating of the panels is 32/16?

    a. 0 inches; the panels may not be used as subfloor sheathing

    b. 16 inches

    c. 20 inches

    d. 24 inches

Reference _____

17. Where Species Group 2 sanded plywood is used as combination subfloor underlayment, what is the minimum required plywood thickness where the joists are spaced at 16 inches on center?

    a. $^1/_2$ inch   b. $^5/_8$ inch

    c. $^3/_4$ inch   d. $^7/_8$ inch

Reference _____

18. Steel floor framing constructed in accordance with the prescriptive provisions of the IRC is limited to buildings a maximum of _____ stories above grade plane with each story having a maximum length perpendicular to the joist span of _____ feet.

    a. 2, 40   b. 2, 60

    c. 3, 40   d. 3, 60

Reference _____

19. Where complying screws are used for the connection of steel floor framing members, screws shall extend through the steel a minimum of _____ .

    a. $^{1}/_{8}$ inch   b. $^{3}/_{8}$ inch

    c. three exposed threads   d. five exposed threads

Reference _____

20. In a steel floor framing system, No. 8 screws spaced at a maximum of _____ on center on the edges and _____ on center at the intermediate supports shall be used to fasten the subfloor to the floor joists.

    a. 6 inches, 10 inches   b. 6 inches, 12 inches

    c. 8 inches, 12 inches   d. 8 inches, 14 inches

Reference _____

21. Cold-formed 33 ksi steel joists are installed as single spans at 16 inches on center in the floor framing system for a dwelling. If the nominal joist size is 800S162-43, the maximum span for a 40 psf live load is _____ .

    a. 10 feet, 5 inches   b. 12 feet, 3 inches

    c. 14 feet, 1 inch   d. 15 feet, 6 inches

Reference _____

22. Where a wood beam does not bear directly on a masonry wall, a 2-inch-thick sill plate must be provided under the beam with a minimum nominal bearing area of _____ square inches.

    a. 32   b. 48

    c. 60   d. 64

Reference _____

23. The maximum fill depth when preparing a site for construction of a concrete slab-on-ground floor is _____ for earth and _____ for clean sand or gravel.

    a. 6 inches, 12 inches   b. 8 inches, 16 inches

    c. 8 inches, 24 inches   d. 12 inches, 24 inches

Reference _____

24. A base course is not required for the prepared subgrade for a concrete floor slab below grade where the soil is classified as _____ .

    a. GM   b. SC

    c. CH   d. OH

Reference _____

25. Where a vapor retarder is required between a concrete floor slab and the prepared subgrade, the joints of the vapor retarder shall be lapped a minimum of _____ .

    a. 2 inches   b. 4 inches

    c. 6 inches   d. 8 inches

Reference _____

26. Where supporting only a light-frame exterior bearing wall and roof, what is the maximum cantilever span for 2 by 10 floor joists spaced at 16 inches on center, provided the roof has a width of 32 feet and the ground snow load is 30 psf?

    a. 18 inches   b. 21 inches

    c. 22 inches   d. 26 inches

Reference _____

27. Where supporting an exterior balcony in an area having a ground snow load of 50 psf, what is the maximum cantilever span for 2 by 12 floor joists spaced at 16 inches on center?

    a. 49 inches   b. 57 inches

    c. 67 inches   d. 72 inches

Reference _____

28. Unless the joists are of sufficient size to carry the load, bearing partitions perpendicular to floor joists shall be offset a maximum of _____ from supporting girders, walls or partitions.

    a. 6 inches

    b. 12 inches

    c. the depth of the floor sheathing

    d. the depth of the floor joists

Reference _____

29. Where an opening in floor framing is framed with a single header joist the same size as the floor joist, the maximum header joist span shall be _____ feet.

    a. 3   b. 4

    c. 6   d. 12

Reference _____

30. If a base course is required for a concrete floor slab installed below grade, the minimum thickness of the base course shall be _____ inches.

    a. 3   b. 4

    c. 6   d. 8

Reference _____

31. Unless supported by other approved means, the ends of each joist, beam or girder shall have a minimum of _____ inches of bearing on masonry or concrete.

    a. $1^{1}/_{2}$ b. 2

    c. 3   d. 4

Reference _____

32. Notches at the ends of solid lumber joists, rafters and beams shall have a maximum depth of _____ .

    a. 2 inches

    b. 3 inches

    c. one-half the depth of the member

    d. one-fourth the depth of the member

Reference _____

33. Holes bored into solid lumber joists, rafter and beams shall be located a minimum of _____ from the top or bottom of the member.

    a. 1 inch

    b. 2 inches

    c. one-sixth the member depth

    d. one-third the member depth

Reference _____

34. Where floor framing members are spaced at 24 inches on center, what is the maximum allowable live load for floor sheathing rated at 32/16 if 1.5 inches of lightweight concrete is placed over the subfloor?

    a. 30 psf

    b. 40 psf

    c. 70 psf

    d. 180 psf

    Reference _____

35. Where reinforcement is provided in concrete slab-on-ground floors, the reinforcement shall be supported to remain within what section of the slab during the concrete placement?

    a. the middle one-third

    b. the middle one-half

    c. the center to the upper one-third

    d. the center to the upper one-fourth

    Reference _____

2012 IRC Chapters 6 and 7
Wall Construction and Wall Covering

OBJECTIVE: To obtain an understanding of the requirements dealing with wall construction and wall covering, including wood wall framing details; headers in both exterior and interior walls; fireblocking; braced wall panels and braced wall lines; cripple walls; steel wall framing; masonry wall construction; exterior windows and exterior doors; gypsum board plaster, ceramic tile and other interior wall finishes; and exterior wall finishes such as siding, plaster and veneer.

REFERENCE: Chapters 6 and 7, 2012 *International Residential Code*

KEY POINTS:
- How is the maximum allowable spacing of wood studs determined? How is a double top plate to be installed? A single top plate?
- What is the appropriate number and type of fasteners for connecting wood structural members? What is the maximum allowable spacing of such fasteners? What alternative fastening methods are acceptable?
- What are the height limitations for 2x4 wood studs? For 2x6 studs?
- Where must a bored hole in a stud be located? What is the maximum permitted size of a bored hole located in a bearing wall? In a nonbearing wall?
- What is the maximum notch size allowed in a bearing stud wall? In a nonbearing wall? How are notches and bored holes in a top plate to be addressed?
- What is the method for determining the maximum header span in an interior bearing wall? In an exterior bearing wall? In a nonbearing wall?
- What is a wood structural panel box header? How is a box header to be constructed?
- Where is fireblocking required? What materials can be used as fireblocking?
- How are foundation cripple studs to be sized? When must cripple walls be sheathed?
- How are braced wall lines to be spaced?
- What intermittent methods are acceptable for the construction of braced wall panels? What percentage of a braced wall line must consist of braced wall panels? How close to the end of a braced wall line is a braced wall panel required? What is minimum required panel length?

KEY POINTS:
(Cont'd)

- What alternate methods of construction are provided for braced wall panels? What are the code benefits of continuous structural panel sheathing?

- How do braced wall panels need to be connected to the floor framing and top plates? What special conditions apply in Seismic Design Categories D_0, D_1 and D_2?

- What methods are acceptable for connecting top plates of exterior braced wall panels to roof framing members? How are braced wall panels required to be supported?

- What is the simplified wall bracing method? When is it permitted to be utilized?

- What are the general provisions for load-bearing steel wall framing? How are such walls to be fastened? How are connections to be made?

- What is the minimum required thickness of masonry walls? Parapet walls?

- How must masonry walls be laterally supported? What is the minimum spacing between lateral supports? How are walls to be anchored at floors and roofs?

- How shall glass masonry be constructed?

- What limitations are placed on insulating concrete form wall construction?

- What is the performance level required for exterior windows and doors? How must such windows and doors be tested and identified? How should they be attached to the structure?

- What are the applicability limits on the use of structural insulated panel (SIP) walls? How are the walls to be connected? What are the limitations for notching and drilling?

- How must gypsum board be applied on an interior wall? How must it be fastened?

- Where is the use of water-resistant gypsum backing board prohibited?

- When is weather-resistant sheathing paper required on an exterior wall?

- What is the criteria for the installation of weep screeds in exterior plaster applications?

- How shall masonry veneer be supported? What method of anchorage is mandated?

- How must fiber cement siding be installed? Vinyl siding? Adhered masonry veneer?

Code Text: *Components of exterior walls shall be fastened in accordance with Tables R602.3(1) through R602.3(4).*

Discussion and Commentary: The fastener schedules provide the minimum nailing requirements for the connection of structural wood framing members, wood structural panels, and other types of wall sheathing materials. For dimensional lumber, the number of nails and maximum nail spacing is specified. The use of staples is permitted in specific locations. Wood structural panels and other sheathing materials are regulated by thickness for nail size, number and spacing. Staples and other alternate attachments are also permitted when in compliance with the code. Although many of the listed building elements are to some degree related to wall construction, a number of the connection locations are specific to floor and roof framing.

TABLE R602.3(1)
FASTENER SCHEDULE FOR STRUCTURAL MEMBERS

ITEM	DESCRIPTION OF BUILDING ELEMENTS	NUMBER AND TYPE OF FASTENER[a, b, c]	SPACING OF FASTENERS
	Roof		
1	Blocking between joists or rafters to top plate, toe nail	3-8d ($2^1/_2$ " × 0.113 ")	—
2	Ceiling joists to plate, toe nail	3-8d ($2^1/_2$ " × 0.113 ")	—
3	Ceiling joists not attached to parallel rafter, laps over partitions, face nail	3-10d	—
4	Collar tie to rafter, face nail or $1^1/_4$ " × 20 gage ridge strap	3-10d (3 " × 0.128 ")	—
5	Rafter or roof truss to plate, toe nail	3-16d box nails ($3^1/_2$ " × 0.135 ") or 3-10d common nails (3 " × 0.148 ")	2 toe nails on one side and 1 toe nail on opposite side of each rafter or truss[j]
6	Roof rafters to ridge, valley or hip rafters: toe nail face nail	4-16d ($3^1/_2$ " × 0.135 ") 3-16d ($3^1/_2$ " × 0.135 ")	—
	Wall		
7	Built-up studs-face nail	10d (3 " × 0.128 ")	24 " o.c.
8	Abutting studs at intersecting wall corners, face nail	16d (3 ½" x 0.135 ")	12 " o.c.
9	Built-up header, two pieces with $1/_2$ " spacer	16d ($3^1/_2$ " × 0.135 ")	16 " o.c. along each edge
10	Continued header, two pieces	16d ($3^1/_2$ " × 0.135 ")	16 " o.c. along each edge
11	Continuous header to stud, toe nail	4-8d ($2^1/_2$ " × 0.113 ")	—
12	Double studs, face nail	10d (3 " × 0.128 ")	24 " o.c.
13	Double top plates, face nail	10d (3 " × 0.128 ")	24 " o.c.
14	Double top plates, minimum 24-inch offset of end joints, face nail in lapped area	8-16d ($3^1/_2$ " × 0.135 ")	—
15	Sole plate to joist or blocking, face nail	16d ($3^1/_2$ " × 0.135 ")	16 " o.c.
16	Sole plate to joist or blocking at braced wall panels	3-16d ($3^1/_2$ " × 0.135 ")	16 " o.c.
		3-8d ($2^1/_2$ " × 0.113 ")	—

(continued)

The code allows the use of built-up members as headers and girders in various applications. The fastening schedules require staggered nailing with minimum 10d nails on built-up members so they may act as a single member. Thus, the loads are properly transferred to all elements of the built-up member.

Code Text: *The size, height and spacing of studs shall be in accordance with Table R602.3(5). See exceptions for limits on use of utility grade studs and studs exceeding 10 feet in height. Wood stud walls shall be capped with a double top plate installed to provide overlapping at corners and intersections with bearing partitions. End joints in top plates shall be offset at least 24 inches (610 mm). Joints in plates need not occur over studs. Plates shall be not less than 2 inches (51 mm) nominal thickness and have a width at least equal to the width of the studs. See exception for use of a single top plate.*

Discussion and Commentary: Unless the studs are supporting two floors and a roof /ceiling assembly or habitable attic, it is possible to use 2-inch by 4-inch studs for the framing of bearing walls up to 10 feet in height. The center-to-center spacing varies based on the loading conditions and the lumber grade. If the stud height exceeds 10 feet, it is likely that larger studs and/or closer spacing will be required.

TABLE R602.3(5)
SIZE, HEIGHT AND SPACING OF WOOD STUDS[a]

STUD SIZE (inches)	BEARING WALLS					NONBEARING WALLS	
	Laterally unsupported stud height[a] (feet)	Maximum spacing when supporting a roof-ceiling assembly or a habitable attic assembly, only (inches)	Maximum spacing when supporting one floor, plus a roof-ceiling assembly or a habitable attic assembly (inches)	Maximum spacing when supporting two floors, plus a roof-ceiling assembly or a habitable attic assembly (inches)	Maximum spacing when supporting one floor height[a] (feet)	Laterally unsupported stud height[a] (feet)	Maximum spacing (inches)
2 × 3[b]	—	—	—	—	—	10	16
2 × 4	10	24[c]	16[c]	—	24	14	24
3 × 4	10	24	24	16	24	14	24
2 × 5	10	24	24	—	24	16	24
2 × 6	10	24	24	16	24	20	24

For SI: 1 inch = 25.4 mm, 1 foot = 304.8 mm, 1 square foot = 0.093 m².

a. Listed heights are distances between points of lateral support placed perpendicular to the plane of the wall. Increases in unsupported height are permitted where justified by analysis.

b. Shall not be used in exterior walls.

c. A habitable attic assembly supported by 2 × 4 studs is limited to a roof span of 32 feet. Where the roof span exceeds 32 feet, the wall studs shall be increased to 2 × 6 or the studs shall be designed in accordance with accepted engineering practice.

The installation of double top plates is mandated for a variety of reasons. They tie the building together at corners and at wall intersections, serve as beams by supporting joists, rafters and trusses that do not bear above studs, and work as chords for floor and roof diaphragms.

Topic: Drilling and Notching

Category: Wall Construction

Reference: IRC R602.6

Subject: Wood Wall Framing

Code Text: *Any stud in an exterior wall or bearing partition may be cut or notched to a depth not exceeding 25 percent of its width. Studs in nonbearing partitions may be notched to a depth not to exceed 40 percent of a single stud width. Any stud may be bored or drilled, provided that the diameter of the resulting hole is no greater than 60 percent of the stud width, the edge of the hole is no closer than ⁵/₈ inch (15.9 mm) to the edge of the stud, and the hole is not located in the same section as a cut or notch. Studs located in exterior walls or bearing partitions drilled over 40 percent and up to 60 percent shall also be doubled with no more than two successive doubled studs bored.* See exceptions for the use of approved stud shoes.

Discussion and Commentary: In order to maintain the structural integrity of stud wall systems, limits are imposed on the notching or drilling of studs. Where a bored hole is oversized, it must be considered a notch for compliance purposes.

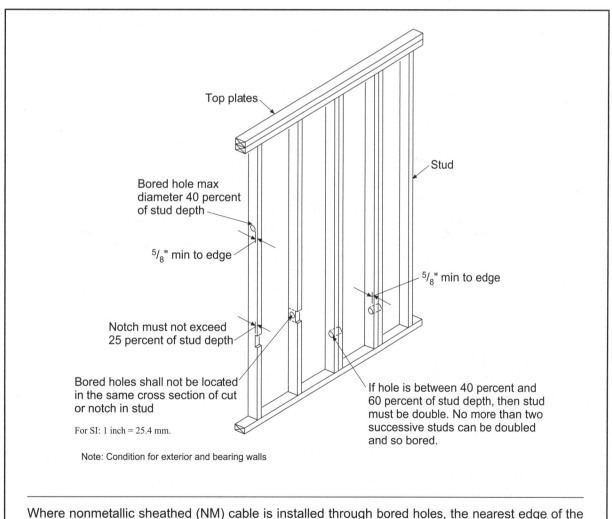

Top plates

Stud

Bored hole max diameter 40 percent of stud depth

⁵/₈" min to edge

⁵/₈" min to edge

Notch must not exceed 25 percent of stud depth

Bored holes shall not be located in the same cross section of cut or notch in stud

For SI: 1 inch = 25.4 mm.

Note: Condition for exterior and bearing walls

If hole is between 40 percent and 60 percent of stud depth, then stud must be double. No more than two successive studs can be doubled and so bored.

Where nonmetallic sheathed (NM) cable is installed through bored holes, the nearest edge of the hole must be located at least 1¼ inches from the edge of the stud. Otherwise, a minimum ¹/₁₆-inch steel plate, shoe or sleeve must be installed to provide the necessary physical protection.

Topic: Notching of Top Plate

Category: Wall Construction

Reference: IRC R602.6.1

Subject: Wood Wall Framing

Code Text: *When piping or ductwork is placed in or partly in an exterior wall or interior load-bearing wall, necessitating cutting, drilling or notching of the top plate by more than 50 percent of its width, a galvanized metal tie of not less than 0.054 inch thick (1.37 mm)(16 ga) and $1^1/_2$ inches (38 mm) wide shall be fastened across and to the plate at each side of the opening with not less than eight 10d (0.148 inch diameter) nails having a minimum length of $1^1/_2$ inches (38 mm) at each side or equivalent.* See exception for conditions where entire side of wall with notch is covered by wood structural panel sheathing.

Discussion and Commentary: It is not uncommon for the top plate to be cut or notched to allow for the placement of plumbing, heating or other pipes in exterior walls and load-bearing interior walls. Where removal of the top plate is extensive (more than one-half of its width), a substantial tie is necessary to provide for plate continuity and retain the structural integrity of the wall system

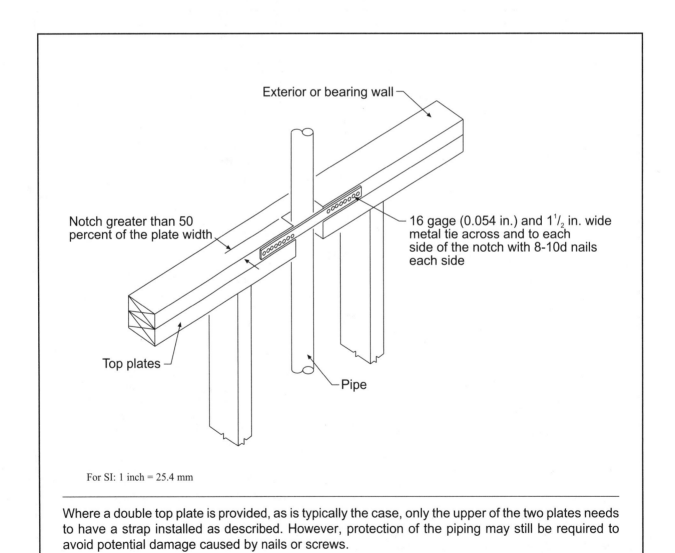

Exterior or bearing wall

Notch greater than 50 percent of the plate width

16 gage (0.054 in.) and $1^1/_2$ in. wide metal tie across and to each side of the notch with 8-10d nails each side

Top plates

Pipe

For SI: 1 inch = 25.4 mm

Where a double top plate is provided, as is typically the case, only the upper of the two plates needs to have a strap installed as described. However, protection of the piping may still be required to avoid potential damage caused by nails or screws.

Topic: Headers   **Category:** Wall Construction
Reference: IRC R602.7, R602.7.3   **Subject:** Wood Wall Framing

Code Text: *For header spans see Tables R502.5(1), R502.5(2) and R602.7.1. Load-bearing headers are not required in interior or exterior nonbearing walls. A single flat 2-inch-by-4-inch (51 mm by 102 mm) member may be used as a header in interior or exterior nonbearing walls for openings up to 8 feet (2438 mm) in width if the vertical distance to the parallel nailing surface above is not more than 24 inches (610 mm).*

Discussion and Commentary: Where it is necessary to transfer floor, ceiling, wall and roof loads to the foundation, headers must be installed over openings created for doors, windows and similar items. The maximum allowable spans are provided for various conditions, with the minimum permitted size of built-up header and supporting jack studs identified. Tables R502.5(1) and R502.5(2) establish maximum header spans for built-up headers consisting of two or more members. Single 2x member headers are also permitted in load-bearing walls under the limitations established in Table R602.7.1.

TABLE R602.7.1
SPANS FOR MINIMUM No.2 GRADE SINGLE HEADER[a, b, c, f]

SINGLE HEADERS SUPPORTING	SIZE	WOOD SPECIES	GROUND SNOW LOAD (psf)								
			≤ 20[d]			30			50		
			Building Width (feet)[e]								
			20	28	36	20	28	36	20	28	36
Roof and ceiling	2 × 8	Spruce-Pine-Fir	4-10	4-2	3-8	4-3	3-8	3-3	3-7	3-0	2-8
		Hem-Fir	5-1	4-4	3-10	4-6	3-10	3-5	3-9	3-2	2-10
		Douglas-Fir or Southern Pine	5-3	4-6	4-0	4-7	3-11	3-6	3-10	3-3	2-11
	2 × 10	Spruce-Pine-Fir	6-2	5-3	4-8	5-5	4-8	4-2	4-6	3-11	3-1
		Hem-Fir	6-6	5-6	4-11	5-8	4-11	4-4	4-9	4-1	3-7
		Douglas-Fir or Southern Pine	6-8	5-8	5-1	5-10	5-0	4-6	4-11	4-2	3-9
	2 × 12	Spruce-Pine-Fir	7-6	6-5	5-9	6-7	5-8	4-5	5-4	3-11	3-1
		Hem-Fir	7-10	6-9	6-0	6-11	5-11	5-3	5-9	4-8	3-8
		Douglas-Fir or Southern Pine	8-1	6-11	6-2	7-2	6-1	5-5	5-11	5-1	4-6
Roof, ceiling and one center-bearing floor	2 × 8	Spruce-Pine-Fir	3-10	3-3	2-11	3-9	3-3	2-11	3-5	2-11	2-7
		Hem-Fir	4-0	3-5	3-1	3-11	3-5	3-0	3-7	3-0	2-8
		Douglas-Fir or Southern Pine	4-1	3-7	3-2	4-1	3-6	3-1	3-8	3-2	2-9
	2 × 10	Spruce-Pine-Fir	4-11	4-2	3-8	4-10	4-1	3-6	4-4	3-7	2-10
		Hem-Fir	5-1	4-5	3-11	5-0	4-4	3-10	4-6	3-11	3-4
		Douglas-Fir or Southern Pine	5-3	4-6	4-1	5-2	4-5	4-0	4-8	4-0	3-7
	2 × 12	Spruce-Pine-Fir	5-8	4-2	3-4	5-5	4-0	3-6	4-9	3-6	2-10
		Hem-Fir	5-11	4-11	3-11	5-10	4-9	4-2	5-5	4-2	3-4
		Douglas-Fir or Southern Pine	6-1	5-3	4-8	6-0	5-2	4-10	5-7	4-10	4-3
Roof, ceiling and one clear span floor	2 × 8	Spruce-Pine-Fir	3-5	2-11	2-7	3-4	2-11	2-7	3-3	2-10	2-6
		Hem-Fir	3-7	3-1	2-9	3-6	3-0	2-8	3-5	2-11	2-7
		Douglas-Fir or Southern Pine	3-8	3-2	2-10	3-7	3-1	2-9	3-6	3-0	2-9
	2 × 10	Spruce-Pine-Fir	4-4	3-7	2-10	4-3	3-6	2-9	4-2	3-4	2-7
		Hem-Fir	4-7	3-11	3-5	4-6	3-10	3-3	4-4	3-9	3-1
		Douglas-Fir or Southern Pine	4-8	4-0	3-7	4-7	4-0	3-6	4-6	3-10	3-5
	2 × 12	Spruce-Pine-Fir	4-11	3-7	2-10	4-9	3-6	2-9	4-6	3-4	2-7
		Hem-Fir	5-6	4-3	3-5	5-6	4-2	3-3	5-4	3-11	3-1
		Douglas-Fir or Southern Pine	5-8	4-11	4-4	5-7	4-10	4-3	5-6	4-8	4-2

For SI: 1 inch=25.4 mm, 1 pound per square foot = 0.0479 kPa.
a. Spans are given in feet and inches.
b. Table is based on a maximum roof-ceiling dead load of 15 psf.
c. The header is permitted to be supported by an approved framing anchor attached to the full-height wall stud and to the header in lieu of the required jack stud.
d. The 20 psf ground snow load condition shall apply only when the roof pitch is 9:12 or greater. In conditions where the ground snow load is 30 psf or less and the roof pitch is less than 9:12, use the 30 psf ground snow load condition.
e. Building width is measured perpendicular to the ridge. For widths between those shown, spans are permitted to be interpolated.
f. The header shall bear on a minimum of one jack stud at each end.

Box headers are permitted when minimum $^{15}/_{32}$-inch wood structural panels are used. Header depths of both 9 inches and 15 inches are addressed, with panels attached to both sides or one side only. Adequate attachment to the framing is critical, with nails spaced at 3 inches on center.

Code Text: *Fireblocking shall be provided in accordance with Section R302.11. In combustible construction, fireblocking shall be provided to cut off all concealed draft openings (both vertical and horizontal) and to form an effective barrier between stories, and between a top story and the roof space. Fireblocking shall be provided in wood-frame construction in the following locations: 1) in concealed spaces of stud walls and partitions, including furred spaces, vertically at the ceiling and floor levels and horizontally at intervals not exceeding 10 feet (3048 mm); 2) at all interconnections between concealed vertical and horizontal spaces; 3) in concealed spaces between stair stringers at the top and bottom of the run; 4) at openings around vents, pipes and ducts at ceiling and floor level; 5) at chimneys and fireplaces per Section R1003.19; and 6) at cornices of a two-family dwelling at the line of dwelling unit separation.*

Discussion and Commentary: Fireblocking is defined as the installation of building materials to resist the free passage of flame to other areas of the building through concealed spaces.

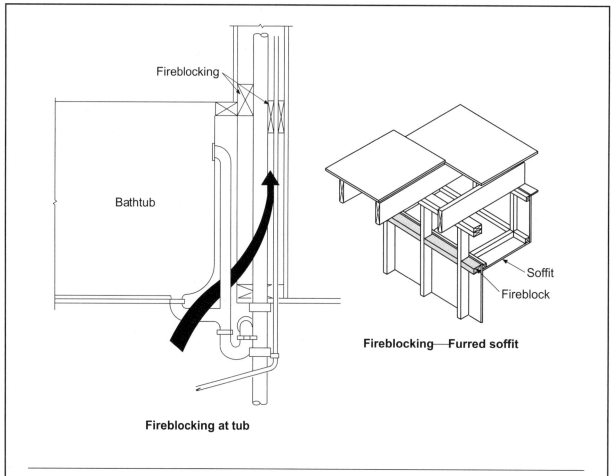

Fireblocking at tub

Fireblocking—Furred soffit

The technique of wood-frame platform construction provides natural fireblocking in many locations to restrict fire movement between horizontal and vertical concealed spaces. Where platform framing and vertical openings occur, a more prescriptive approach is necessary.

Topic: Braced Wall Panel Location	**Category:** Wall Construction
Reference: IRC R602.10.2	**Subject:** Wood Wall Framing

Code Text: *Braced wall panels shall be full-height sections of wall that shall have no vertical or horizontal offsets. Braced wall panels shall be constructed and placed along a braced wall line in accordance with Section R602.10 and the bracing methods specified in Section R602.10.4. A braced wall panel shall begin within 10 feet (3810 mm) from each end of a braced wall line as determined in Section R602.10.1.1. The distance between adjacent edges of braced wall panels along a braced wall line shall be no greater than 20 feet (6096 mm) as shown in Figure R602.10.2.2. See exceptions for buildings located in Seismic Design Categories D_0, D_1 and D_2.*

Discussion and Commentary: Braced wall lines are the lateral-resisting elements in conventional construction that are similar to shear walls in engineered structures. In order to properly transfer the lateral loads through the floor or roof diaphragms to the braced wall panels in the braced wall lines, the structure must be adequately connected.

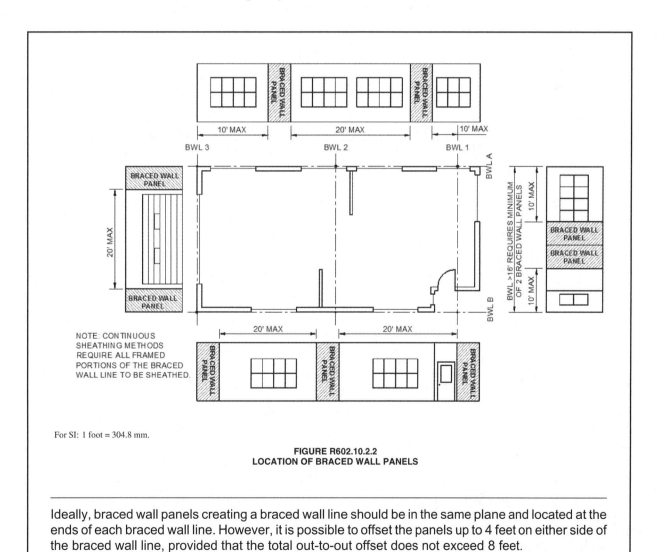

For SI: 1 foot = 304.8 mm.

FIGURE R602.10.2.2
LOCATION OF BRACED WALL PANELS

Ideally, braced wall panels creating a braced wall line should be in the same plane and located at the ends of each braced wall line. However, it is possible to offset the panels up to 4 feet on either side of the braced wall line, provided that the total out-to-out offset does not exceed 8 feet.

Code Text: *Intermittent and continuously sheathed braced wall panels shall be constructed in accordance with Section R602.10.4 and the methods listed in Table R602.10.4. Continuous sheathing methods require structural panel sheathing to be used on all sheathable surfaces on one side of a braced wall line including areas above and below openings and gable end walls and shall meet the requirements of Section R602.10.7. Braced wall panels shall have gypsum wall board installed on the side of the wall opposite the bracing material.* See exceptions for various construction methods.

Discussion and Commentary: Intermittent braced wall panels are often utilized as isolated elements of a braced wall line, as opposed to the use of continuous sheathing methods. Both types of bracing methods are described in Table R602.10.4 through a tabular format with a description, illustrative icon and connection criteria.

TABLE R602.10.4
BRACING METHODS

METHODS, MATERIAL		MINIMUM THICKNESS	FIGURE	CONNECTION CRITERIA[a]	
				Fasteners	Spacing
Intermittent Bracing Method	**LIB** Let-in-bracing	1 × 4 wood or approved metal straps at 45° to 60° angles for maximum 16″ stud spacing		Wood: 2-8d common nails or 3-8d ($2^1/_2$″ long x 0.113″ dia.) nails	Wood: per stud and top and bottom plates
				Metal strap: per manufacturer	Metal: per manufacturer
	DWB Diagonal wood boards	$^3/_4$″(1″ nominal) for maximum 24″ stud spacing		2-8d ($2^1/_2$″ long × 0.113″ dia.) nails or 2 - $1^3/_4$″ long staples	Per stud
	WSP Wood structural panel (See Section R604)	$^3/_8$″		Exterior sheathing per Table R602.3(3)	6″ edges 12″ field
				Interior sheathing per Table R602.3(1) or R602.3(2)	Varies by fastener
	BV-WSP[c] Wood Structural Panels with Stone or Masonry Veneer (See Section R602.10.6.5)	$^7/_{16}$″	See Figure R602.10.6.5	8d common ($2^1/_2$″ × 0.131) nails	4″ at panel edges 12″ at intermediate supports 4″ at braced wall panel end posts
	SFB Structural fiberboard sheathing	$^1/_2$″ or $^{25}/_{32}$″ for maximum 16″ stud spacing		$1^1/_2$″ long × 0.12″ dia. (for $^1/_2$″ thick sheathing) $1^3/_4$″ long × 0.12″ dia. (for $^{25}/_{32}$″ thick sheathing) galvanized roofing nails or 8d common ($2^1/_2$″ long × 0.131″ dia.) nails	3″ edges 6″ field
	GB Gypsum board	$^1/_2$″		Nails or screws per Table R602.3(1) for exterior locations	For all braced wall panel locations: 7″ edges (including top and bottom plates) 7″ field
				Nails or screws per Table R702.3.5 for interior locations	

(continued)

The gypsum board often required on the side of the wall opposite the bracing materials provides additional stiffness and resistance to lateral loads. Without the installation of gypsum board on the opposite side of the braced wall panel, it is possible that the braced wall line will have insufficient load-resisting capacity.

Topic: Structural Framing	**Category:** Wall Construction
Reference: IRC R603.2, R603.3	**Subject:** Steel Wall Framing

Code Text: *Load-bearing cold-formed steel wall framing members shall comply with Figure R603.2(1) and with the dimensional and minimum thickness requirements specified in Tables R603.2(1) and R603.2(2). All exterior cold-formed steel framed walls and interior load-bearing cold-formed steel framed walls shall be constructed in accordance with the provisions of Section R603.3.*

Discussion and Commentary: The use of load-bearing cold-formed steel members is regulated prescriptively not only for wall framing but also for floor and roof system framing. In order to verify that the steel members are in compliance with appropriate materials standards, the framing members must have a legible label, stencil, stamp or embossment with the manufacturer's identification as well as the minimum steel thickness, coating designation and yield strength.

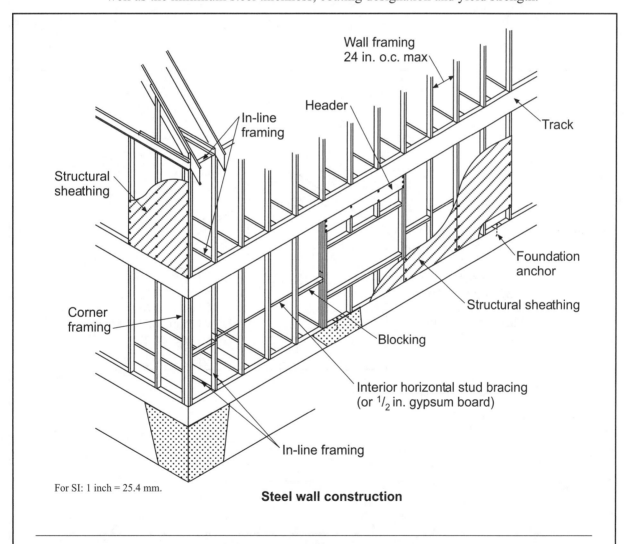

For SI: 1 inch = 25.4 mm.

Steel wall construction

A variety of tables, figures and details are provided in the code to assist the user in understanding the construction techniques and structural limitations of cold-formed steel-framed structures. When in compliance with the prescriptive provisions, the building is considered structurally sound.

Topic: Masonry Thickness

Category: Wall Construction

Reference: IRC R606.2.1

Subject: General Masonry Construction

Code Text: *The minimum thickness of masonry bearing walls more than one story high shall be 8 inches (203 mm). Solid masonry walls of one-story dwellings and garages shall not be less than 6 inches (152 mm) in thickness when not greater than 9 feet (2743 mm) in height, provided that when gable construction is used, an additional 6 feet (1829 mm) is permitted to the peak of the gable. Masonry walls shall be laterally supported in either the horizontal or vertical direction at intervals as required by Section R606.9.*

Discussion and Commentary: The minimum required width of masonry units is based directly on the height of the wall, both in number of stories and vertical dimension. Lateral support may be provided horizontally at complying intervals along the length of masonry walls, determined by the walls' length-to-thickness ratio, or vertically at the floors and/or the roof.

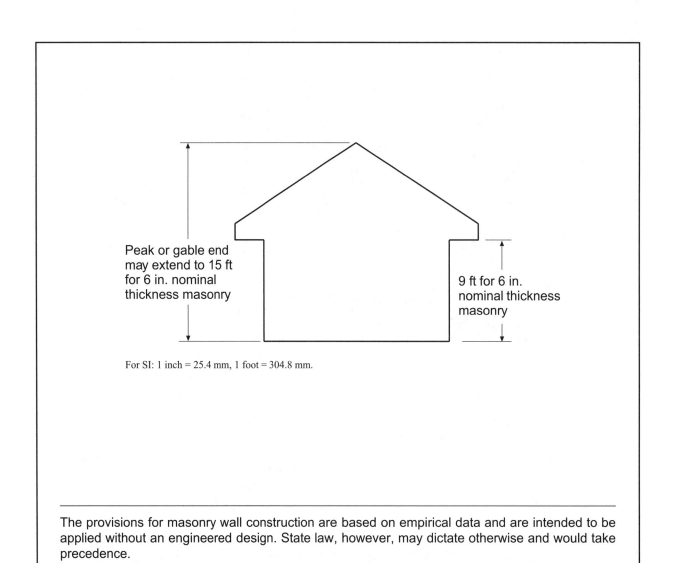

Peak or gable end may extend to 15 ft for 6 in. nominal thickness masonry

9 ft for 6 in. nominal thickness masonry

For SI: 1 inch = 25.4 mm, 1 foot = 304.8 mm.

The provisions for masonry wall construction are based on empirical data and are intended to be applied without an engineered design. State law, however, may dictate otherwise and would take precedence.

Code Text: *Masonry walls shall be laterally supported in either the horizontal or the vertical direction. The maximum spacing between lateral supports shall not exceed the distances in Table R606.9. Lateral support shall be provided by cross walls, pilasters, buttresses or structural frame members when the limiting distance is taken horizontally, or by floors or roofs when the limiting distance is taken vertically.*

Discussion and Commentary: The limitations on the maximum unsupported height or length of masonry walls specified in Table R606.9 provide reasonable performance. At the base of the wall, footings are a lateral support point. Thus, the unsupported height from the footing to the anchorage point at the floor or roof is the unsupported height, which must be limited to the values in Table R606.9.

TABLE R606.9
SPACING OF LATERAL SUPPORT FOR MASONRY WALLS

CONSTRUCTION	MAXIMUM WALL LENGTH TO THICKNESS OR WALL HEIGHT TO THICKNESS[a,b]
Bearing walls:	
Solid or solid grouted	20
All other	18
Nonbearing walls:	
Exterior	18
Interior	36

For SI: 1 foot = 304.8 mm.

a. Except for cavity walls and cantilevered walls, the thickness of a wall shall be its nominal thickness measured perpendicular to the face of the wall. For cavity walls, the thickness shall be determined as the sum of the nominal thicknesses of the individual wythes. For cantilever walls, except for parapets, the ratio of height to nominal thickness shall not exceed 6 for solid masonry, or 4 for hollow masonry. For parapets, see Section R606.2.4.

b. An additional unsupported height of 6 feet is permitted for gable end walls.

In lieu of unsupported dimensional limitations being measured vertically from footing to supporting floor or roof, the span limitations may be met with the use of pilasters, columns, piers, cross walls or similar elements whose relative stiffness is greater than the wall.

Code Text: *Masonry walls in Seismic Design Categories A, B and C shall be anchored to roof structures with metal strap anchors spaced in accordance with the manufacturer's instructions, $^{1}/_{2}$-inch (13 mm) bolts spaced not more than 6 feet (1829 mm) on center, or other approved anchors. Anchors shall be embedded at least 16 inches (406 mm) into the masonry, or be hooked or welded to bond beam reinforcement placed not less than 6 inches (152 mm) from the top of the wall.*

Discussion and Commentary: All anticipated lateral forces must have a path of transfer from the roof and floor diaphragms to the masonry walls for transfer to the building's foundation. It is necessary to provide adequate connections to maintain the required load path. Where the building is located in one of the higher seismic design categories, additional conditions are mandated.

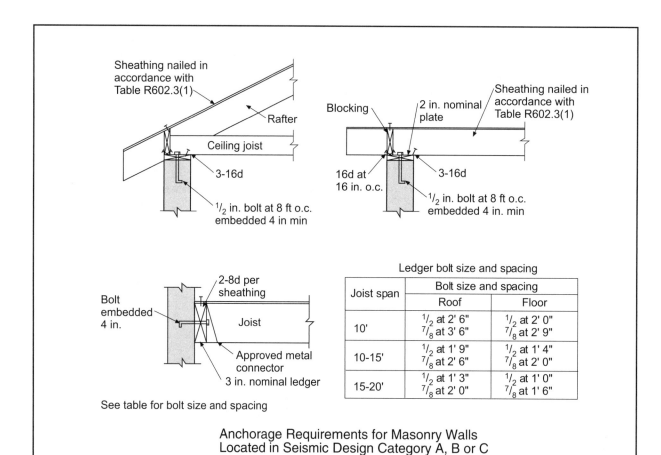

Ledger bolt size and spacing

| Joist span | Bolt size and spacing | |
	Roof	Floor
10'	$^{1}/_{2}$ at 2' 6" $^{7}/_{8}$ at 3' 6"	$^{1}/_{2}$ at 2' 0" $^{7}/_{8}$ at 2' 9"
10-15'	$^{1}/_{2}$ at 1' 9" $^{7}/_{8}$ at 2' 6"	$^{1}/_{2}$ at 1' 4" $^{7}/_{8}$ at 2' 0"
15-20'	$^{1}/_{2}$ at 1' 3" $^{7}/_{8}$ at 2' 0"	$^{1}/_{2}$ at 1' 0" $^{7}/_{8}$ at 1' 6"

Anchorage Requirements for Masonry Walls
Located in Seismic Design Category A, B or C
and where Wind Loads are Less than 30 psf

For SI: 1 inch = 25.4 mm, 1 foot = 304.8 mm.

Where the floor diaphragm is connected to a masonry wall, a ledger is used. The ledger fasteners transfer shear forces when loading is in the plane of the wall and prevent separation of the wall and flooring system when the forces occur out of plane.

Topic: Performance and Anchorage	**Category:** Wall Construction
Reference: IRC R612.2, R612.7.1	**Subject:** Exterior Windows and Glass Doors

Code Text: *Exterior windows and doors shall be designed to resist the design wind loads specified in Table R301.2(2) adjusted for height and exposure per Table R301.2(3). Window and glass door assemblies shall be anchored in accordance with the published manufacturer's recommendations to achieve the design pressure specified. Substitute anchoring systems used for substrates not specified by the fenestration manufacturer shall provide equal or greater anchoring performance as demonstrated by accepted engineering practice.*

Discussion and Commentary: Glass doors and windows are components of the exterior wall. As such, they are regulated through the specification of performance criteria for such components, as well as their supporting elements, in order to protect against glass breakage due to high wind pressure.

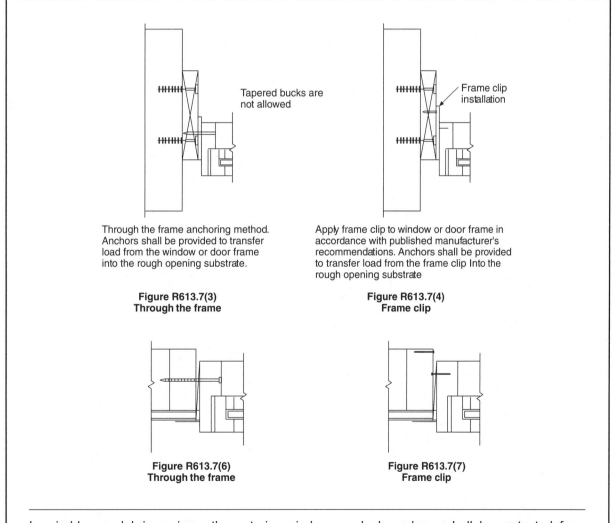

Tapered bucks are not allowed

Frame clip installation

Through the frame anchoring method. Anchors shall be provided to transfer load from the window or door frame into the rough opening substrate.

Apply frame clip to window or door frame in accordance with published manufacturer's recommendations. Anchors shall be provided to transfer load from the frame clip Into the rough opening substrate

Figure R613.7(3)
Through the frame

Figure R613.7(4)
Frame clip

Figure R613.7(6)
Through the frame

Figure R613.7(7)
Frame clip

In wind-borne debris regions, the exterior windows and glass doors shall be protected from wind-borne debris in accordance with Section R301.2.1.2. The installation of impact-resistant glazing or precut wood structural panels are options.

Code Text: *Maximum spacing of supports and the size and spacing of fasteners used to attach gypsum board shall comply with Table R702.3.5. Gypsum board shall be applied at right angles or parallel to framing members. All edges and ends of gypsum board shall occur on the framing members, except those edges and ends that are perpendicular to the framing members. Interior gypsum board shall not be installed where it is directly exposed to the weather or to water.*

Discussion and Commentary: Table R702.3.5 is a comprehensive table identifying the minimum thickness and fastening requirements for gypsum wallboard. The wallboard thickness ($^3/_8$-inch, $^1/_2$-inch, or $^5/_8$-inch), location of the wallboard (wall or ceiling), orientation of the wallboard to the framing members (parallel or perpendicular), spacing of framing members (16 inches or 24 inches on center), type of fasteners (nails or screws) and use of adhesive (with or without) are all set forth in the table.

TABLE R702.3.5
MINIMUM THICKNESS AND APPLICATION OF GYPSUM BOARD

THICKNESS OF GYPSUM BOARD (inches)	APPLICATION	ORIENTATION OF GYPSUM BOARD TO FRAMING	MAXIMUM SPACING OF FRAMING MEMBERS (inches o.c.)	MAXIMUM SPACING OF FASTENERS (inches) Nails[a]	Screws[b]	SIZE OF NAILS FOR APPLICATION TO WOOD FRAMING[c]
Application without adhesive						
$^3/_8$	Ceiling[d]	Perpendicular	16	7	12	13 gage, $1^1/_4$″ long, $^{19}/_{64}$″ head; 0.098″ diameter, $1^1/_4$″ long, annular-ringed; or 4d cooler nail, 0.080″ diameter, $1^3/_8$″ long, $^7/_{32}$″ head.
	Wall	Either direction	16	8	16	
$^1/_2$	Ceiling	Either direction	16	7	12	13 gage, $1^3/_8$″ long, $^{19}/_{64}$″ head; 0.098″ diameter, $1^1/_4$″ long, annular-ringed; 5d cooler nail, 0.086″ diameter, $1^5/_8$″ long, $^{15}/_{64}$″ head; or gypsum board nail, 0.086″ diameter, $1^5/_8$″ long, $^9/_{32}$″ head.
	Ceiling[d]	Perpendicular	24	7	12	
	Wall	Either direction	24	8	12	
	Wall	Either direction	16	8	16	
$^5/_8$	Ceiling	Either direction	16	7	12	13 gage, $1^5/_8$″ long, $^{19}/_{64}$″ head; 0.098″ diameter, $1^3/_8$″ long, annular-ringed; 6d cooler nail, 0.092″ diameter, $1^7/_8$″ long, $^1/_4$″ head; or gypsum board nail, 0.0915″ diameter, $1^7/_8$″ long, $^{19}/_{64}$″ head.
	Ceiling[e]	Perpendicular	24	7	12	
	Wall	Either direction	24	8	12	
	Wall	Either direction	16	8	16	
Application with adhesive						
$^3/_8$	Ceiling[d]	Perpendicular	16	16	16	Same as above for $^3/_8$″ gypsum board
	Wall	Either direction	16	16	24	
$^1/_2$ or $^5/_8$	Ceiling	Either direction	16	16	16	Same as above for $^1/_2$″ and $^5/_8$″ gypsum board, respectively
	Ceiling[d]	Perpendicular	24	12	16	
	Wall	Either direction	24	16	24	
Two $^3/_8$ layers	Ceiling	Perpendicular	16	16	16	Base ply nailed as above for $^1/_2$″ gypsum board; face ply installed with adhesive
	Wall	Either direction	24	24	24	

For SI: 1 inch = 25.4 mm.

a. For application without adhesive, a pair of nails spaced not less than 2 inches apart or more than $2^1/_2$ inches apart may be used with the pair of nails spaced 12 inches on center.

b. Screws shall be Type S or W per ASTM C 1002 and shall be sufficiently long to penetrate wood framing not less than $^5/_8$ inch and metal framing not less than $^3/_8$ inch.

c. Where metal framing is used with a clinching design to receive nails by two edges of metal, the nails shall be not less than $^5/_8$ inch longer than the gypsum board thickness and shall have ringed shanks. Where the metal framing has a nailing groove formed to receive the nails, the nails shall have barbed shanks or be 5d, $13^1/_2$ gage, $1^5/_8$ inches long, $^{15}/_{64}$-inch head for $^1/_2$-inch gypsum board; and 6d, 13 gage, $1^7/_8$ inches long, $^{15}/_{64}$-inch head for $^5/_8$-inch gypsum board.

d. Three-eighths-inch-thick single-ply gypsum board shall not be used on a ceiling where a water-based textured finish is to be applied, or where it will be required to support insulation above a ceiling. On ceiling applications to receive a water-based texture material, either hand or spray applied, the gypsum board shall be applied perpendicular to framing. When applying a water-based texture material, the minimum gypsum board thickness shall be increased from $^3/_8$ inch to $^1/_2$ inch for 16-inch on center framing, and from $^1/_2$ inch to $^5/_8$ inch for 24-inch on center framing or $^1/_2$-inch sag-resistant gypsum ceiling board shall be used.

e. Type X gypsum board for garage ceilings beneath habitable rooms shall be installed perpendicular to the ceiling framing and shall be fastened at maximum 6 inches o.c. by minimum $1^7/_8$ inches 6d coated nails or equivalent drywall screws.

Where screws are used for attaching gypsum board to wood framing, they shall be either Type W or Type S and must penetrate the wood at least $^5/_8$ inch. Type S screws are to be used where attachment is made to light-gage steel framing, with a minimum penetration of $^3/_8$ inch.

Code Text: *One layer of No. 15 asphalt felt, free from holes and breaks, complying with ASTM D 226 for Type 1 felt or other approved water-resistive barrier shall be applied over studs or sheathing of all exterior walls. Such felt or material shall be applied horizontally, with the upper layer lapped over the lower layer not less than 2 inches (51 mm). Where joints occur, felt shall be lapped not less than 6 inches (152 mm).* See three exceptions identifying conditions under which the water-resistive barrier may be omitted.

Discussion and Commentary: Structural members are adversely affected by moisture, particularly under cyclic conditions of wetting and drying. Due to the potential for water accumulation within the wall cavity, a water-resistant membrane is mandated behind any exterior siding or veneer.

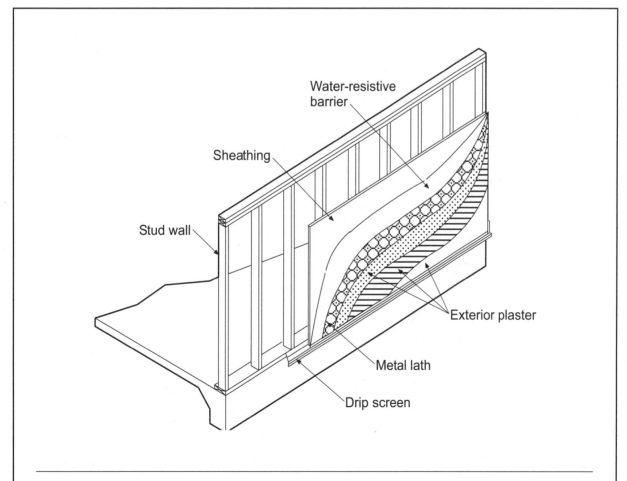

Under certain conditions, the value of asphalt felt or a similar weather-resistant barrier is questionable or its use unnecessary. It is permissible to eliminate the membrane in detached accessory buildings or where the siding, finish materials or lath provides the needed protection.

Code Text: *Veneer ties, if strand wire, shall not be less in thickness than No. 9 U.S. gage [(0.148 in.) (4 mm)] wire and shall have a hook embedded in the mortar joint, or if sheet metal, shall be not less than No. 22 U.S. gage by [(0.0299 in.)(0.76 mm)] $^7/_8$ inch (22 mm) corrugated. Each tie shall support not more than 2.67 square feet (0.25 m^2) of wall area and shall be spaced not more than 32 inches (813 mm) on center horizontally and 24 inches (635 mm) on center vertically.* See exception for a reduction in support area for high-wind and high-seismic areas.

Discussion and Commentary: Two types of ties are approved for the attachment of masonry veneer to wood construction: corrugated sheet metal ties and metal strand wire ties. The clearance between the veneer and the wood backing is limited and varies based upon which type of tie is used.

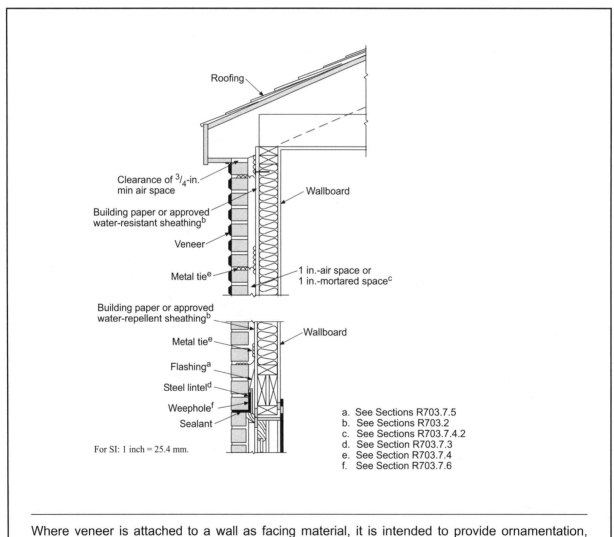

Roofing

Clearance of $^3/_4$-in. min air space

Building paper or approved water-resistant sheathing[b]

Veneer

Metal tie[e]

Wallboard

1 in.-air space or 1 in.-mortared space[c]

Building paper or approved water-repellent sheathing[b]

Metal tie[e]

Flashing[a]

Steel lintel[d]

Weephole[f]

Sealant

Wallboard

a. See Sections R703.7.5
b. See Sections R703.2
c. See Sections R703.7.4.2
d. See Section R703.7.3
e. See Section R703.7.4
f. See Section R703.7.6

For SI: 1 inch = 25.4 mm.

Where veneer is attached to a wall as facing material, it is intended to provide ornamentation, protection, insulation or a combination of these features. It does not provide any structural strength to the wall, nor is it intended to carry any load other than its own weight.

Quiz

Study Session 7
IRC Chapters 6 and 7

1. What is the maximum center-to-center stud spacing permitted for a 2-inch by 6-inch, 8-foot-high wood stud bearing wall supporting two floors, roof and ceiling?

   a. 8 inches

   b. 12 inches

   c. 16 inches

   d. 24 inches

   Reference _____

2. An exterior bearing wall consists of 2-inch by 4-inch wood studs at 12 inches on center. If located in Seismic Design Category C and exposed to a wind speed of 90 mph, what is the maximum stud height when the wall supports a roof only?

   a. 10 feet

   b. 12 feet

   c. 14 feet

   d. 16 feet

   Reference _____

3. The minimum offset for end joints in a double top plate shall be _____ .

   a. 24 inches

   b. 36 inches

   c. 48 inches

   d. 60 inches

   Reference _____

4. A single top plate is permitted in a wood stud bearing wall where the rafters or joists are located within _____ of the center of the studs.

    a. 0 inches; no tolerance is permitted

    b. 1 inch

    c. $1^1/_2$ inches

    d. 5 inches

Reference _____

5. Where the wall sheathing is used to resist wind pressures, what is the maximum stud spacing permitted for $^3/_8$-inch wood structural panel wall sheathing with a span rating of 24/0.

    a. 12 inches o.c.   b. 16 inches o.c.

    c. 20 inches o.c.   d. 24 inches o.c.

Reference _____

6. In a nonbearing partition, a wood stud may be notched a maximum of _____.

    a. $^5/_8$ inch

    b. $1^3/_8$ inches

    c. 40 percent of the stud width

    d. 60 percent of the stud width

Reference _____

7. A bored hole shall be located a minimum of _____ from the edge of a wood stud.

    a. $^3/_8$ inch   b. $^1/_2$ inch

    c. $^5/_8$ inch   d. 1 inch

Reference _____

8. In a nonbearing interior partition, what is the maximum diameter permitted for a bored hole in a wood stud?

    a. $1^1/_2$ inches

    b. 25 percent of the stud depth

    c. 40 percent of the stud depth

    d. 60 percent of the stud depth

    Reference _____

9. A 15-inch-deep box header in an exterior wall is constructed with wood structural panels on both sides. What is the maximum allowable header span for a condition where the header supports a clear-span roof truss with a span of 26 feet?

    a. 4 feet   b. 5 feet

    c. 7 feet   d. 8 feet

    Reference _____

10. Which of the following materials is not specifically identified by the IRC as a fireblocking material?

    a. $^1/_4$-inch cement-based millboard

    b. $^1/_2$-inch gypsum board

    c. $^{15}/_{32}$-inch wood structural panel

    d. $^3/_4$-inch particleboard

    Reference _____

11. Where unfaced fiberglass batt insulation is used as a fireblocking material in the wall cavity of a wood stud wall system, the insulation shall be installed with a minimum vertical height of _____ .

    a. 16 inches   b. 3 feet

    c. 4 feet   d. the entire stud space

    Reference _____

12. A foundation cripple wall shall be considered an additional story for stud sizing requirements where the wall height exceeds _____ .

    a. 14 inches   b. 30 inches

    c. 36 inches   d. 48 inches

    Reference _____

13. The distance between adjacent edges of braced wall panels along a braced wall line must be a maximum of _____ feet.

    a. 16   b. 20

    c. 25   d. 35

Reference _____

14. Where located in Seismic Design Category B, braced wall panels shall begin a maximum of _____ from each end of a braced wall line.

    a. 4 feet   b. 8 feet

    c. 10 feet   d. 12 feet, 6 inches

Reference _____

15. In steel-framed wall construction, rafters may be offset a maximum of _____ from the centerline of the load-bearing steel studs.

    a. 0 inches; no tolerance is permitted

    b. $^3/_4$ inch

    c. 1 inch

    d. 5 inches

Reference _____

16. Web holes in load-bearing steel wall framing members shall have a minimum distance of _____ between the edge of the bearing surface and the edge of the hole.

    a. $1^1/_2$ inches   b. 3 inches

    c. 4 inches   d. 10 inches

Reference _____

17. All exterior walls parallel to a braced wall line must be offset a maximum of _____ feet from the designated braced wall line location.

    a. 2   b. 4

    c. 5   d. 6

Reference _____

18. Where ledgers are used at the connection between a masonry wall and a wood floor system having floor joists spanning 16 feet, $^1/_2$-inch ledger bolts shall be located at a maximum of _____ on center, provided the building is in Seismic Design Category A, B or C and the wind loads are less than 30 psf.

    a. 1 foot, 0 inches   b. 1 foot, 3 inches

    c. 1 foot, 9 inches   d. 2 feet, 0 inches

Reference _____

19. When $^1/_2$-inch gypsum board is used as an interior wall covering and installed perpendicular to framing members at 16 inches on center, the maximum spacing of nails is _____ on center where adhesive is used.

    a. 7 inches   b. 8 inches

    c. 12 inches   d. 16 inches

Reference _____

20. Screws for attaching gypsum board to wood framing shall penetrate the wood a minimum of _____ .

    a. $^1/_4$ inch   b. $^3/_8$ inch

    c. $^1/_2$ inch   d. $^5/_8$ inch

Reference _____

21. One-half-inch thick water-resistant gypsum backing board is permitted to be used on a ceiling where the framing members are spaced a maximum of _____ on center.

    a. 12 inches   b. 16 inches

    c. 20 inches   d. 24 inches

Reference _____

22. Where $^3/_8$-inch particleboard is used as an exterior wall covering, what type of fasteners are required if the particleboard is attached directly to the studs?

    a. 0.120 nail, 2 inches long

    b. 6d box nail

    c. 8d box nail

    d. direct attachment to the studs is prohibited

Reference _____

23. In Seismic Design Category D_1, metal ties for anchoring masonry veneer to a supporting wall shall support a maximum of _____ of wall area.

    a. 2 square feet   b. $2^2/_3$ square feet

    c. $3^1/_2$ square feet   d. $4^1/_2$ square feet

Reference _____

24. An Exterior Insulation Finish System (EIFS) shall terminate a minimum of _____ above the finished ground level.

    a. 1 inch   b. 2 inches

    c. 6 inches   d. 8 inches

Reference _____

25. What is the minimum required size of a steel angle spanning 8 feet used as a lintel supporting one story of masonry veneer above?

    a. 3 x 3 x $^1/_4$ b. 4 x 3 x $^1/_4$

    c. 5 x $3^1/_2$ x $^5/_{16}$ d. 6 x $3^1/_2$ x $^5/_{16}$

Reference _____

26. Utility grade studs, where used in loadbearing walls supporting only a roof and a ceiling, shall have a maximum height of _____ feet.

    a. 8   b. 10

    c. 12   d. 14

Reference _____

27. Double top plates shall be face-nailed together in the required lapped area with _____ fasteners.

    a. four 16d   b. six 10d

    c six 16d   d. eight 16d

Reference _____

28. $^{15}/_{32}$-inch wood structural panels attached to wall framing with 15 gage staples shall be fastened at a maximum of _____ inches at the panel edges and _____ inches at the intermediate supports.

    a. 3, 6   b. 4, 8

    c. 5, 10   d. 6, 12

Reference _____

29. Where the top plate of an interior load-bearing wall is notched by more than _____ of its width to accommodate piping, a complying metal tie shall be installed.

    a. 25 percent   b. $33^{1}/_{3}$ percent

    c. 40 percent   d. 50 percent

Reference _____

30. A weep screed installed on exterior stud walls in an exterior plaster application must be placed a minimum of _____ inch(es) above paved areas.

    a. 1   b. 2

    c. 4   d. 6

Reference _____

31. Where an exterior wall top plate is notched to the extent that a metal tie is required across the opening, the tie shall be fastened to the plate at each side of the opening with a minimum of _____10d nails at each side, or equivalent.

    a. two   b. four

    c. six   d. eight

Reference _____

32. For a building located in Seismic Design Category D$_0$, plate washers used in the connection of braced wall line sills to a concrete foundation shall be a minimum of _____ in size.

    a. 0.188 inch by 2 inches by 2 inches

    b. 0.229 inch by 2 inches by 2 inches

    c. 0.229 inch by 3 inches by 3 inches

    d. 0.375 inch by 3 inches by 3 inches

Reference _____

33. A rectangular penetration in a structural insulated wall panel (SIP) shall have a maximum dimension of _____ inches.

    a. 6   b. 8

    c. 12   d. 16

Reference _____

34. End-jointed lumber used in an assembly required by the IBC to have a fire resistance rating must have the designation _____ included in its grade mark.

    a. Fire Assembly Certified (FAC)

    b. Fire Resistant Rated (FRR)

    c. Heat and Fire Resistant (HFR)

    d. Heat Resistant Adhesive (HRA)

Reference _____

35. The water-resistive vapor permeable barrier required to be applied over wood-based wall sheathing shall have a performance level equivalent to that of _____ layer(s) of Grade _____ paper.

    a. one, A   b. one, B

    c. two, C   d. two, D

Reference _____

Study Session

8

2012 IRC Chapters 8 and 9
Roof/Ceiling Construction and Roof Assemblies

OBJECTIVE: To gain an understanding of the requirements for ceiling construction, roof construction and roof coverings, including ceiling joist and rafter sizing; cutting and notching of framing members; framing at openings; purlins; steel roof framing; roof ventilation and attic access; and roof covering materials such as asphalt shingles, concrete or clay tiles, wood shingles and wood shakes.

REFERENCE: Chapters 8 and 9, 2012 *International Residential Code*

KEY POINTS:
- How must roof drainage water be discharged?
- How must wood framing members be identified? What special concerns must be addressed when using fire-retardant-treated wood?
- What are the acceptable framing methods at the roof ridge? At valleys and hips? Under what condition are ridges, valleys and hips required to be designed as beams?
- How are rafters to be tied together? At what maximum intervals must collar ties be located?
- How is the maximum allowable ceiling joist span determined? Roof rafter span? How does the location of rafter ties affect the maximum allowable rafter span? Where are collar ties to be located?
- What methods are prescribed for framing around ceiling and roof openings? When is lateral support mandated?
- What is the purpose of a purlin system? How are purlins sized? What is the maximum spacing of purlin braces? The maximum unbraced length?
- How is the maximum allowable span for roof sheathing determined? How must roof sheathing be installed?
- How are trusses to be connected to the wall plates to address uplift pressures?
- What are the general limitations for the prescriptive use of steel roof framing systems? What are the general requirements?
- When is roof ventilation required? What is minimum required ventilating area? How can this area be reduced?

KEY POINTS: • Under what circumstances are unvented conditioned attic assemblies permitted?

(Cont'd) • When is attic access required? What is the minimum size needed for an attic access opening? How much headroom clearance must be provided over the access opening?

• What is Class A, Class B or Class C roofing? When is such roofing required? What specific materials are considered Class A roof coverings?

• Where is flashing required? How is roof drainage to be addressed?

• How must roofing materials be identified?

• What are the installation requirements for asphalt shingles? Clay and concrete tile? Wood shingles? Wood shakes?

• What other types of roof coverings are regulated by the code?

• What level of work is considered reroofing? What is a roof repair? Roof replacement? When must the existing roof coverings be removed prior to any reroofing work?

Code Text: *Rafters shall be framed to ridge board or to each other with a gusset plate as a tie. Ridge board shall be at least 1-inch (25 mm) nominal thickness and not less in depth than the cut end of the rafter. At all valleys and hips there shall be a valley or hip rafter not less than 2-inch (51 mm) nominal thickness and not less in depth than the cut end of the rafter.*

Discussion and Commentary: For the roof loads to be carried to the exterior walls, a continuous tie must be created at the roof ridge. This is typically accomplished by installing the rafters in direct opposition to each other and nailing them directly to a ridge board having a depth at least that of the rafter ends. Another option is the use of a gusset plate or similar element that ties the roof system together at the ridge line. Continuity from rafter to rafter is the key requirement.

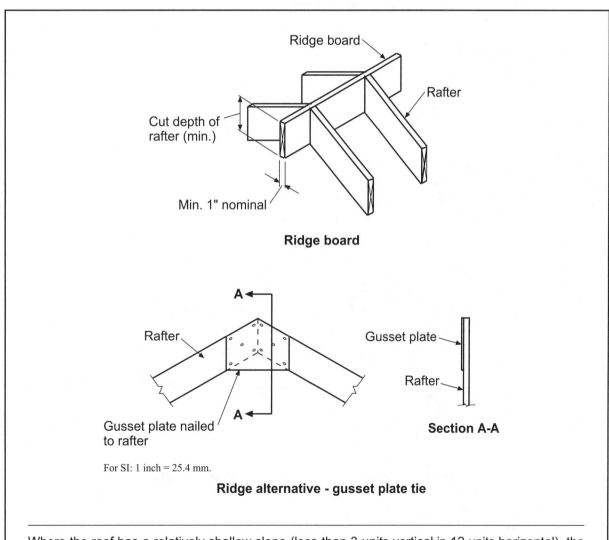

Ridge board

Ridge alternative - gusset plate tie

For SI: 1 inch = 25.4 mm.

Where the roof has a relatively shallow slope (less than 3 units vertical in 12 units horizontal), the ridge must be designed as a beam in order to carry the necessary tributary loading. Positive bearing for the rafters must be provided unless an approved alternative method is used.

Code Text: *Ceiling joists and rafters shall be nailed to each other in accordance with Table R802.5.1(9), and the rafter shall be nailed to the top wall plate in accordance with Table R602.3(1). Ceiling joists shall be continuous or securely joined in accordance with Table R802.5.1(9) where they meet over interior partitions and are nailed to adjacent rafters to provide a continuous tie across the building when such joists are parallel to the rafters.*

Discussion and Commentary: Where ceiling joists are located parallel to roof rafters, they must be tied together. Adequate attachment is mandated to transfer the thrust from the rafters to the joists. Where the ceiling joists do not run parallel to the rafters, it is necessary to use other means, such as minimum 2-inch by 4-inch rafter ties located near the top plate. Collar ties or ridge straps are also required in order to resist wind uplift. The ties shall be at least 1-inch by 4-inch members spaced at a maximum of 4 feet on center.

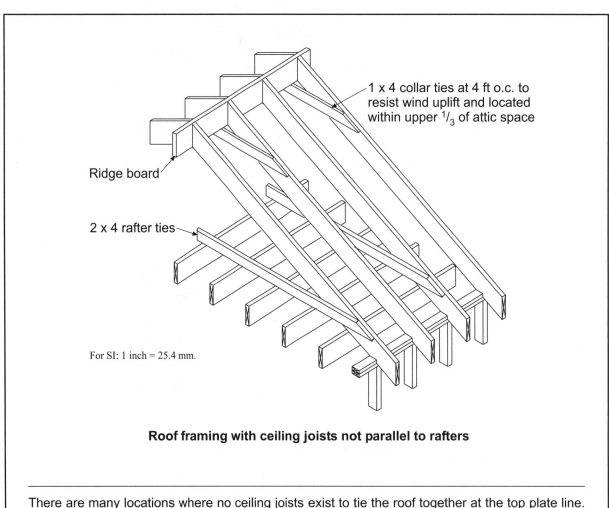

1 x 4 collar ties at 4 ft o.c. to resist wind uplift and located within upper $^1/_3$ of attic space

Ridge board

2 x 4 rafter ties

For SI: 1 inch = 25.4 mm.

Roof framing with ceiling joists not parallel to rafters

There are many locations where no ceiling joists exist to tie the roof together at the top plate line. Under such conditions, the ridge must be designed to act as a girder and support the necessary tributary roof load. Both the girder and its supporting elements must be designed.

Code Text: *Spans for ceiling joists shall be in accordance with Tables R802.4(1) and R802.4(2).*

Discussion and Commentary: Two tables are provided for the more common ceiling joist conditions. The maximum allowable spans are provided for uninhabitable attics with no storage (10 psf live load) and limited storage (20 psf). Table R301.5 indicates that the 10 psf criterion is to be used only where the attic height is limited to 42 inches. It is assumed that higher attics create conditions that allow for some degree of storage, thus mandating a higher design load. Criteria for truss roof systems is also provided. Similar to the listings for floor joists and rafter spans, four common lumber species are listed, along with a variety of lumber grades. Four center-to-center spacing conditions are included, as are four sizes of dimension lumber. The combination of these criteria provides all the information necessary to determine the maximum allowable ceiling joist spans.

TABLE R802.4(1)
CEILING JOIST SPANS FOR COMMON LUMBER SPECIES
(Uninhabitable attics without storage, live load = 10 psf, L/Δ = 240)

CEILING JOIST SPACING (inches)	SPECIES AND GRADE		DEAD LOAD = 5 psf			
			2 × 4	2 × 6	2 × 8	2 × 10
			Maximum ceiling joist spans			
			(feet - inches)	(feet - inches)	(feet - inches)	(feet - inches)
12	Douglas fir-larch	SS	13-2	20-8	Note a	Note a
	Douglas fir-larch	#1	12-8	19-11	Note a	Note a
	Douglas fir-larch	#2	12-5	19-6	25-8	Note a
	Douglas fir-larch	#3	10-10	15-10	20-1	24-6
	Hem-fir	SS	12-5	19-6	25-8	Note a
	Hem-fir	#1	12-2	19-1	25-2	Note a
	Hem-fir	#2	11-7	18-2	24-0	Note a
	Hem-fir	#3	10-10	15-10	20-1	24-6
	Southern pine	SS	12-11	20-3	Note a	Note a
	Southern pine	#1	12-8	19-11	Note a	Note a
	Southern pine	#2	12-5	19-6	25-8	Note a
	Southern pine	#3	11-6	17-0	21-8	25-7
	Spruce-pine-fir	SS	12-2	19-1	25-2	Note a
	Spruce-pine-fir	#1	11-10	18-8	24-7	Note a
	Spruce-pine-fir	#2	11-10	18-8	24-7	Note a
	Spruce-pine-fir	#3	10-10	15-10	20-1	24-6
16	Douglas fir-larch	SS	11-11	18-9	24-8	Note a
	Douglas fir-larch	#1	11-6	18-1	23-10	Note a
	Douglas fir-larch	#2	11-3	17-8	23-0	Note a
	Douglas fir-larch	#3	9-5	13-9	17-5	21-3
	Hem-fir	SS	11-3	17-8	23-4	Note a
	Hem-fir	#1	11-0	17-4	22-10	Note a
	Hem-fir	#2	10-6	16-6	21-9	Note a
	Hem-fir	#3	9-5	13-9	17-5	21-3
	Southern pine	SS	11-9	18-5	24-3	Note a
	Southern pine	#1	11-6	18-1	23-1	Note a
	Southern pine	#2	11-3	17-8	23-4	Note a
	Southern pine	#3	10-0	14-9	18-9	22-2
	Spruce-pine-fir	SS	11-0	17-4	22-10	Note a
	Spruce-pine-fir	#1	10-9	16-11	22-4	Note a
	Spruce-pine-fir	#2	10-9	16-11	22-4	Note a
	Spruce-pine-fir	#3	9-5	13-9	17-5	21-3

(continued)

In a limited number of applications, the span tables may not address the loading conditions under consideration. Therefore, it may be necessary to refer to another source to determine the maximum allowable joist spans. AF&PA provides span tables for such situations.

Code Text: *Spans for rafters shall be in accordance with Tables R802.5.1(1) through R802.5.1(8). The span of each rafter shall be measured along the horizontal projection of the rafter.*

Discussion and Commentary: Eight span tables are available in the code to determine the maximum allowable span of rafters in light-frame wood construction. The tables are specific for snow loads of 30 psf, 50 psf and 70 psf, as well as a live load of 20 psf. Each of these loading conditions includes two different deflection criteria. Where a ceiling is attached to the rafters, such as a cathedral ceiling, the deflection is limited to L/240. Where there is no ceiling material supported by the rafters, a maximum deflection of L/180 is permitted. The tables' format is similar to that for floor joists and ceiling joists.

TABLE R802.5.1(1)—continued
RAFTER SPANS FOR COMMON LUMBER SPECIES
(Roof live load=20 psf, ceiling not attached to rafters, L/Δ = 180)

RAFTER SPACING (inches)	SPECIES AND GRADE		DEAD LOAD = 10 psf					DEAD LOAD = 20 psf				
			2 × 4	2 × 6	2 × 8	2 × 10	2 × 12	2 × 4	2 × 6	2 × 8	2 × 10	2 × 12
			Maximum rafter spans[a]									
			(feet - inches)	(feet - inches)	(feet - inches)	(feet - inches)	(feet - inches)	(feet - inches)	(feet - inches)	(feet - inches)	(feet - inches)	(feet - inches)
24	Douglas fir-larch	SS	9-1	14-4	18-10	23-4	Note b	8-11	13-1	16-7	20-3	23-5
	Douglas fir-larch	#1	8-7	12-6	15-10	19-5	22-6	7-5	10-10	13-9	16-9	19-6
	Douglas fir-larch	#2	8-0	11-9	14-10	18-2	21-0	6-11	10-2	12-10	15-8	18-3
	Douglas fir-larch	#3	6-1	8-10	11-3	13-8	15-11	5-3	7-8	9-9	11-10	13-9
	Hem-fir	SS	8-7	13-6	17-10	22-9	Note b	8-7	12-10	16-3	19-10	23-0
	Hem-fir	#1	8-4	12-3	15-6	18-11	21-11	7-3	10-7	13-5	16-4	19-0
	Hem-fir	#2	7-11	11-7	14-8	17-10	20-9	6-10	10-0	12-8	15-6	17-11
	Hem-fir	#3	6-1	8-10	11-3	13-8	15-11	5-3	7-8	9-9	11-10	13-9
	Southern pine	SS	8-11	14-1	18-6	23-8	Note b	8-11	14-1	18-6	22-11	Note b
	Southern pine	#1	8-9	13-9	17-9	21-1	25-2	8-3	12-3	15-4	18-3	21-9
	Southern pine	#2	8-7	12-3	15-10	18-11	22-2	7-5	10-8	13-9	16-5	19-3
	Southern pine	#3	6-5	9-6	12-1	14-4	17-1	5-7	8-3	10-6	12-5	14-9
	Spruce-pine-fir	SS	8-5	13-3	17-5	21-8	25-2	8-4	12-2	15-4	18-9	21-9
	Spruce-pine-fir	#1	8-0	11-9	14-10	18-2	21-0	6-11	10-2	12-10	15-8	18-3
	Spruce-pine-fir	#2	8-0	11-9	14-10	18-2	21-0	6-11	10-2	12-10	15-8	18-3
	Spruce-pine-fir	#3	6-1	8-10	11-3	13-8	15-11	5-3	7-8	9-9	11-10	13-9

Check sources for availability of lumber in lengths greater than 20 feet.

For SI: 1 inch = 25.4 mm, 1 foot = 304.8 mm, 1 pound per square foot = 0.0479 kPa.

a. The tabulated rafter spans assume that ceiling joists are located at the bottom of the attic space or that some other method of resisting the outward push of the rafters on the bearing walls, such as rafter ties, is provided at that location. When ceiling joists or rafter ties are located higher in the attic space, the rafter spans shall be multiplied by the factors given below:

H_C/H_R	Rafter Span Adjustment Factor
1/3	0.67
1/4	0.76
1/5	0.83
1/6	0.90
1/7.5 or less	1.00

where:

H_C = Height of ceiling joists or rafter ties measured vertically above the top of the rafter support walls.

H_R = Height of roof ridge measured vertically above the top of the rafter support walls.

b. Span exceeds 26 feet in length.

It is important that the horizontal thrust created by roof loads be resisted by the connection of the rafters to ceiling joists or rafter ties. This connection must occur very near the top plate line. Where the horizontal ties are higher in the space, a reduction in the rafter span is required.

Topic: Purlins

Category: Roof-Ceiling Construction

Reference: IRC R802.5.1

Subject: Wood Roof Framing

Code Text: *Installation of purlins to reduce the span of rafters is permitted as shown in Figure R802.5.1. Purlins shall be sized no less than the required size of the rafters that they support. Purlins shall be continuous and shall be supported by 2-inch by 4-inch (51 mm by 102 mm) braces installed to bearing walls at a slope not less than 45 degrees from the horizontal. The braces shall be spaced not more than 4 feet (1219 mm) on center and the unbraced length of braces shall not exceed 8 feet (2438 mm).*

Discussion and Commentary: Purlins provide an effective method for reducing rafter size by reducing the rafter span. The span is measured horizontally from the exterior wall to the purlin and from the purlin to the ridge, with the controlling span determined by the greater length.

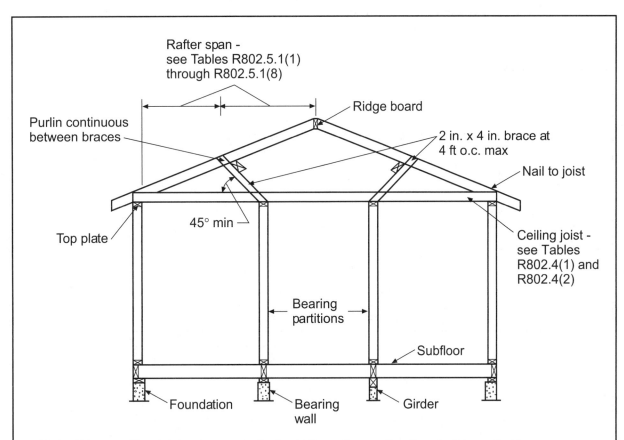

Note: Where ceiling joists run perpendicular to the rafters, rafter ties shall be nailed to the rafter near the plate line and spaced not more than 4 ft on center.

For SI: 1 inch = 25.4 mm, 1 foot = 304.8 mm, 1 degree = 0.01745 rad.

Braced rafter construction

It is critical that purlin braces be supported by bearing walls. The use of 2-inch by 4-inch braces is based on a maximum span of 4 feet between such purlin supports. Although there is no specific distance limit between bearing points at the bearing wall, the 45-degree criteria must be met.

Topic: Roof Tie-Down

Category: Roof-Ceiling Construction

Reference: IRC R802.11.1

Subject: Wood Roof Framing

Code Text: *Where the uplift force does not exceed 200 pounds, rafters and trusses spaced not more than 24 inches (610 mm) on center shall be permitted to be attached to their supporting wall assemblies in accordance with Table R602.3(1). Where the basic wind speed does not exceed 90 mph, the wind exposure category is B, the roof pitch is 5:12 or greater, and the roof span is 32 feet (9754 mm) or less, rafters and trusses spaced not more than 24 inches (610 mm) on center shall be permitted to be attached to their supporting wall assemblies in accordance with Table R602.3(1).*

Discussion and Commentary: Because roof uplift caused by high winds can be a significant factor in roof-system damage, the code requires roof-to-wall connections capable of resisting this force. In addition, it is necessary to maintain the integrity of the continuous load path by providing a complying means to transmit the uplift forces from the rafter or truss ties to the foundation.

TABLE R802.11
RAFTER OR TRUSS UPLIFT CONNECTION FORCES FROM WIND (POUNDS PER CONNECTION)[a, b, c, d, e, f, g, h]

RAFTER OR TRUSS SPACING	ROOF SPAN (feet)	EXPOSURE B							
		Basic Wind Speed (mph)							
		85		90		100		110	
		Roof Pitch		Roof Pitch		Roof Pitch		Roof Pitch	
		< 5:12	≥ 5:12	< 5:12	≥ 5:12	< 5:12	≥ 5:12	< 5:12	≥ 5:12
12" o.c.	12	47	41	62	54	93	81	127	110
	18	59	51	78	68	119	104	165	144
	24	70	61	93	81	145	126	202	176
	28	77	67	104	90	163	142	227	197
	32	85	74	115	100	180	157	252	219
	36	93	81	126	110	198	172	277	241
	42	105	91	143	124	225	196	315	274
	48	116	101	159	138	251	218	353	307
16" o.c.	12	63	55	83	72	124	108	169	147
	18	78	68	103	90	159	138	219	191
	24	93	81	124	108	193	168	269	234
	28	102	89	138	120	217	189	302	263
	32	113	98	153	133	239	208	335	291
	36	124	108	168	146	264	230	369	321
	42	139	121	190	165	299	260	420	365

(continued)

Trusses must be attached to supporting wall assemblies with connections that are capable of resisting the uplift forces specified on the Truss Design Drawings. In addition to accepted engineering practice, the use of Table R802.11, where applicable, is permitted for the determination of uplift forces.

Topic: Structural Framing **Category:** Roof-Ceiling Construction
Reference: IRC R804.2, R804.3 **Subject:** Steel Roof Framing

Code Text: *Load-bearing cold-formed steel roof framing members shall comply with Figure R804.2(1) and the dimensional and minimum thickness requirements specified in Tables R804.2(1) and R804.2(2). Cold-formed steel roof systems constructed in accordance with the provisions of Section R804.3 shall consist of both ceiling joists and rafters in accordance with Figure R804.3 and fastened in accordance with Table R804.3, and hip framing in accordance with Section R804.3.3.*

Discussion and Commentary: The conventional use of steel roof framing members is regulated in much the same manner as steel floor and wall framing systems. It is important that roof rafters and ceiling joists bear directly above load-bearing studs, with a maximum tolerance of $^3/_4$ inch between the centerline of the stud and the roof joist or rafter.

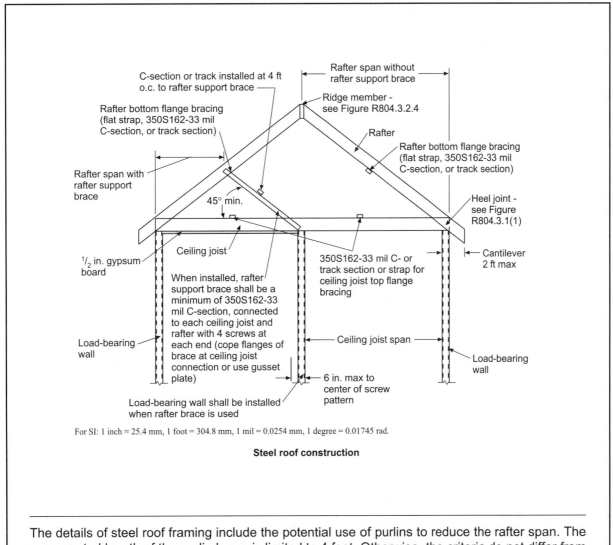

For SI: 1 inch = 25.4 mm, 1 foot = 304.8 mm, 1 mil = 0.0254 mm, 1 degree = 0.01745 rad.

Steel roof construction

The details of steel roof framing include the potential use of purlins to reduce the rafter span. The unsupported length of the purlin brace is limited to 4 feet. Otherwise, the criteria do not differ from those for light-frame wood construction.

Code Text: *Enclosed attics and enclosed rafter spaces formed where ceilings are applied directly to the underside of roof rafters shall have cross ventilation for each separate space by ventilating openings protected against the entrance of rain or snow.* See exception where ventilation deemed unnecessary due to atmospheric or climatic conditions. *The minimum net free ventilating area shall be $^1/_{150}$ of the area of the vented space.* See two methods that will allow for a ventilating area reduction to $^1/_{300}$ of the area of the space ventilated. *Where eave or cornice vents are installed, insulation shall not block the free flow of air. A minimum of a 1-inch (25 mm) space shall be provided between the insulation and the roof sheathing and at the location of the vent.*

Discussion and Commentary: Large amounts of water vapor can migrate into the attic and condense on wood roof components. As the wetting and drying cycle continues, rotting and decay are possible. If the attic or enclosed rafter spaces are properly ventilated, water will not accumulate on the building components.

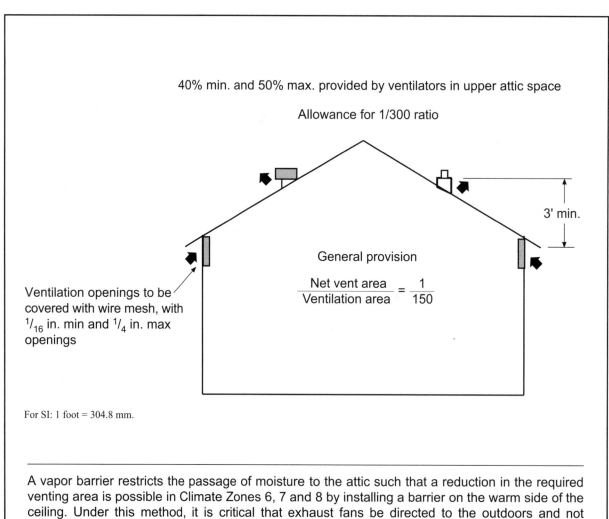

40% min. and 50% max. provided by ventilators in upper attic space

Allowance for 1/300 ratio

3' min.

General provision

$$\frac{\text{Net vent area}}{\text{Ventilation area}} = \frac{1}{150}$$

Ventilation openings to be covered with wire mesh, with $^1/_{16}$ in. min and $^1/_4$ in. max openings

For SI: 1 foot = 304.8 mm.

A vapor barrier restricts the passage of moisture to the attic such that a reduction in the required venting area is possible in Climate Zones 6, 7 and 8 by installing a barrier on the warm side of the ceiling. Under this method, it is critical that exhaust fans be directed to the outdoors and not terminate in the attic.

Code Text: *Unvented attic assemblies (spaces between the ceiling joists of the top story and the roof rafters) and unvented enclosed rafter assemblies (spaces between ceilings that are applied directly to the underside of roof framing members/rafters and the structural roof sheathing at the top of the roof framing members/rafters) shall be permitted if all of the following conditions are met:* See five conditions that address 1) where the unvented attic space is fully within the building thermal envelope, 2) where wood shakes or shingles are installed, 3) the prohibition of Class I vapor retarders, 4) the application of air-impermeable insulation and 5) special criteria.

Discussion and Commentary: In recent years, much research and experimentation has occurred regarding the concept of conditioned attic spaces. The allowance for such a practice, limited by a number of conditions set forth in the code, is based on the positive results that have occurred. Different requirements are in place to address the climatic region where the building is located.

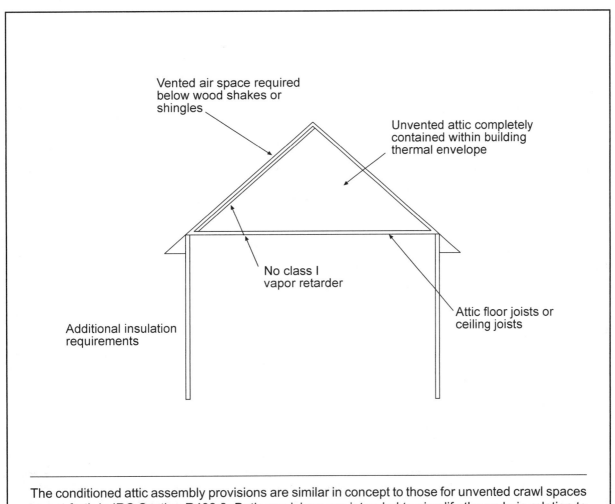

The conditioned attic assembly provisions are similar in concept to those for unvented crawl spaces as set forth in IRC Section R408.3. Both provisions are intended to simplify the code in relation to energy-efficiency requirements.

Code Text: *Buildings with combustible ceiling or roof construction shall have an attic access opening to attic areas that exceed 30 square feet (2.8 m²) and have a vertical height of 30 inches (762 mm) or greater. The rough-framed opening shall not be less than 22 inches by 30 inches (559 mm by 762 mm) and shall be located in a hallway or other readily accessible location. When located in a wall, the opening shall be a minimum of 22 inches wide by 30 inches high. When the access is located in a ceiling, minimum unobstructed headroom in the attic space shall be 30 inches (762 mm) at some point above the access measured vertically from the bottom of ceiling framing members.*

Discussion and Commentary: During the life of a dwelling structure, it is quite probable that the attic will need to be accessed for a variety of reasons. Therefore, an adequately sized access opening must be provided with adequate headroom above the opening.

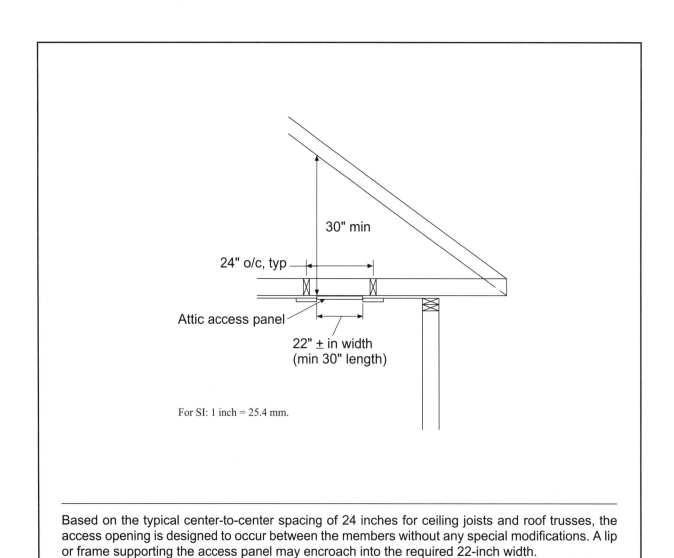

30" min

24" o/c, typ

Attic access panel

22" ± in width
(min 30" length)

For SI: 1 inch = 25.4 mm.

Based on the typical center-to-center spacing of 24 inches for ceiling joists and roof trusses, the access opening is designed to occur between the members without any special modifications. A lip or frame supporting the access panel may encroach into the required 22-inch width.

Code Text: *Roofs shall be covered with materials as set forth in Sections R904 and R905. Class A, B or C roofing shall be installed in areas designated by law as requiring their use or when the edge of the roof is less than 3 feet (914 mm) from a lot line. Classes A, B and C roofing required by Section R902.1 to be listed shall be tested in accordance with UL 790 or ASTM E 108.* See three exceptions recognizing concrete decks, clay tile, concrete tile, copper sheets and similar noncombustible materials as Class A roof assemblies.

Discussion and Commentary: The general scope of Chapter 9 includes the use of roof coverings as a weather-resistant material for protection of the building's interior. To a limited degree, the roof covering materials must also be resistant to external fire conditions. Because the primary consideration is that of fire spread from building to building, the requirements for roof covering effective against fire exposure are limited to those structures located in very close proximity to adjoining lot lines.

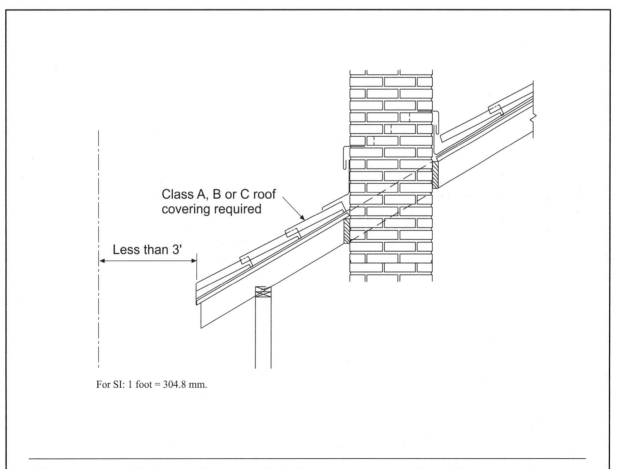

Class A, B or C roof covering required

Less than 3'

For SI: 1 foot = 304.8 mm.

Although not specifically tested and classified, slate, clay and concrete roof tiles, exposed concrete roof decks, and metal and copper sheets and shingles are considered equivalent to Class A roofing. Otherwise, a listed Class A, B or C roof covering must be used when the roof is located within 3 feet of the lot line.

Code Text: *Unless roofs are sloped to drain over roof edges, roof drains shall be installed at each low point of the roof. Where roof drains are required, secondary emergency overflow roof drains or scuppers shall be provided where the roof perimeter construction extends above the roof in such a manner that water will be entrapped if the primary drains allow buildup for any reason.*

Discussion and Commentary: Although roofs on most buildings regulated by the IRC are distinctly pitched and simply drain roof water over the edge (often to a gutter and downspout system), occasionally a flat roof system is provided. In such situations, it is necessary to install roof drains as well as overflow drains. The overflow drains shall be installed with the inlet flow line located 2 inches above the low point of the roof. Overflow drains shall not be connected to roof drain lines and must discharge to a location approved by the building official.

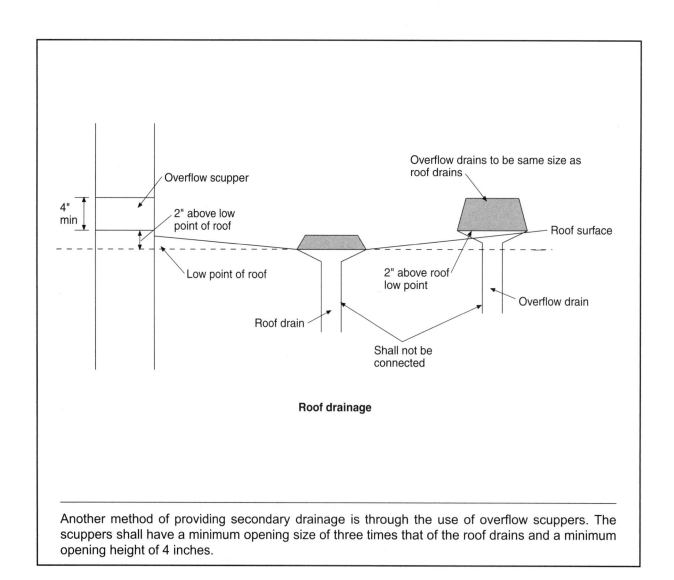

Roof drainage

Another method of providing secondary drainage is through the use of overflow scuppers. The scuppers shall have a minimum opening size of three times that of the roof drains and a minimum opening height of 4 inches.

Topic: Asphalt Shingles

Category: Roof Assemblies

Reference: IRC R905.2

Subject: Roof Covering Requirements

Code Text: *Asphalt shingles shall only be used on roof slopes of two units vertical in 12 units horizontal (2:12) or greater. For roof slopes from . . . 2:12 up to . . . 4:12, double underlayment application is required in accordance with Section R905.2.7. Fasteners for asphalt shingles shall be . . . of a length to penetrate through the roofing materials and a minimum of $^3/_4$ inch (19 mm) into the roof sheathing. Where the roof sheathing is less than $^3/_4$ inch (19 mm) thick, the fasteners shall penetrate through the sheathing. Asphalt shingles shall have the minimum number of fasteners required by the manufacturer, but not less than four fasteners per strip shingle or two fasteners per individual shingle.*

Discussion and Commentary: A very common roofing material, asphalt shingles are composed of organic or glass felt coated with mineral granules. For wind-resistance purposes, asphalt shingles must be classified in accordance with the appropriate basic wind speed, and the packaging shall be labeled to indicate compliance.

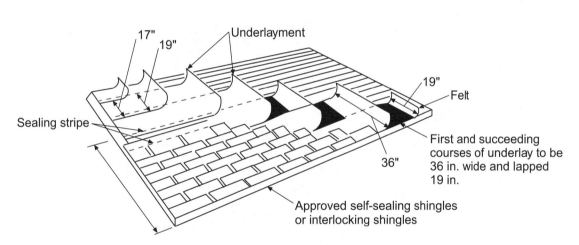

Note: In areas where there has been a history of ice forming along the eaves causing a backup of water, special methods, rather than normal underlayment, shall extend up from eaves for enough to overlie a point 24 in. inside the wall line of the building.

source NRCA

For SI: 1 inch = 25.4 mm, °C = [(°F)-32/1.8].

Application of asphalt shingle on slopes between 2:12 and 4:12

Where asphalt shingles are installed, they are regulated for sheathing, roof slope, underlayment, fasteners and attachment. As with other types of roof covering materials, they must also be installed in conformance with the manufacturer's installation instructions.

Topic: Ice Barriers	Category: Roof Assemblies
Reference: IRC R905.2.7.1	Subject: Roof Covering Requirements

Code Text: *In areas where there has been a history of ice forming along the eaves causing a backup of water as designated in Table R301.2(1), an ice barrier that consists of at least two layers of underlayment cemented together or of a self-adhering polymer modified bitumen sheet, shall be used in lieu of normal underlayment and extend from the lowest edges of all roof surfaces to a point at least 24 inches (610 mm) inside the exterior wall line of the building. See exception for detached accessory dwellings that contain no conditioned floor area.*

Discussion and Commentary: Ice dams can form along a roof eave because rain and snow continually freezes and thaws, or frozen slush backs up in gutters. Therefore, the underlayment must be modified to prevent ice dams from forcing water under the roofing where it could cause damage to the residence's ceilings, walls and insulation. Beyond the 24-inch point, such special underlayment is deemed unnecessary as the interior of building provides adequate warmth to prevent ice dams from forming above the heated space.

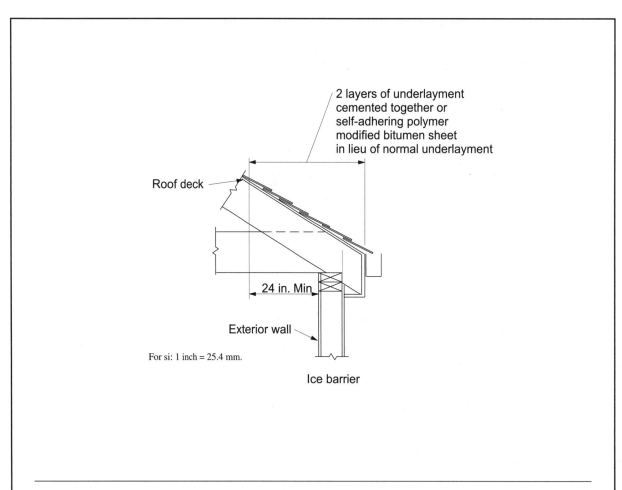

2 layers of underlayment cemented together or self-adhering polymer modified bitumen sheet in lieu of normal underlayment

Roof deck

24 in. Min.

Exterior wall

For si: 1 inch = 25.4 mm.

Ice barrier

The requirement for an ice barrier is not limited to the application of asphalt shingles. Similar requirements are applicable to roof coverings of mineral-surfaced roll roofing, slate and slate-type shingles, wood shingles and wood shakes.

Code Text: *Clay and concrete tile shall be installed on roof slopes of two and one-half units vertical in 12 units horizontal (2^1/$_2$:12) or greater. For roof slopes from . . . 2^1/$_2$:12 to . . . 4:12, double underlayment application is required in accordance with Section R905.3.3. Nails shall be . . . of sufficient length to penetrate the deck a minimum of ^{3}/$_4$ inch (19.1 mm) or through the thickness of the deck, whichever is less.*

Discussion and Commentary: Tile has been used as a roofing material for many centuries. It not only provides for weather protection but is also resistant to exterior fire conditions. Both flat tile and roll tile, the two common configurations, can be interlocked through the use of ribs located along the tile edges.

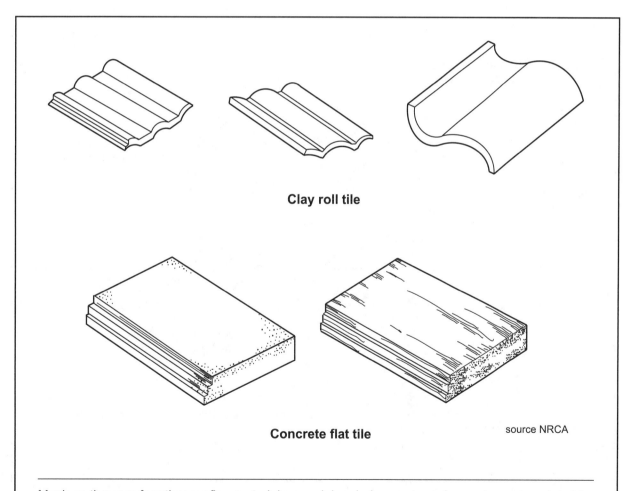

Clay roll tile

Concrete flat tile   source NRCA

Much as they are for other roofing materials, special underlayment requirements are mandated for tile roofs having low slopes. In addition, the underlayment application is more highly regulated for structures located in areas subject to high winds.

Topic: Wood Shingles

Category: Roof Assemblies

Reference: IRC R905.7

Subject: Roof Covering Requirements

Code Text: *Wood shingles shall be installed on slopes of three units vertical in 12 units horizontal (25-percent slope) or greater. Weather exposure for wood shingles shall not exceed those set in Table R905.7.5. Fasteners . . . shall be corrosion-resistant with a minimum penetration of $^1/_2$ inch (12.7 mm) into the sheathing. Wood shingles shall be attached to the roof with two fasteners per shingle, positioned no more than $^3/_4$ inch (19 mm) from each edge and no more than 1 inch (25 mm) above the exposure line.*

Discussion and Commentary: Wood shingles are wood roofing materials that are sawed in a manner to produce a uniform butt thickness. There are three different grades of wood shingles, with each grade available in three different lengths. The application method is specifically described so that following it will eliminate the potential for water intrusion.

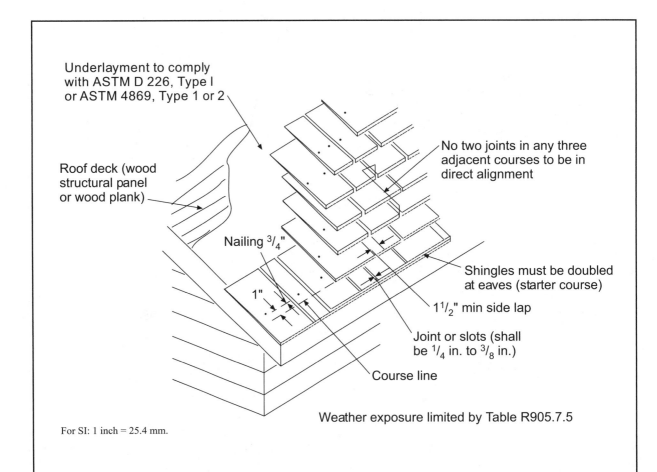

Underlayment to comply with ASTM D 226, Type I or ASTM 4869, Type 1 or 2

Roof deck (wood structural panel or wood plank)

Nailing $^3/_4$"

1"

No two joints in any three adjacent courses to be in direct alignment

Shingles must be doubled at eaves (starter course)

$1^1/_2$" min side lap

Joint or slots (shall be $^1/_4$ in. to $^3/_8$ in.)

Course line

Weather exposure limited by Table R905.7.5

For SI: 1 inch = 25.4 mm.

When fastening wood shingles to either spaced or solid sheathing, it is critical that care be taken so that the shingles are not split. Unless the sheathing is less than $^1/_2$-inch thick, the fasteners must penetrate into the sheathing a minimum of $^1/_2$ inch.

Code Text: *Wood shakes shall only be used on slopes of three units vertical in 12 units horizontal (25-percent slope) or greater. Fasteners for wood shakes shall be corrosion-resistant, with a minimum penetration of $^1/_2$ inch (12.7 mm) into the sheathing. Wood shakes shall be attached to the roof with two fasteners per shake, positioned no more than 1 inch (25 mm) from each edge and no more than 2 inches (51 mm) above the exposure line. Shakes shall be interlaid with 18-inch-wide (457 mm) strips of not less than No. 30 felt shingled between each course in such a manner that no felt is exposed to the weather.*

Discussion and Commentary: Wood shakes are roofing materials that are split from logs, creating a more inconsistent surface finish than that of wood shingles. Shakes are either 18 inches or 24 inches long and graded as No. 1 or No. 2.

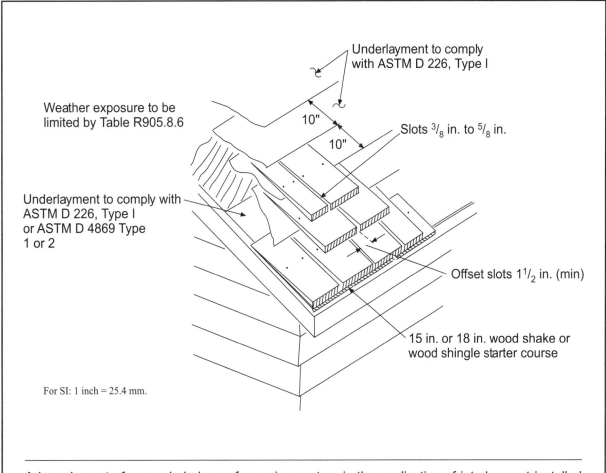

Weather exposure to be limited by Table R905.8.6

Underlayment to comply with ASTM D 226, Type I or ASTM D 4869 Type 1 or 2

Underlayment to comply with ASTM D 226, Type I

10"

10"

Slots $^3/_8$ in. to $^5/_8$ in.

Offset slots $1^1/_2$ in. (min)

15 in. or 18 in. wood shake or wood shingle starter course

For SI: 1 inch = 25.4 mm.

A key element of a wood shake roof covering system is the application of interlayment installed shingle fashion between each course of shakes. By positioning the lower edge of the minimum No. 30 felt above the butt end of the shake, the felt extends twice the required weather exposure.

Quiz

Study Session 8
IRC Chapters 8 and 9

1. On a site having expansive soil, roof drainage water shall discharge a minimum of _____ from the foundation walls or to an approved drainage system.

   a. 2 feet

   b. 5 feet

   c. 4 feet

   d. 10 feet

   Reference _____

2. Where the roof pitch is less than _____ , the structural members that support rafters and ceiling joists (such as ridge beams, hips, and valleys) shall be designed as beams.

   a. 3:12

   b. 4:12

   c. 5:12

   d. 6:12

   Reference _____

3. The maximum spacing of collar ties shall be _____ on center.

   a. 16 inches

   b. 4 feet

   c. 5 feet

   d. 8 feet

   Reference _____

4. Ends of ceiling joists shall be lapped a minimum of _____ unless butted and toenailed to the supporting member.

    a. $1^1/_2$ inches   b. 3 inches

    c. 4 inches   d. 6 inches

    Reference _____

5. Where SPF #2 ceiling joists create an uninhabitable attic without storage and are spaced 24 inches on center, what is the maximum allowable span when 2-inch by 6-inch members are used?

    a. 11 feet, 2 inches   b. 14 feet, 5 inches

    c. 14 feet, 9 inches   d. 15 feet, 11 inches

    Reference _____

6. A roof system is subjected to a 50 psf ground snow load and creates a dead load of 20 psf. The ceiling is not attached to the rafters. Assuming that rafter ties are provided at the top plate line, what is the maximum span of 2-inch by 8-inch Hem-Fir #1 rafters spaced at 24 inches on center?

    a. 9 feet, 9 inches   b. 10 feet, 6 inches

    c. 11 feet, 6 inches   d. 11 feet, 10 inches

    Reference _____

7. Purlins shall be supported by braces having a maximum unbraced length of _____ and spaced at a maximum of _____ on center.

    a. 8 feet, 4 feet   b. 8 feet, 6 feet

    c. 12 feet, 4 feet   d. 12 feet, 6 feet

    Reference _____

8. The ends of a ceiling joist shall have a minimum of _____ bearing on wood and a minimum of _____ on metal.

    a. $1^1/_2$ inches, $1^1/_2$ inches   b. $1^1/_2$ inches, 3 inches

    c. 3 inches, $1^1/_2$ inches   d. 3 inches, 3 inches

    Reference _____

9. Notches in solid sawn lumber ceiling joists shall not be located in the middle _____ of the span and are limited in depth to _____ the depth of the joist.

    a. one-fourth, one-sixth
    b. one-fourth, one-third

    c. one-third, one-sixth
    d. one-third, one-third

Reference _____

10. An opening in a wood framed roof system may be framed with a single header and single trimmer joists provided the header joist is located a maximum of _____ from the trimmer joist bearing.

    a. 3 feet
    b. 4 feet

    c. 6 feet
    d. 12 feet

Reference _____

11. Where $^{15}/_{32}$-inch wood structural panels having a span rating of 32/16 are used as roof sheathing, what is the maximum span without edge support?

    a. 16 inches
    b. 24 inches

    c. 28 inches
    d. 32 inches

Reference _____

12. A dwelling is located in an area having a 110 mph wind speed and Exposure B. The ground snow load is 20 psf. What is the maximum allowable rafter span for 800S162-54 33 ksi steel rafters installed at 24 inches on center with a roof slope of 12:12?

    a. 10 feet, 7 inches
    b. 12 feet, 6 inches

    c. 15 feet, 8 inches
    d. 17 feet, 9 inches

Reference _____

13. In a steel-framed roof system, eave overhangs shall not exceed _____ in horizontal projection.

    a. 12 inches
    b. 18 inches

    c. 24 inches
    d. 30 inches

Reference _____

14. Openings provided for roof ventilation, where covered with corrosion-resistant wire cloth screening, shall have openings a minimum of _____ and a maximum of _____ .

    a. $^1/_{16}$ inch, $^1/_4$ inch   b. $^1/_8$ inch, $^1/_4$ inch

    c. $^1/_4$ inch, $^3/_8$ inch   d. $^1/_4$ inch, $^1/_2$ inch

Reference _____

15. A required attic access opening shall have a minimum rough opening size of _____ .

    a. 18 inches by 24 inches   b. 18 inches by 30 inches

    c. 22 inches by 30 inches   d. 24 inches by 36 inches

Reference _____

16. A net free cross-ventilating area of 1/300 is permitted for roof ventilation in buildings in Climate Zones 6, 7 and 8, provided a minimum _____ vapor barrier is installed on the warm-in-winter side of the ceiling.

    a. Class I only   b. Class II only

    c. Class III   d. Class I or II

Reference _____

17. Fire-retardant-treated wood, when used in roof construction, shall have a maximum listed flame spread index of _____ .

    a. 25   b. 75

    c. 100   d. 200

Reference _____

18. In the attachment of asphalt shingles, special methods of fastening as established by the manufacturer are required where the roof slope of _____ is exceeded.

    a. 12 units vertical in 12 units horizontal

    b. 15 units vertical in 12 units horizontal

    c. 18 units vertical in 12 units horizontal

    d. 21 units vertical in 12 units horizontal

Reference _____

19. Double underlayment is required for an asphalt shingle application where the roof has a maximum slope of _____ .

      a. 2:12   b. $2^{1}/_{2}$:12

      c. 3:12   d. 4:12

Reference _____

20. In an asphalt shingle roof application, step flashing on a sidewall shall be a minimum of _____ inches high and _____ inches wide.

      a. 3, 6   b. 4, 4

      c. 6, 6   d. 6, 8

Reference _____

21. Nails used to attach concrete roof tiles to the roof deck shall penetrate the deck a minimum of _____ or through the thickness of the deck, whichever is less.

      a. $^{1}/_{2}$ inch   b. $^{5}/_{8}$ inch

      c. $^{3}/_{4}$ inch   d. $^{7}/_{8}$ inch

Reference _____

22. Where the cantilevered portion of a rafter is notched, a minimum of _____ inches of the member depth must remain.

      a. 3   b. $3^{1}/_{2}$

      c. 4   d. 5

Reference _____

23. Where No. 2 wood shingles of naturally durable wood are installed on a roof having a 5:12 pitch, the maximum weather exposure for 16-inch shingles shall be _____ .

      a. $3^{1}/_{2}$ inches   b. 4 inches

      c. $5^{1}/_{2}$ inches   d. $6^{1}/_{2}$ inches

Reference _____

24. For preservative-treated tapersawn shakes installed on a roof, the spacing between wood shakes shall be a minimum of _____ and a maximum of _____.

    a. $^1/_8$ inch, $^3/_8$ inch   b. $^1/_8$ inch, $^5/_8$ inch

    c. $^1/_4$ inch, $^3/_8$ inch   d. $^3/_8$ inch, $^5/_8$ inch

    Reference _____

25. For a wood shake roof system, sheet metal roof valley flashing shall extend a minimum of _____ from the centerline of the valley in each direction.

    a. 4 inches   b. 7 inches

    c. 11 inches   d. 12 inches

    Reference _____

26. A notch located in the end of a solid lumber ceiling joist shall have a maximum depth of _____ .

    a. 1 inch   b. 2 inches

    c. one-third the joist depth   d. one-fourth the joist depth

    Reference _____

27. In the framing of an opening in ceiling construction, approved hangers are not required for the header joist to trimmer joist connections where the header joist span is a maximum of _____ feet.

    a. 4   b. 6

    c. 8   d. 12

    Reference _____

28. Where eave or cornice vents are installed, a minimum _____ air space shall be provided between the insulation and the roof sheathing at the vent location.

    a. $^1/_2$-inch   b. 1-inch

    c. 2-inch   d. 3-inch

    Reference _____

29. Where the attic height is 30 inches or greater, an access opening is not required in attics of combustible construction where the attic space is a maximum of _____ square feet in area.

    a. 30   b. 50

    c. 100   d. 120

Reference _____

30. Overflow scuppers utilized for secondary roof drainage shall have a minimum size_____ of the roof drain.

    a. equal to that   b. twice the size

    c. three times the size   d. four times the size

Reference _____

31. A ridge board shall be a minimum of _____ -inch in nominal thickness and have a minimum depth equal to _____.

    a. 1, the cut end of the rafter

    b. 1, the nominal depth of the rafter

    c. 2, the actual depth of the rafter

    d. 2, the cut end of the rafter

Reference _____

32. What is the minimum total net free ventilating area required for a 1,500-square-foot attic space where 75 percent of the required ventilation area is provided by ventilators located in the upper portion of the attic space?

    a. 5 square feet   b. 7.5 square feet

    c. 10 square feet   d. 15 square feet

Reference _____

33. Where an unvented conditioned attic assembly is utilized for a residence, the use of wood shakes on the roof would require a minimum continuous _____ -inch vented air space between the shakes and the roofing felt..

    a. $^1/_4$ b. $^1/_2$

    c. $^3/_4$ d. 1

Reference _____

34. Where required on an asphalt-shingled roof, an ice barrier shall extend from the lowest edges of the roof surfaces to a minimum of _____ inches inside the exterior wall line of the building.

    a. 8   b. 12

    c. 16   d. 24

Reference _____

35. Drip edges provided at eaves and gables of asphalt shingle roofs shall extend a minimum of _____ inch below the roof sheathing.

    a. 0.25   b. 0.50

    c. 0.75   d. 1.00

Reference _____

Study Session

9

2012 IRC Chapters 10 and 11
Chimneys/Fireplaces and Energy Efficiency

OBJECTIVE: To obtain an understanding of the provisions for chimneys and fireplaces, as well as the requirements for energy-efficient design and construction.

REFERENCE: Chapters 10 and 11, 2012 *International Building Code*

KEY POINTS:
- How must a masonry fireplace be supported and reinforced?
- What is the minimum thickness of masonry fireplace walls? What is the maximum width of joints between firebricks? How are steel fireplace units regulated?
- What is the minimum required depth of a firebox in a masonry fireplace? In a Rumsford fireplace?
- Of what material must a lintel be constructed? What is the minimum required bearing length at each end of the lintel? Where is the fireplace throat to be located?
- What construction materials are permitted for hearths and hearth extensions? What is the minimum thickness of a hearth? A hearth extension? What are the minimum dimensions of a hearth extension?
- What is a masonry heater? How must it be supported and reinforced?
- When is reinforcement mandated in the construction of a masonry chimney? How must such chimneys be supported?
- Where are masonry chimneys prohibited from changing size or shape? How shall corbeling be done? Offsets?
- At what minimum height is a masonry chimney required to terminate?
- What are the requirements for spark arresters installed on masonry chimneys?
- What are the various methods and materials used as flue linings? How are multiple flues to be constructed? What methods are provided for determining the minimum required flue area?
- When is a chimney cleanout opening required? Where must it be located?
- How far must combustible material be located from a masonry chimney located in the interior of a building? Located entirely outside a building?
- How is a factory-built chimney regulated?
- When is an exterior air supply required?

KEY POINTS:
(Cont'd)

- What is the relationship between the energy provisions of the IRC and the residential energy provisions in the *International Energy Conservation Code*?

- What is a low-energy building? What energy provisions of the IRC are not applicable to such buildings?

- What is a climate zone and how is it determined? How does the climate zone relate to the energy efficiency requirements?

- How must insulation be labeled? When is a certificate of compliance permitted? How shall the insulation information be permanently identified in an attic area?

- What are the prescriptive requirements regulating the building thermal envelope? What specific insulation requirements must be met?

- How is the performance of fenestration regulated? Replacement fenestration? Recessed light fixtures?

- What methods are established to limit air leakage of the thermal envelope? What manner of testing is required? How are fireplaces to be regulated? Fenestration? Recessed lighting?

- What type of thermostat is required? How many thermostats must be provided?

- How are ducts to be insulated and sealed?

- What criteria are established when using simulated energy performance analysis? What documentation is required?

Code Text: *Footings for masonry fireplaces and their chimneys shall be constructed of concrete or solid masonry at least 12 inches (305 mm) thick and shall extend at least 6 inches (152 mm) beyond the face of the fireplace or foundation wall on all sides. Footings shall be founded on natural, undisturbed earth or engineered fill below frost depth. In areas not subjected to freezing, footings shall be at least 12 inches (305 mm) below finished grade.*

Discussion and Commentary: Because of the extensive weight of a masonry fireplace and chimney, the foundation is likely to support a greater load than the foundation for the surrounding construction. The minimum size mandated is expected to carry such a load. The relationship between the building foundation and the fireplace/chimney foundation must be considered in order to avoid detrimental settling.

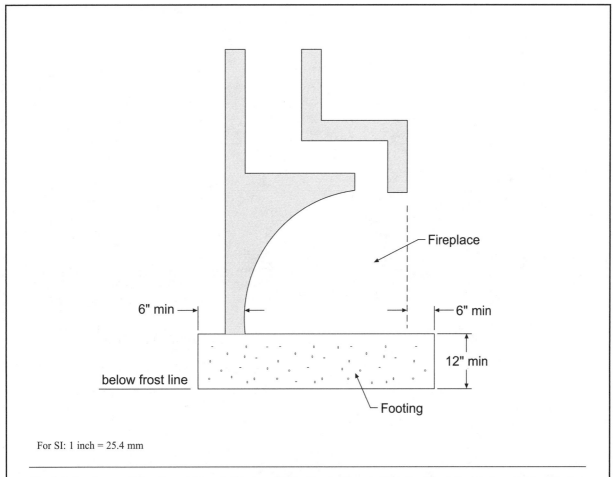

For SI: 1 inch = 25.4 mm

Similar to other conventional foundation systems, the bottom of the footing must extend to below the frost line to eliminate the potential for movement due to freeze/thaw conditions. It is critical that the footing and foundation provide a solid base to avoid any possible settlement.

Topic: Firebox Walls

Category: Chimneys and Fireplaces

Reference: IRC R1001.5

Subject: Masonry Fireplaces

Code Text: *Masonry fireboxes shall be constructed of solid masonry units, hollow masonry units grouted solid, stone or concrete. When a lining of firebrick at least 2 inches (51 mm) in thickness or other approved lining is provided, the total minimum thickness of back and side walls shall each be 8 inches (203 mm) of solid masonry, including the lining. The width of joints between firebricks shall not be greater than 3/4 inch (6 mm). When no lining is provided, the total minimum thickness of back and side walls shall be 10 inches (254 mm) of solid masonry.*

Discussion and Commentary: A variety of noncombustible materials is acceptable for the construction of masonry fireplaces. Where an approved lining is provided, the total masonry thickness is reduced from that needed where no lining is installed. Care should be taken to limit the joint width between firebricks in order to reduce the potential for joint deterioration.

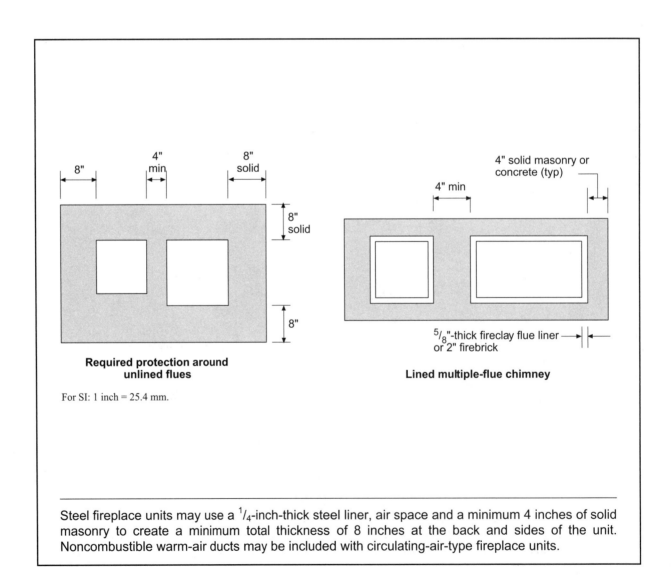

Required protection around unlined flues

Lined multiple-flue chimney

For SI: 1 inch = 25.4 mm.

Steel fireplace units may use a 1/4-inch-thick steel liner, air space and a minimum 4 inches of solid masonry to create a minimum total thickness of 8 inches at the back and sides of the unit. Noncombustible warm-air ducts may be included with circulating-air-type fireplace units.

Code Text: *The firebox of a concrete or masonry fireplace shall have a minimum depth of 20 inches (508 mm). The throat shall not be less than 8 inches (203 mm) above the fireplace opening. The throat opening shall not be less than 4 inches (102 mm) in depth. The cross-sectional area of the passageway above the firebox, including the throat, damper and smoke chamber, shall not be less than the cross-sectional area of the flue. See exception for Rumford fireplaces.*

Discussion and Commentary: To provide for positive drafting, the size of the firebox is regulated, as is the location and depth of the throat opening. These criteria are based on many years of trial and error in the construction of fireplaces and chimneys.

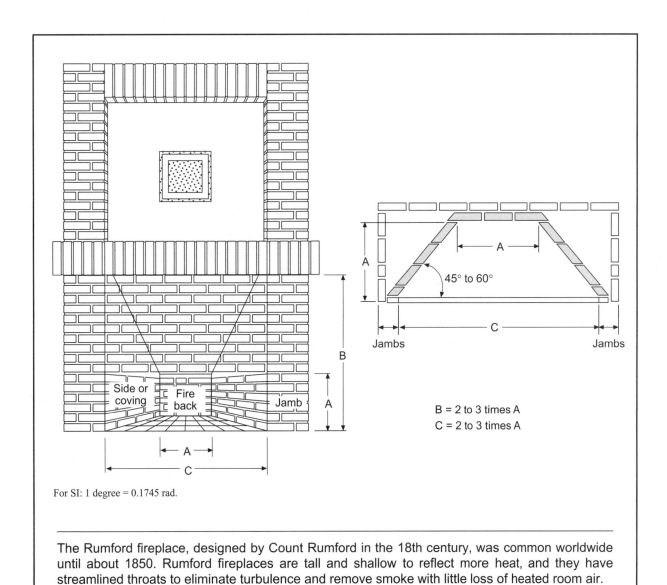

B = 2 to 3 times A
C = 2 to 3 times A

For SI: 1 degree = 0.1745 rad.

The Rumford fireplace, designed by Count Rumford in the 18th century, was common worldwide until about 1850. Rumford fireplaces are tall and shallow to reflect more heat, and they have streamlined throats to eliminate turbulence and remove smoke with little loss of heated room air.

Topic: Hearth and Hearth Extension

Category: Chimneys and Fireplaces

Reference: IRC R1001.9, R1001.10

Subject: Masonry Fireplaces

Code Text: *The minimum thickness of fireplace hearths shall be 4 inches (102 mm). The minimum thickness of hearth extensions shall be 2 inches (51 mm). See exception for permitted reduction in hearth extension thickness. Hearth extensions shall extend at least 16 inches (406 mm) in front of and at least 8 inches (203 mm) beyond each side of the fireplace opening. Where the fireplace opening is 6 square feet (0.6 m²) or larger, the hearth extension shall extend at least 20 inches (508 mm) in front of and at least 12 inches (305 mm) beyond each side of the fireplace opening.*

Discussion and Commentary: The hearth is the floor of the firebox, whereas the hearth extension is the projection of the hearth beyond the firebox opening. The purpose of the hearth extension is to provide a noncombustible area adjacent to the opening where sparks, embers and ashes from the firebox may fall without the risk of igniting of combustible flooring materials.

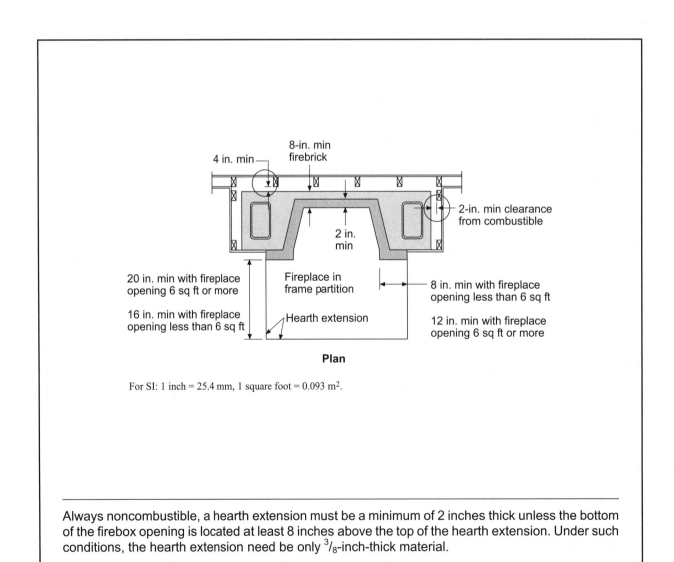

Plan

For SI: 1 inch = 25.4 mm, 1 square foot = 0.093 m².

Always noncombustible, a hearth extension must be a minimum of 2 inches thick unless the bottom of the firebox opening is located at least 8 inches above the top of the hearth extension. Under such conditions, the hearth extension need be only ³/₈-inch-thick material.

Code Text: *All wood beams, joists, studs and other combustible material shall have a clearance of not less than 2 inches (51 mm) from the front faces and sides of masonry fireplaces and not less than 4 inches (102 mm) from the back faces of masonry fireplaces.* See four exceptions for permitted reductions in clearance.

Discussion and Commentary: Radiant heat transfer through the materials used to construct a masonry fireplace and/or chimney necessitates a minimum separation between the masonry and combustible materials, such as wood floor, wall or roof framing. Radiant heat transfer, and to some degree direct flame impingement, are also of considerable concern at the firebox opening. Therefore, a minimum distance is mandated between the opening and any adjacent combustible material.

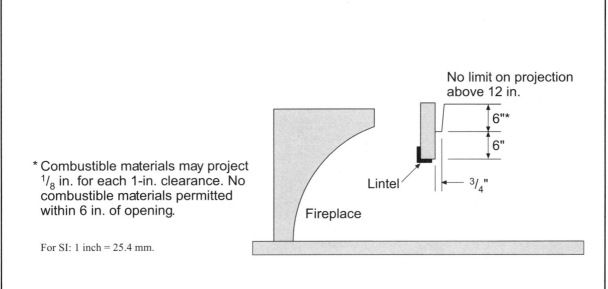

No limit on projection above 12 in.

6"*

6"

* Combustible materials may project $^1/_8$ in. for each 1-in. clearance. No combustible materials permitted within 6 in. of opening.

Lintel

$^3/_4$"

Fireplace

For SI: 1 inch = 25.4 mm.

Combustible woodwork, including trim or mantels, may be placed directly on the front of a masonry fireplace, provided such materials are not located within 6 inches of the fireplace opening. The projection of any combustible materials is strictly regulated where located within 12 inches of the opening, with a limitation of $^1/_8$ inch of projection for each 1-inch distance from the opening.

Code Text: *Masonry chimneys shall not be corbelled more than one-half of the chimney's wall thickness from a wall or foundation, nor shall a chimney be corbelled from a wall or foundation that is less than 12 inches (305 mm) thick unless it projects equally on each side of the wall (exception for second story of two-story dwelling). The projection of a single course shall not exceed one-half the unit height or one-third the unit bed depth, whichever is less.*

Discussion and Commentary: Corbeling occurs where masonry projects from a chimney surface in small increments as the chimney extends vertically. Two conditions are regulated: the maximum projection of a single course, and the limitations for the entire corbelled wall. A single course projection is limited to one-half of the individual unit height, and in no case more than one-third of the individual unit depth. If the projection exceeds these limitations, it is possible that the unit could fail due to its inability to carry the intended load. In addition, masonry walls are typically not permitted to be corbeled more than 50 percent of the chimney's wall thickness unless the projection occurs equally on each side of the wall. An unbalanced condition could potentially lead to cracking, or under extreme conditions, overturning of the wall.

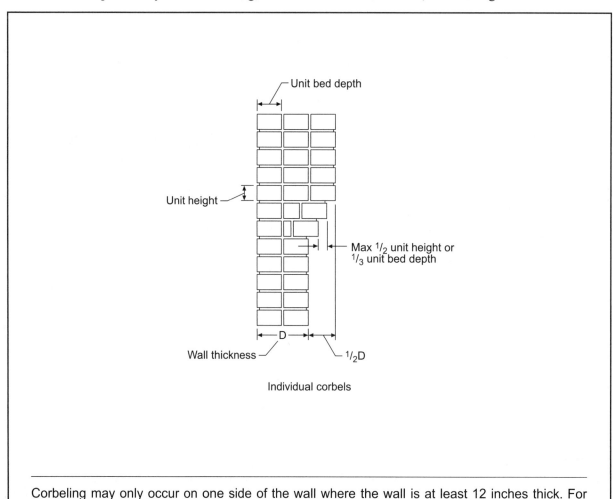

Individual corbels

Corbeling may only occur on one side of the wall where the wall is at least 12 inches thick. For thinner walls, the corbelling must project equally on both sides.

Code Text: *Chimneys shall extend at least 2 feet (610 mm) higher than any portion of a building within 10 feet (3048 mm), but shall not be less than 3 feet (914 mm) above the highest point where the chimney passes through the roof. Masonry chimney walls shall be constructed of solid masonry units or hollow masonry units grouted solid with not less than a 4-inch (102 mm) nominal thickness. Masonry chimneys shall be lined.*

Discussion and Commentary: A chimney must terminate some distance above a roof or other surface in order to provide for the necessary upward draft in the chimney. The minimum extensions mandated by the code, based on experience, were determined considering the possibility of downward drafts creating eddying, which neutralizes the required upward draft in the chimney as well as flue gas temperature.

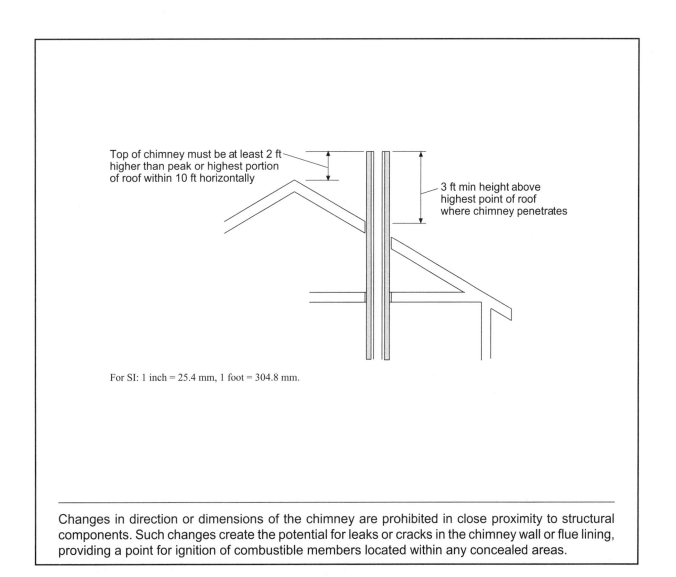

Top of chimney must be at least 2 ft higher than peak or highest portion of roof within 10 ft horizontally

3 ft min height above highest point of roof where chimney penetrates

For SI: 1 inch = 25.4 mm, 1 foot = 304.8 mm.

Changes in direction or dimensions of the chimney are prohibited in close proximity to structural components. Such changes create the potential for leaks or cracks in the chimney wall or flue lining, providing a point for ignition of combustible members located within any concealed areas.

Code Text: *Cleanout openings shall be provided within 6 inches (152 mm) of the base of each flue within every masonry chimney.* See exception for chimney flues where cleaning is possible through the fireplace opening. *The upper edge of the cleanout shall be located at least 6 inches (152 mm) below the lowest chimney inlet opening. The height of the opening shall be at least 6 inches (152 mm). The cleanout shall be provided with a noncombustible cover.*

Discussion and Commentary: The chimney interior can become coated with carbonaceous material that attacks the masonry and mortar, resulting in deposits that fall to the bottom of the chimney. Chimney cleaning also results in the accumulation of deposits at the chimney base. The deposits can be removed as necessary through the cleanout opening at the base of the chimney.

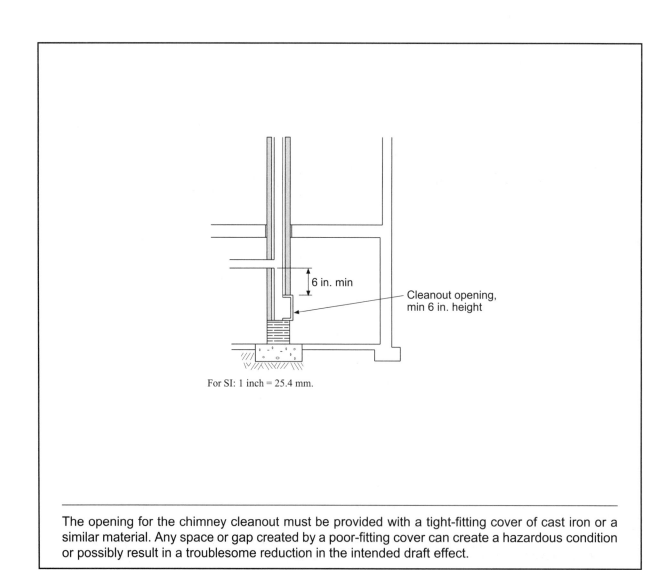

For SI: 1 inch = 25.4 mm.

The opening for the chimney cleanout must be provided with a tight-fitting cover of cast iron or a similar material. Any space or gap created by a poor-fitting cover can create a hazardous condition or possibly result in a troublesome reduction in the intended draft effect.

Topic: Chimney Clearances

Category: Chimneys and Fireplaces

Reference: IRC R1003.18

Subject: Masonry Chimneys

Code Text: *Any portion of a masonry chimney located in the interior of the building or within the exterior wall of the building shall have a minimum air space clearance to combustibles of 2 inches (51 mm). Chimneys located entirely outside the exterior walls of the building, including chimneys that pass through the soffit or cornice, shall have a minimum air space clearance of 1 inch (25 mm).* See exceptions for alternate chimney designs or systems.

Discussion and Commentary: On account of the hot gases and products of combustion, the chimney walls become quite hot. As the heat is transferred through the masonry material, any combustible material in close proximity to the heated walls may reach the point of ignition. Thus, an air space of limited depth must be created around the chimney to help in the dissipation of heat.

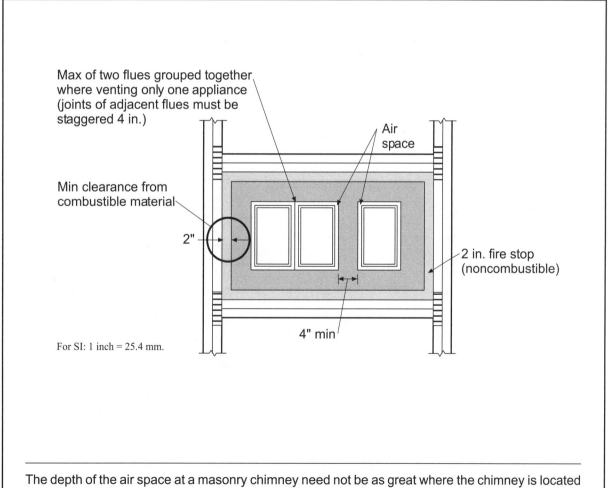

For SI: 1 inch = 25.4 mm.

The depth of the air space at a masonry chimney need not be as great where the chimney is located entirely outside of the exterior walls of the dwelling. The exterior surface will remain cooler because of its direct exposure to the ambient outdoor temperature.

Code Text: *Chimneys shall be provided with crickets when the dimension parallel to the ridgeline is greater than 30 inches (762 mm) and does not intersect the ridgeline. The intersection of the cricket and the chimney shall be flashed and counterflashed in the same manner as normal roof-chimney intersections. Crickets shall be constructed in compliance with Figure R1003.20 and Table R1003.20.*

Discussion and Commentary: The potential for water intrusion at chimney openings in roofs is of concern, particularly where the length of the chimney wall that runs perpendicular to the roof slope is sizable. By providing a cricket, positive water flow away from the roof opening can be ensured.

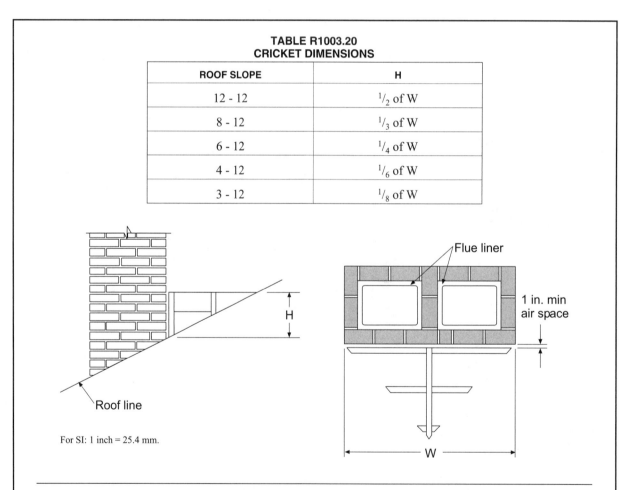

TABLE R1003.20
CRICKET DIMENSIONS

ROOF SLOPE	H
12 - 12	$^1/_2$ of W
8 - 12	$^1/_3$ of W
6 - 12	$^1/_4$ of W
4 - 12	$^1/_6$ of W
3 - 12	$^1/_8$ of W

For SI: 1 inch = 25.4 mm.

The use of flashing and counterflashing at a roof opening such as a chimney is based on their ability to prevent moisture from breaching the weather protective membrane. Where corrosion-resistant metal is used as flashing, it must be minimum No. 26 galvanized sheet steel.

Code Text: *The exterior air intake for factory-built and masonry fireplaces regulated by the IRC shall be capable of supplying all combustion air from the exterior of the dwelling or from spaces within the dwelling ventilated with outside air such as non-mechanically ventilated crawl or attic spaces. The exterior air intake shall not be located within the garage or basement of the dwelling nor shall the air intake be located at an elevation higher than the firebox. The exterior air intake shall be covered with a corrosion-resistant screen of $^1/_4$-inch (6 mm) mesh.*

Discussion and Commentary: Unless it is possible to mechanically ventilate and control the indoor pressure of the room in which a fireplace is located so that the pressure remains neutral or positive, both factory-built and masonry fireplaces must have an exterior air supply. The presence of a complying exterior air intake allows for an adequate supply of combustion air to the fireplace.

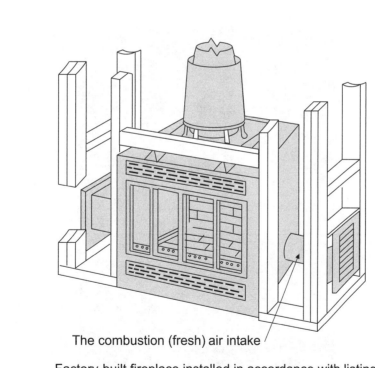

The combustion (fresh) air intake

Factory-built fireplace installed in accordance with listing and manufacturer's instructions

Combustion air ducts are regulated for clearance, size and location. The outlet may be located in the back or sides of the firebox chamber, or as an alternative, located on or near the floor within a distance of 24 inches of the firebox opening.

Code Text: Chapter *11 regulates the energy efficiency for the design and construction of buildings regulated by* the IRC. ***Note:*** *The text of the Sections N1101.2 through N1105 is extracted from the 2012 edition of the* International Energy Conservation Code—Residential Provisions *and has been editorially revised to conform to the scope and application of* the IRC. *The section numbers appearing in parenthesis after each section number are the section numbers of the corresponding text in the* International Energy Conservation Code—Residential Provisions.

Discussion and Commentary: The energy efficiency provisions of the IRC enable the effective use of energy in new dwelling construction by regulating the building envelope, mechanical systems, electrical systems and service water heating systems.

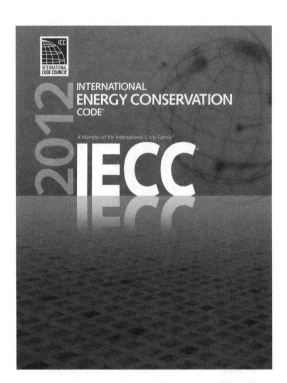

In the IECC, residential buildings are defined as detached one- and two-family dwellings and multiple single-family dwellings (townhouses), and Group R-2, R-3 and R-4 buildings three stories or less in height above grade. To provide consistency and coordination, IRC Chapter 11 includes the IECC residential provisions applicable to residential construction that fall under the scope of the IRC.

Code Text: *Climate zones from Figure N1101.10 or Table N1101.10 shall be used in determining the applicable requirements in Sections N1101 through N1105. Warm humid counties are identified in Table 1101.10 by an asterisk. The climate zone for any location outside the United States shall be determined by applying Table N1101.10.2(1) and then Table N1101.10.2(2).*

Discussion and Commentary: The IRC approach to energy efficiency compliance utilizes climate zones that are assigned to geographical areas of the United States based on multiple climate variables that address both cooling and heating considerations. The climate zone for the building under consideration is the basis for numerous requirements.

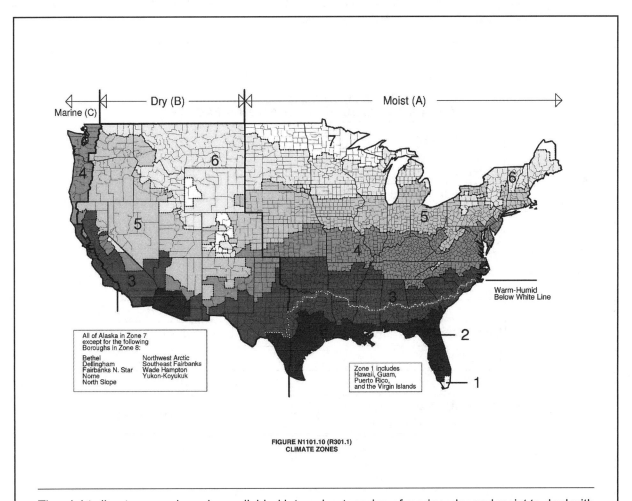

FIGURE N1101.10 (R301.1)
CLIMATE ZONES

The eight climate zones have been divided into subcategories of marine, dry and moist to deal with varying levels of humidity. In addition, the southeastern portion of the country is designated as warm-humid, which results in specialized requirements.

Topic: Insulation and Fenestration Products **Category:** Energy Efficiency
Reference: IRC N1101.12.1, N1101.12.3   **Subject:** General Provisions

Code Text: *An R-value identification mark shall be applied by the manufacturer to each piece of building thermal envelope insulation 12 inches (305 mm) or greater in width. Alternately, the insulation installers shall provide a certification listing the type, manufacturer and R-value of insulation installed in each element of the building thermal envelope. U-factors of fenestration products (windows, door and skylights) shall be determined in accordance with NFRC 100 by an accredited, independent laboratory, and labeled and certified by the manufacturer. Products lacking such a labeled U-factor shall be assigned a default U-factor from Tables N1101.12.3(1) or N1101.12.3(2).*

Discussion and Commentary: The IRC requires certification of the installed density and *R*-value of blown-in or sprayed insulation in the wall cavities. Where installed in the attic, the certification must include both the initial installed thickness and the settled thickness, the coverage area and the number of bags of insulating material installed. Markers that indicate in 1-inch-high numbers the installed thickness of the insulation must be provided for every 300 square feet of attic area, attached to the trusses, rafters or joists.

TABLE N1101.12.3(1) [R303.1.3(1)]
DEFAULT GLAZED FENESTRATION *U*-FACTOR

FRAME TYPE	SINGLE PANE	DOUBLE PANE	SKYLIGHT	
			Single	Double
Metal	1.20	0.80	2.00	1.30
Metal with Thermal Break	1.10	0.65	1.90	1.10
Nonmetal or Metal Clad	0.95	0.55	1.75	1.05
Glazed Block	0.60			

TABLE N1101.12.3(2) [R303.1.3(2)]
DEFAULT DOOR *U*-FACTORS

DOOR TYPE	*U*-FACTOR
Uninsulated Metal	1.20
Insulated Metal	0.60
Wood	0.50
Insulated, nonmetal edge, max 45% glazing, any glazing double pane	0.35

Fenestration includes skylights, roof windows, vertical windows, opaque doors, glazed doors, glass block and combination opaque/glazed doors. The U-factor for fenestration products shall be established in accordance with National Fenestration Rating Council (NFRC) 100.

Code Text: *Projects shall comply with sections identified as "mandatory"* (such as N1102.4 on air leakage) *and with either sections identified as "prescriptive"* (such as N1102.1 on building thermal envelope) *or the performance approach in Section N1105* (simulated energy performance analysis).

Discussion and Commentary: Chapter 11 provides three methods for demonstrating compliance with the insulation and glazing requirements of the code. Section N1102.1 contains two prescriptive methods of compliance. The code user can choose to comply with the insulation R-value requirements, glazing U-factor requirements and SHGC requirements by using Table N1102.1.1. The code user can also choose to perform a prescriptive trade-off approach by calculating the U-factor of each assembly (e.g., wall assembly, floor assembly, etc.) and multiplying this by the area (ft^2) of the assembly to determine a UA. Table N1102.3 is provided for this approach. In addition, the code user can use the Simulated Performance Alternative (Performance) approach to demonstrate compliance. Computer modeling is used to determine the cost of energy to operate a residence annually. Mandatory requirements must be met as well, which include air sealing, vapor retarders, duct insulation and duct sealing.

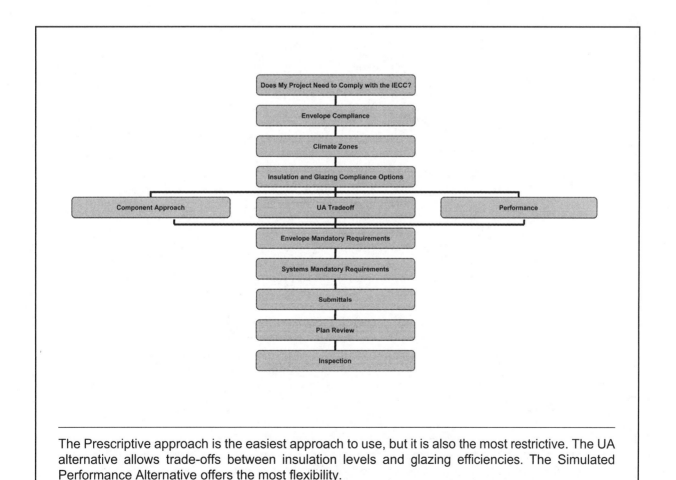

The Prescriptive approach is the easiest approach to use, but it is also the most restrictive. The UA alternative allows trade-offs between insulation levels and glazing efficiencies. The Simulated Performance Alternative offers the most flexibility.

Code Text: *A permanent certificate shall be completed and posted on or in the electrical distribution panel by the builder or registered design professional. The certificate shall list the predominant R-values of insulation installed in or on ceiling/roof, walls, foundation (slab, basement wall, crawlspace wall and/or floor) and ducts outside conditioned spaces; U-factors for fenestration and the solar heat gain coefficient (SHGC) of fenestration; and the results from any required duct system and building envelope air leakage testing done on the building.*

Discussion and Commentary: The requirement mandating that energy-related information be provided on a certificate, and that certificate be posted at the electrical panelboard, is intended to encourage an increased awareness of the energy-efficiency ratings for various building elements. The energy efficiency of the building as a system is a function of many elements considered as parts of a whole, and it is almost always impossible to have a proper identification and analysis of a building's energy efficiency once the structure is completed.

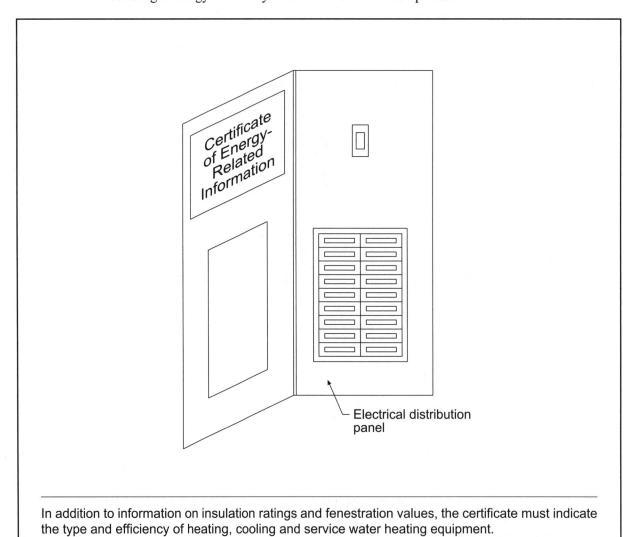

Certificate of Energy-Related Information

Electrical distribution panel

In addition to information on insulation ratings and fenestration values, the certificate must indicate the type and efficiency of heating, cooling and service water heating equipment.

Code Text: *The building thermal envelope shall meet the requirements of Table N1102.1.1 based on the climate zone specified in Table N1101.10.*

Discussion and Commentary: The insulation and fenestration requirements are established based upon two criteria: the type of building component and the climate zone in which the building is located. In addition to the maximum U-factor fenestration and skylight ratings, minimum insulation *R*-values are established for ceilings, wood frame walls, mass walls, floors, basement walls, slabs and crawl space walls.

TABLE N1102.1.1 (R402.1.1)
INSULATION AND FENESTRATION REQUIREMENTS BY COMPONENT[a]

CLIMATE ZONE	FENESTRATION U-FACTOR[b]	SKYLIGHT[b] U-FACTOR	GLAZED FENESTRATION SHGC[b, e]	CEILING R-VALUE	WOOD FRAME WALL R-VALUE	MASS WALL R-VALUE[i]	FLOOR R-VALUE	BASEMENT[c] WALL R-VALUE	SLAB[d] R-VALUE & DEPTH	CRAWL SPACE[c] WALL R-VALUE
1	NR	0.75	0.25	30	13	3/4	13	0	0	0
2	0.40	0.65	0.25	38	13	4/6	13	0	0	0
3	0.35	0.55	0.25	38	20 or 13 + 5[h]	8/13	19	5/13[f]	0	5/13
4 except Marine	0.35	0.55	0.40	49	20 or 13 + 5[h]	8/13	19	10 /13	10, 2 ft	10/13
5 and Marine 4	0.32	0.55	NR	49	20 or 13 + 5[h]	13/17	30[g]	15/19	10, 2 ft	15/19
6	0.32	0.55	NR	49	20 + 5 or 13 + 10[h]	15/20	30[g]	15/19	10, 4 ft	15/19
7 and 8	0.32	0.55	NR	49	20 + 5 or 13 + 10[h]	19/21	38[g]	15/19	10, 4 ft	15/19

For SI: 1 foot = 304.8 mm.

a. *R*-values are minimums. *U*-factors and SHGC are maximums. When insulation is installed in a cavity which is less than the label or design thickness of the insulation, the installed *R*-value of the insulation shall not be less than the *R*-value specified in the table.

b. The fenestration *U*-factor column excludes skylights. The SHGC column applies to all glazed fenestration.
   Exception: Skylights may be excluded from glazed fenestration SHGC requirements in Climate Zones 1 through 3 where the SHGC for such skylights does not exceed 0.30.

c. "15/19" means R-15 continuous insulation on the interior or exterior of the home or R-19 cavity insulation at the interior of the basement wall. "15/19" shall be permitted to be met with R-13 cavity insulation on the interior of the basement wall plus R-5 continuous insulation on the interior or exterior of the home. "10/13" means R-10 continuous insulation on the interior or exterior of the home or R-13 cavity insulation at the interior of the basement wall.

d. R-5 shall be added to the required slab edge *R*-values for heated slabs. Insulation depth shall be the depth of the footing or 2 feet, whichever is less in Zones 1 through 3 for heated slabs.

e. There are no SHGC requirements in the Marine Zone.

f. Basement wall insulation is not required in warm-humid locations as defined by Figure N1101.10 and Table N1101.10.

g. Or insulation sufficient to fill the framing cavity, R-19 minimum.

h. First value is cavity insulation, second is continuous insulation or insulated siding, so "13 + 5" means R-13 cavity insulation plus R-5 continuous insulation or insulated siding. If structural sheathing covers 40 percent or less of the exterior, continuous insulation *R*-value shall be permitted to be reduced by no more than R-3 in the locations where structural sheathing is used – to maintain a consistent total sheathing thickness.

i. The second *R*-value applies when more than half the insulation is on the interior of the mass wall.

A total UA (sum of U-factor times assembly area) method allows for comparing the entire building thermal envelope UA to the sum of individual U-factors. If the total UA is equivalent or less than the sum of the individual U-factors, then the building is considered in compliance.

Code Text: *The building thermal envelope shall be constructed to limit air leakage. The sealing methods between dissimilar materials shall allow for differential expansion and contraction. The components of the building thermal envelope as listed in Table N1102.4.1.1 shall be installed in accordance with the manufacturer's instructions and the criteria listed in Table N1102.4.1.1, as applicable. Where required by the building official, an approve third party shall inspect all components and verify compliance. The building or dwelling unit shall be tested and verified as having an air leakage rate of not exceeding five air changes per hour in Zones 1 and 2, and three air changes per hour in Zones 3 through 8. Testing shall be conducted with a blower door at a pressure of 0.2 inches w.g. (50 Pascals).*

Discussion and Commentary: Sealing the building envelope is critical to good thermal performance for the building. It will prevent warm, conditioned air from leaking out around doors, windows and other cracks during the heating season, thereby reducing the cost of heating the residence. During the hot summer months proper sealing will stop hot air from entering the residence, helping to reduce the air conditioning load.

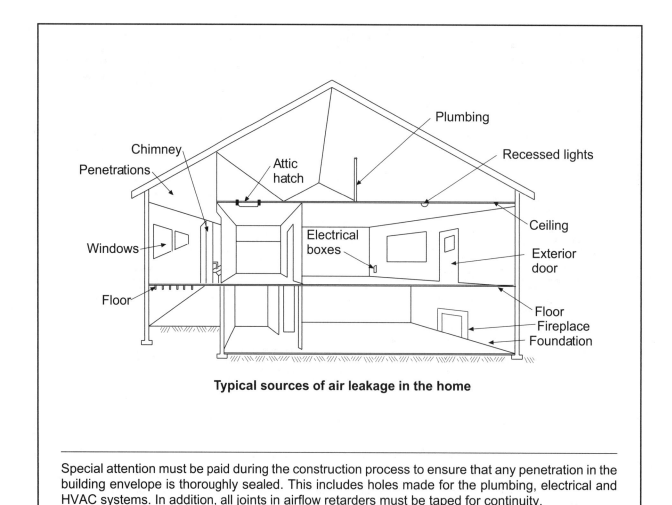

Typical sources of air leakage in the home

Special attention must be paid during the construction process to ensure that any penetration in the building envelope is thoroughly sealed. This includes holes made for the plumbing, electrical and HVAC systems. In addition, all joints in airflow retarders must be taped for continuity.

Quiz

Study Session 9
IRC Chapters 10 and 11

1. The footing for a masonry fireplace and chimney shall be a minimum of _____ thick.

   a 6 inches   b. 8 inches

   c. 10 inches   d. 12 inches

   Reference _____

2. A masonry chimney wall shall not change size or shape for a minimum of _____ above or below where the chimney passes through floor, ceiling or roof components.

   a. 6 inches   b. 12 inches

   c. 24 inches   d. 30 inches

   Reference _____

3. In all cases, a masonry chimney shall extend a minimum of _____ above the highest point where the chimney passes through the roof.

   a. 12 inches   b. 18 inches

   c. 24 inches   d. 36 inches

   Reference _____

4.   A clay flue liner for a masonry chimney shall be carried vertically, with a maximum permitted slope of _____ from the vertical.

     a.   15 degrees   b. $22^1/_2$ degrees

     c.   30 degrees   d.   45 degrees

Reference _____

5.   When two or more flues are located in the same masonry chimney, they shall be separated by masonry wythes having a minimum thickness of _____.

     a.   2 inches   b.   4 inches

     c.   6 inches   d.   8 inches

Reference _____

6.   Unless cleaning is possible through the fireplace opening, a cleanout opening shall be provided within _____ of the base of a flue in a masonry chimney.

     a.   6 inches   b.   8 inches

     c.   12 inches   d.   18 inches

Reference _____

7.   Where a masonry chimney is constructed totally outside the exterior walls of a dwelling, a minimum air space clearance to combustibles of _____ is required.

     a.   0 inches; no clearance is required

     b. $^1/_2$ inch

     c.   1 inch

     d.   2 inches

Reference _____

8.   Foundations for masonry fireplaces and their chimneys shall extend a minimum of _____ beyond the face of the fireplace or support walls on all sides.

     a.   4 inches   b.   6 inches

     c.   8 inches   d.   12 inches

Reference _____

9. Where a 60-inch-wide masonry chimney and fireplace are located in Seismic Design Category D$_1$, the chimney shall be reinforced with _____ No. 4 continuous vertical reinforcing bars.

    a. 0; no reinforcement is required

    b. 2

    c. 4

    d. 6

Reference _____

10. What is the minimum required thickness of the back and side walls of a masonry fireplace constructed without a firebrick lining?

    a. 6 inches   b. 8 inches

    c. 10 inches   d. 12 inches

Reference _____

11. Where firebricks are used as lining for a masonry fireplace, the maximum joint spacing shall be _____ between firebricks.

    a. $^1/_8$ inch   b. $^1/_4$ inch

    c. $^3/_8$ inch   d. $^1/_2$ inch

Reference _____

12. The damper or throat for a masonry fireplace shall be located a minimum of _____ above the lintel supporting masonry over the fireplace opening.

    a. 4 inches   b. 6 inches

    c. 8 inches   d. 12 inches

Reference _____

13. What is the minimum required thickness for the hearth of a masonry fireplace?

    a. $^3/_8$ inch   b. $^3/_4$ inch

    c. 2 inches   d. 4 inches

Reference _____

14. What is the minimum required size of a hearth extension for a masonry fireplace having a 30-inch-high by 42-inch-wide fireplace opening?

   a. 16 inches by 42 inches
   b. 16 inches by 58 inches
   c. 20 inches by 58 inches
   d. 20 inches by 66 inches

   Reference _____

15. A minimum $^3/_8$-inch-thick noncombustible hearth extension is permitted for a masonry fireplace where the bottom of the firebox opening is a minimum of _____ inches above the top of the hearth extension.

   a   6
   b. 8
   c. 12
   d. 15

   Reference _____

16. Wood studs shall have a minimum clearance of _____ from the back face of a masonry fireplace.

   a   0 inches; no clearance is required
   b. 1 inch
   c. 2 inches
   d. 4 inches

   Reference _____

17. When masonry chimneys are constructed as part of masonry walls, combustible materials in contact with the masonry wall must be located a minimum of _____ inches from the inside surface of the nearest flue lining.

   a. 4
   b. 6
   c. 12
   d. 30

   Reference _____

18. Combustible trim projecting a distance of 1 inch beyond the fireplace opening of a masonry fireplace shall be located a minimum of _____ from the opening.

   a. 2 inches
   b. 6 inches
   c. 8 inches
   d. 12 inches

   Reference _____

19. Factory-built fireplaces shall be tested in accordance with _____ .

    a. ASTM C 199   b. ASTM C 315

    c. UL 103   d. UL 127

Reference _____

20. Unless the room is mechanically ventilated and controlled so that the indoor pressure is neutral or positive, a combustion air passageway having a minimum size of _____ square inches and a maximum size of _____ square inches shall be provided as an exterior air supply for a factory-built or masonry fireplace.

    a. 6, 25   b. 6, 55

    c. 25, 55   d. 25, 75

Reference _____

21. Unless a certification is provided by the insulation installers, an *R*-value identification mark shall be applied by the manufacturer to each piece of building thermal envelope insulation that is a minimum of _____ inches wide.

    a. 12   b. 15

    c. 18   d. 22

Reference _____

22. Exposed insulation applied to the exterior of basement walls shall be protected with a covering that extends a minimum of _____ inches below grade.

    a. 6   b. 12

    c. 18   d. 24

Reference _____

23. In Climate Zone 2, what is the minimum *R*-value required for a wood-frame exterior wall when using the prescriptive method for the building thermal envelope?

    a. 11   b. 13

    c. 15   d. 19

Reference _____

24. Other than site-built elements, windows, skylights and sliding glass doors shall have a maximum air infiltration rate of _____ cfm per square foot.

    a. 0.05   b. 0.1

    c. 0.3   d. 1.0

Reference _____

25. Unless located completely inside the building thermal envelope, supply ducts in attics shall be insulated to a minimum of _____ when using the prescriptive method.

    a. R-6   b. R-8

    c. R-11   d. R-13

Reference _____

26. Concrete foundations for masonry chimneys shall extend a minimum of _____ inches beyond each side of the exterior dimensions of the chimney.

    a. 3   b. 6

    c. 8   d. 12

Reference _____

27. The net free area of a spark arrester for a masonry chimney shall be a minimum of _____ time(s) the net free area of the outlet of the chimney flue it serves.

    a. one   b. two

    c. three   d. four

Reference _____

28. Where a 54-inch-wide masonry chimney is constructed above a roof having a slope of 8:12, a required cricket shall have a minimum height of _____ inches.

    a. 9   b. $13^1/_2$

    c. 18   d. 27

Reference _____

29. Except for Rumford fireplaces, the firebox of a masonry fireplace shall have a minimum depth of _____ inches.

    a. 12   b. 18

    c. 20   d. 24

Reference _____

30. In general, a minimum clearance of _____ inches is required between the outside surface of a masonry heater and combustible materials.

    a. 6   b. 12

    c. 18   d. 36

Reference _____

31. Where a steel fireplace unit incorporating a steel firebox lining is installed with solid masonry to form a masonry fireplace, it shall be constructed with minimum _____ -inch steel.

    a. $^1/_8$ b. $^3/_{16}$

    c. $^1/_4$ d. $^5/_{16}$

Reference _____

32. A Rumsford fireplace shall have a minimum fireplace depth of _____ inches where the fireplace opening is 42 inches in width.

    a. 12   b. 14

    c. 15   d. 20

Reference _____

33. Where the hearth extension for a masonry fireplace is located at the same level as the bottom of the firebox opening, the minimum required thickness of the hearth extension is _____ inches.

    a. $1^1/_2$ b. 2

    c. 3   d. 4

Reference _____

34. Unlisted combustion air ducts installed to serve a factory-built or masonry fireplace shall be installed with a minimum _____ -inch clearance to combustibles for those portions of the duct within _____ feet of the duct outlet.

    a. 1, 5   b. 1, 10

    c. 2, 5   d. 2, 10

Reference _____

35. The thickness of blown-in cellulose roof/ceiling insulation shall be provided on markers installed throughout the attic space at a maximum rate of one marker for every _____ square feet maximum.

    a. 100   b. 200

    c. 300   d. 500

Reference _____

Study Session

10

2012 IRC Chapters 13 and 14

General Mechanical System Requirements and Heating/Cooling Equipment

OBJECTIVE: To gain an understanding of the general provisions for the installation of mechanical appliances, equipment and systems, including requirements regulating heating and cooling equipment.

REFERENCE: Chapters 13 and 14, 2012 *International Residential Code*

KEY POINTS:
- What information must appear on the factory-applied nameplate affixed to an appliance?
- How much working space is needed in front of the control side of an appliance in order to service the appliance? How much of a reduction is allowed in front of a room heater?
- Where a furnace or air handler is installed in an alcove or compartment, what is the minimum total width of the enclosed area? What is the minimum working clearance required along the sides of the unit? Along the back? On the top? How much front clearance is mandated where the firebox is open to the atmosphere?
- How large a passageway must be provided to an appliance installed in a compartment, alcove, basement or similar space? How must access to an attic appliance be addressed? For appliances installed in under-floor spaces?
- What minimum clearance is required between the ground and the appliance where the appliance is suspended from the floor? How far must the excavation extend beyond the appliance enclosure?
- What is the minimum clearance needed between an appliance and unprotected combustible construction? How may such clearances be reduced? What is the minimum required clearance for solid fuel appliances?
- In what seismic design categories must water heaters be anchored or strapped to resist horizontal displacement due to earthquake forces? What prescriptive method is set forth?
- How high above a floor in a garage is an appliance with an ignition source required to be located?
- What special provisions apply to hydrogen-generating and refueling operations?
- How shall piping used as a portion of a mechanical system be protected where installed in concealed locations through holes and notches in framing members?

- How are radiant heating systems required to be installed? Duct heaters?
- What minimum clearance from a wall is mandated for a floor furnace? What is the minimum clearance from doors, draperies or similar combustible objects? What is the minimum required clearance from the furnace register to a combustible projecting object above?
- Where are vented wall furnaces prohibited from being located? What clearances are mandated between the furnace air inlet or outlet and the swing of a door?
- How must vented room heaters be installed? What minimum clearance is required on all sides of the appliance?
- How is condensate from refrigeration cooling equipment to be addressed? What methods are described for an auxiliary or secondary drain system? What is the minimum drain pipe size?
- What limitations are mandated for the use of fireplace stoves?

Code Text: *Furnaces and air handlers within compartments or alcoves shall have a minimum working space clearance of 3 inches (76 mm) along the sides, back and top with a total width of the enclosing space being at least 12 inches (305 mm) wider than the furnace or air handler. Furnaces having a firebox open to the atmosphere shall have at least a 6-inch (152 mm) working space along the front combustion chamber side. See exception for replacement appliances.*

Discussion and Commentary: A minimum amount of clearance is desired around a furnace or air handler to allow for access to the unit, provide space to facilitate the removal of the unit, create an air space for ventilating and cooling the appliance, and protect any surrounding combustible materials. The increased clearance at the front protects against flame roll-out and provides free movement of combustion air.

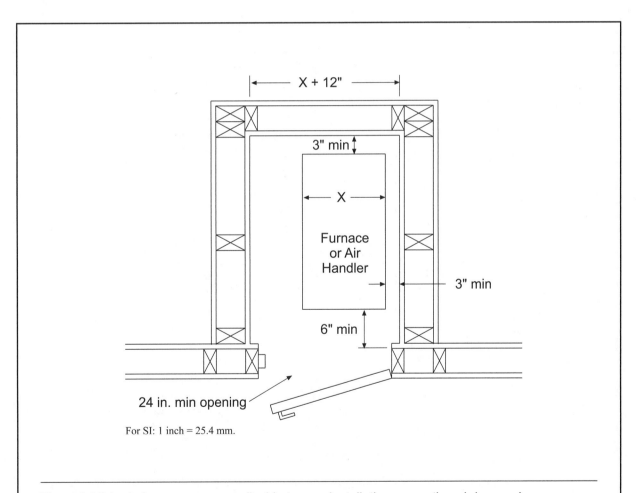

24 in. min opening

For SI: 1 inch = 25.4 mm.

The established clearances are applicable to new installations even though lesser clearances are permitted by the manufacturer. Replacement appliances installed in existing compartments may have lesser clearances if in conformance with the manufacturer's instructions.

Code Text: *Attics containing appliances shall be provided with an opening and a clear and unobstructed passageway large enough to allow removal of the largest appliance, but not less than 30 inches (762 mm) high and 22 inches (559 mm) wide and not more than 20 feet (6096 mm) long measured along the centerline of the passageway from the opening to the appliance.* See exceptions to 1) eliminate the passageway and level service space where appliance can be serviced and removed through the required opening, and 2) permit passageway length of 50 feet maximum where minimum 6-foot by 22-inch unobstructed passageway is provided.

Discussion and Commentary: By placing the mechanical equipment in the attic, it is possible to better utilize the building's floor area. However, it is important that such equipment be accessible for service or removal. In addition to a properly sized access opening, a continuous passageway of solid flooring must be provided where needed to reach the appliance. A minimum 30-inch by 30-inch service area is required in front of any appliance access.

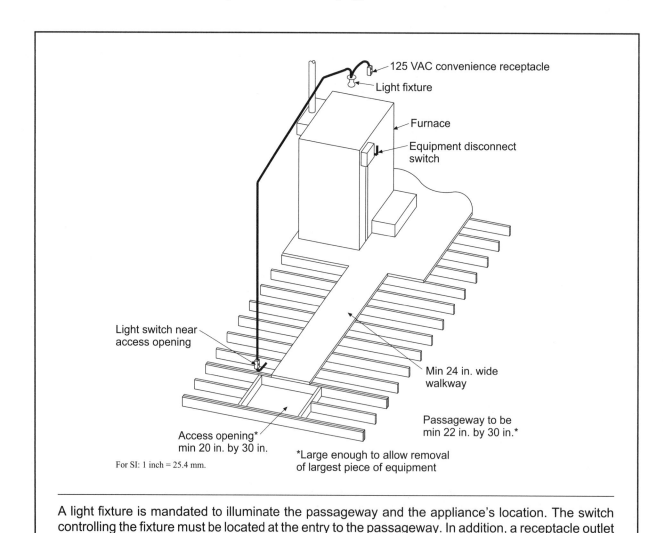

125 VAC convenience receptacle

Light fixture

Furnace

Equipment disconnect switch

Light switch near access opening

Min 24 in. wide walkway

Passageway to be min 22 in. by 30 in.*

Access opening* min 20 in. by 30 in.

For SI: 1 inch = 25.4 mm.

*Large enough to allow removal of largest piece of equipment

A light fixture is mandated to illuminate the passageway and the appliance's location. The switch controlling the fixture must be located at the entry to the passageway. In addition, a receptacle outlet shall be located at or near the appliance for use during maintenance or repair.

Category: Mechanical Systems
Subject: Appliance Access

Code Text: *Underfloor spaces containing appliances shall be provided with an unobstructed passageway large enough to remove the largest appliance, but not less than 30 inches (762 mm) high and 22 inches (559 mm) wide, nor more than 20 feet (6096 mm) long measured along the centerline of the passageway from the opening to the appliance. If the depth of the passageway or the service space exceeds 12 inches (305 mm) below the adjoining grade, the walls shall be lined with concrete or masonry extending 4 inches (102 mm) above the adjoining grade in accordance with Chapter 4 (Foundations). See exceptions to 1) eliminate the passageway and level service space where appliance can be serviced and removed through the required opening, and 2) permit unlimited passageway length where minimum 6-foot high by 22-inch high unobstructed passageway is provided.*

Discussion and Commentary: The requirements for access to an underfloor furnace are similar to those for furnaces located in attics and other concealed areas. A switched light fixture and receptacle outlet are mandated to allow for safe and convenient appliance repair and maintenance.

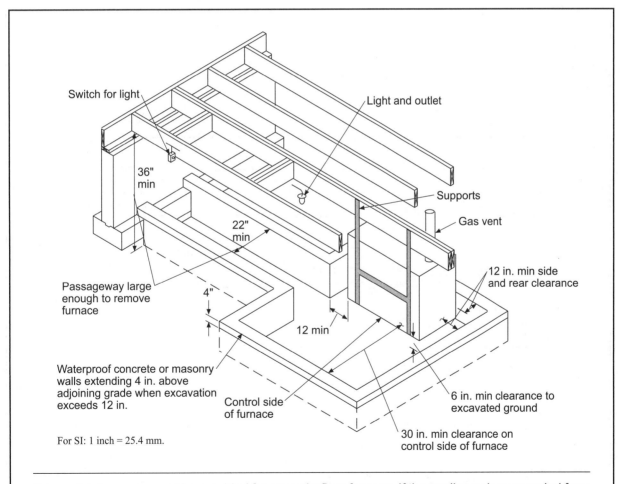

Switch for light

Light and outlet

36" min

22" min

Supports

Gas vent

Passageway large enough to remove furnace

4"

12 in. min side and rear clearance

Waterproof concrete or masonry walls extending 4 in. above adjoining grade when excavation exceeds 12 in.

Control side of furnace

12 min

6 in. min clearance to excavated ground

For SI: 1 inch = 25.4 mm.

30 in. min clearance on control side of furnace

Adequate clearance must be provided for an underfloor furnace. If the appliance is suspended from the floor, it must be at least 6 inches above the ground. Where the underfloor space is excavated, clearances of 12 inches on the sides and 30 inches on the control side are required.

Code Text: *Appliances shall be installed with the clearances from unprotected combustible materials as indicated on the appliance label and in the manufacturer's installation instructions.*

Discussion and Commentary: The clearances established by the manufacturer's installation instructions are assumed to be from unprotected combustible construction. These clearances create the necessary airspace for reducing the conduction of heat that might ignite adjacent combustible materials. In addition, convection cooling provided by the airspace assists in the proper operation of the appliance. It is important to note that the provisions and guidelines are not intended to apply to the installation of unlisted appliances.

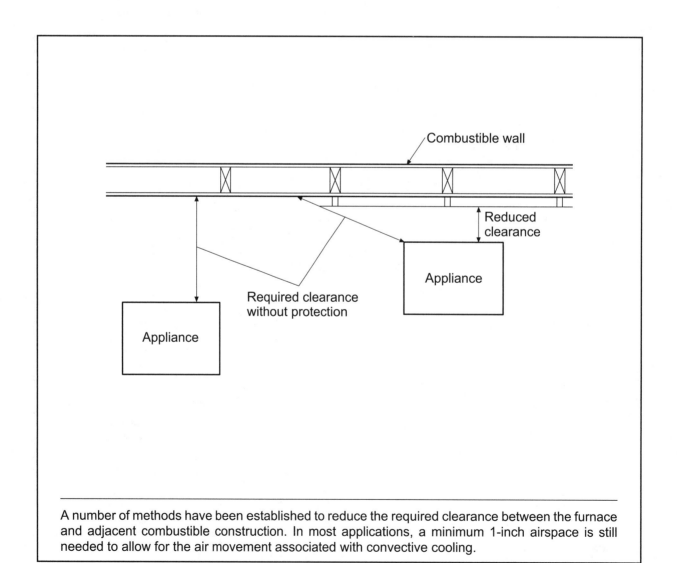

A number of methods have been established to reduce the required clearance between the furnace and adjacent combustible construction. In most applications, a minimum 1-inch airspace is still needed to allow for the air movement associated with convective cooling.

Code Text: *Reduction of clearances shall be in accordance with the appliance manufacturer's instructions and Table M1306.2. Forms of protection with ventilated air space shall conform to the following requirements: 1) not less than 1-inch (25 mm) air space shall be provided between the protection and combustible wall surface; 2) air circulation shall be provided by having edges of the wall protection open at least 1 inch (25 mm); 3) if the wall protection is mounted on a single flat wall away from corners, air circulation shall be provided by having the bottom and top edges, or the side and top edges open at least 1 inch (25 mm); and 4) wall protection covering two walls in a corner shall be open at the bottom and top edges at least 1 inch (25 mm).*

Discussion and Commentary: The allowances for reduced clearances require that a 1-inch air space be maintained to allow the air circulation needed to keep the temperature rise within acceptable limits.

TABLE M1306.2
REDUCTION OF CLEARANCES WITH SPECIFIED FORMS OF PROTECTION[a, c, d, e, f, g, h, i, j, k, l]

TYPE OF PROTECTION APPLIED TO AND COVERING ALL SURFACES OF COMBUSTIBLE MATERIAL WITHIN THE DISTANCE SPECIFIED AS THE REQUIRED CLEARANCE WITH NO PROTECTION (See Figures M1306.1 and M1306.2)	WHERE THE REQUIRED CLEARANCE WITH NO PROTECTION FROM APPLIANCE, VENT CONNECTOR, OR SINGLE WALL METAL PIPE IS:									
	36 inches		18 inches		12 inches		9 inches		6 inches	
	Allowable clearances with specified protection (Inches)[b] Use column 1 for clearances above an appliance or horizontal connector. Use column 2 for clearances from an appliance, vertical connector and single-wall metal pipe.									
	Above column 1	Sides and rear column 2	Above column 1	Sides and rear column 2	Above column 1	Sides and rear column 2	Above column 1	Sides and rear column 2	Above column 1	Sides and rear column 2
3½-inch thick masonry wall without ventilated air space	—	24	—	12	—	9	—	6	—	5
½-in. insulation board over 1-inch glass fiber or mineral wool batts	24	18	12	9	9	6	6	5	4	3
Galvanized sheet steel having a minimum thickness of 0.0236-inch (No. 24 gage) over 1-inch glass fiber or mineral wool batts reinforced with wire or rear face with a ventilated air space	18	12	9	6	6	4	5	3	3	3
3½-inch thick masonry wall with ventilated air space	—	12	—	6	—	6	—	6	—	6
Galvanized sheet steel having a minimum thickness of 0.0236-inch (No. 24 gage) with a ventilated air space 1-inch off the combustible assembly	18	12	9	6	6	4	5	3	3	2
½-inch thick insulation board with ventilated air space	18	12	9	6	6	4	5	3	3	3
Galvanized sheet steel having a minimum thickness of 0.0236-inch (No. 24 gage) with ventilated air space over 24 gage sheet steel with a ventilated space	18	12	9	6	6	4	5	3	3	3
1-inch glass fiber or mineral wool batts sandwiched between two sheets of galvanized sheet steel having a minimum thickness of 0.0236-inch (No. 24 gage) with a ventilated air space	18	12	9	6	6	4	5	3	3	3

For SI: 1 inch = 25.4 mm, 1 pound per cubic foot = 16.019 kg/m³, °C = [(°F)-32/1.8], 1 Btu/(h × ft² × °F/in.) = 0.001442299 (W/cm² × °C/cm).

a. Reduction of clearances from combustible materials shall not interfere with combustion air, draft hood clearance and relief, and accessibility of servicing.

b. Clearances shall be measured from the surface of the heat producing appliance or equipment to the outer surface of the combustible material or combustible assembly.

c. Spacers and ties shall be of noncombustible material. No spacer or tie shall be used directly opposite appliance or connector.

d. Where all clearance reduction systems use a ventilated air space, adequate provision for air circulation shall be provided as described. (See Figures M1306.1 and M1306.2.)

e. There shall be at least 1 inch between clearance reduction systems and combustible walls and ceilings for reduction systems using ventilated air space.

f. If a wall protector is mounted on a single flat wall away from corners, adequate air circulation shall be permitted to be provided by leaving only the bottom and top edges or only the side and top edges open with at least a 1-inch air gap.

g. Mineral wool and glass fiber batts (blanket or board) shall have a minimum density of 8 pounds per cubic foot and a minimum melting point of 1,500°F.

h. Insulation material used as part of a clearance reduction system shall have a thermal conductivity of 1.0 Btu per square foot per hour °F or less. Insulation board shall be formed of noncombustible material.

i. There shall be at least 1 inch between the appliance and the protector. In no case shall the clearance between the appliance and the combustible surface be reduced below that allowed in this table.

j. All clearances and thicknesses are minimum; larger clearances and thicknesses are acceptable.

k. Listed single-wall connectors shall be permitted to be installed in accordance with the terms of their listing and the manufacturer's instructions.

l. For limitations on clearance reduction for solid-fuel-burning appliances see Section M1306.2.1.

Because solid-fuel-burning appliances can produce radiant high-intensity heat and have wide variations in heat output, there are restrictions on the use of Table M1306.2 for reducing clearances to combustible construction. A minimum of 12 inches must always be maintained.

Code Text: *Appliances designed to be fixed in position shall be fastened or anchored in an approved manner. In Seismic Design Categories D_1 and D_2, water heaters shall be anchored or strapped to resist horizontal displacement caused by earthquake motion. Strapping shall be at points within the upper one-third and lower one-third of the appliance's vertical dimensions. At the lower point, the strapping shall maintain a minimum distance of 4 inches (102 mm) above the controls.*

Discussion and Commentary: In those areas determined to have a high earthquake risk, it is important that any water heater be fastened in place to avoid damage. The strapping should be adequately attached to the studs or other framing members. It must be located an adequate distance from the control panel to allow access and use.

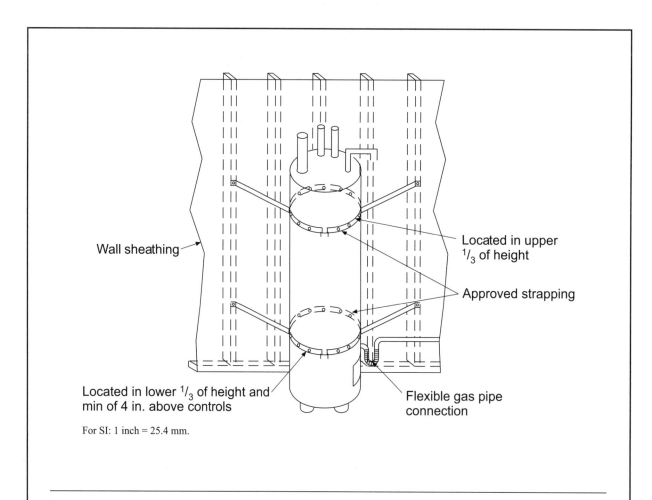

Wall sheathing

Located in upper $^1/_3$ of height

Approved strapping

Located in lower $^1/_3$ of height and min of 4 in. above controls

Flexible gas pipe connection

For SI: 1 inch = 25.4 mm.

Large water heaters weigh several hundred pounds; if they are not secured, the water, gas and electrical connections may be damaged or severed in an earthquake, creating a hazardous situation. In addition to strapping, approved flexible connectors should be used.

Topic: Elevation of Ignition Source   **Category:** Mechanical Systems
Reference: IRC M1307.3, M1307.3.1   **Subject:** Appliance Installation

Code Text: *Appliances having an ignition source shall be elevated such that the source of ignition is not less than 18 inches (457 mm) above the floor in garages.* See exception for appliances that are listed as flammable vapor ignition resistant. *For the purpose of* Section M1307.3, *rooms or spaces that are not part of the living space of a dwelling unit and that communicate with a private garage through openings shall be considered to be part of the garage. Appliances located in a garage or carport shall be protected from impact by automobiles.*

Discussion and Commentary: Gasoline leakage or spillage in a garage is a possible danger. Gasoline fumes will evaporate from liquid puddles at the floor level. Any potential ignition source must be elevated to keep open flame, spark-producing elements and heating elements above the gasoline fume level.

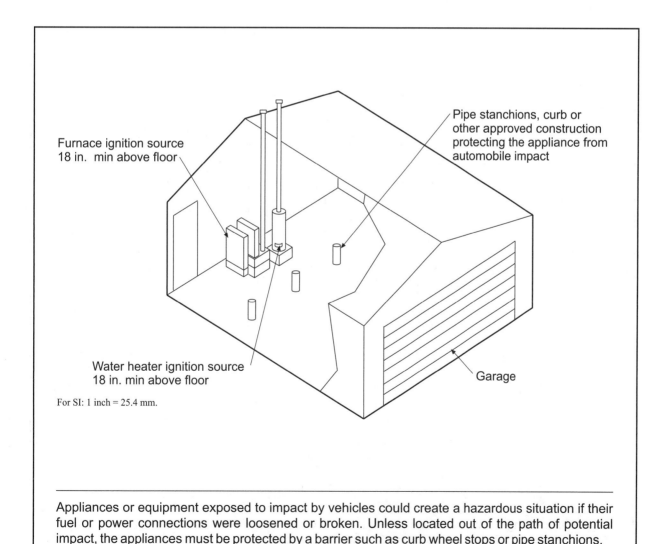

Furnace ignition source 18 in. min above floor

Pipe stanchions, curb or other approved construction protecting the appliance from automobile impact

Water heater ignition source 18 in. min above floor

Garage

For SI: 1 inch = 25.4 mm.

Appliances or equipment exposed to impact by vehicles could create a hazardous situation if their fuel or power connections were loosened or broken. Unless located out of the path of potential impact, the appliances must be protected by a barrier such as curb wheel stops or pipe stanchions.

Code Text: *In concealed locations where piping, other than cast-iron or galvanized steel, is installed through holes or notches in studs, joists, rafters or similar members less than 1.5 inches (38 mm) from the nearest edge of the member, the pipe shall be protected by shield plates. Protective shield plates shall be a minimum thickness of 0.0575-inch (1.463 mm) (No. 16 gage), shall cover the area of the pipe where the member is notched or bored, and shall extend a minimum of 2 inches (51 mm) above sole plates and below top plates.*

Discussion and Commentary: It is critical that piping passing through joists, studs, rafters and other structural members be protected from damage, typically caused by wallboard fasteners and/or wood structural panels. Therefore, steel shield plates are to be utilized where the distance from the member surface to the notch or bored hole exceeds a specified distance.

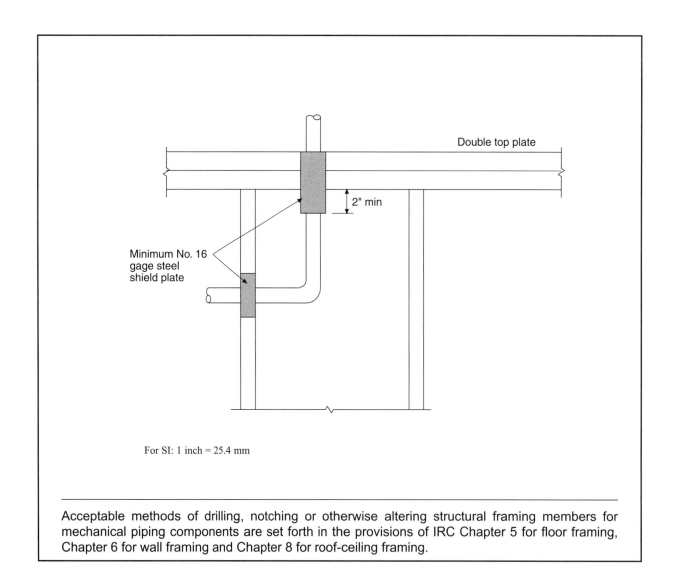

Double top plate

2" min

Minimum No. 16 gage steel shield plate

For SI: 1 inch = 25.4 mm

Acceptable methods of drilling, notching or otherwise altering structural framing members for mechanical piping components are set forth in the provisions of IRC Chapter 5 for floor framing, Chapter 6 for wall framing and Chapter 8 for roof-ceiling framing.

Code Text: *Supports and foundations for the outdoor unit of a heat pump shall be raised at least 3 inches (76 mm) above the ground to permit free drainage of defrost water, and shall conform to the manufacturer's installation instructions.*

Discussion and Commentary: All heat pumps installed outdoors are subject to corrosion and other environmental effects. The installation of such units on the ground could subject them to additional and unnecessary effects of drainage water and possible ponded water. Therefore, all heat pumps must be raised to a minimum height of 3 inches above the surrounding finished grade.

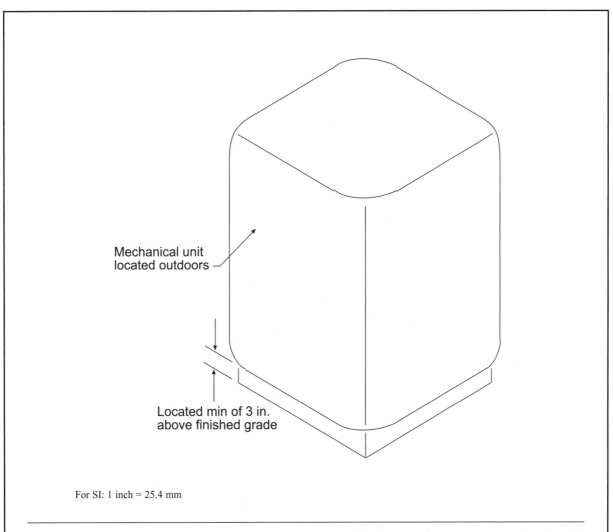

Mechanical unit located outdoors

Located min of 3 in. above finished grade

For SI: 1 inch = 25.4 mm

The protection afforded by raising the heat pump a small distance above the ground surface is primarily necessary because of the unit's production of defrost water, whereas the presence of storm surface drainage is also responsible for damage to any unit placed outdoors.

Code Text: *Floor registers of floor furnaces shall be installed not less than 6 inches (152 mm) from a wall. Wall registers of floor furnaces shall be installed not less than 6 inches (152 mm) from the adjoining wall at inside corners. The furnace register shall be located not less than 12 inches (305 mm) from doors in any position, draperies or similar combustible objects. The furnace register shall be located at least 5 feet (1524 mm) below any projecting combustible materials. The floor furnace burner assembly shall not project into an occupied under-floor area. The floor furnace shall not be installed in concrete floor construction built on grade. The floor furnace shall not be installed where a door can swing within 12 inches (305 mm) of the grille opening.*

Discussion and Commentary: Flat floor furnaces and wall register floor furnaces are designed to be built directly into a wood-framed floor. However, there are clearances necessary because of the heat generated in the combustion chamber and heat coming from the register.

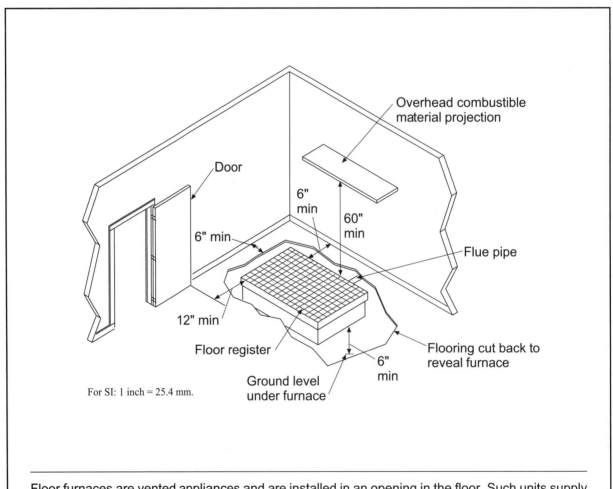

For SI: 1 inch = 25.4 mm.

Floor furnaces are vented appliances and are installed in an opening in the floor. Such units supply heat to the room by gravity convection and direct radiation and typically serve as the sole source of space heating. They are sometimes used in cottages and small homes.

Code Text: *Floor furnaces shall be installed not closer than 6 inches (152 mm) to the ground. Clearance may be reduced to 2 inches (51 mm), provided that the lower 6 inches (152 mm) of the furnace is sealed to prevent water entry. Where excavation is required for a floor furnace installation, the excavation shall extend 30 inches (762 mm) beyond the control side of the floor furnace and 12 inches (305 mm) beyond the remaining sides.*

Discussion and Commentary: Floor furnaces must have supports independent of the floor register. A 6-inch clearance must also be maintained from the ground to protect against moisture, rust and corrosion of the furnace casing. This may be reduced to 2 inches if the unit is sealed against flooding. Floor furnaces are not permitted to be supported on the ground.

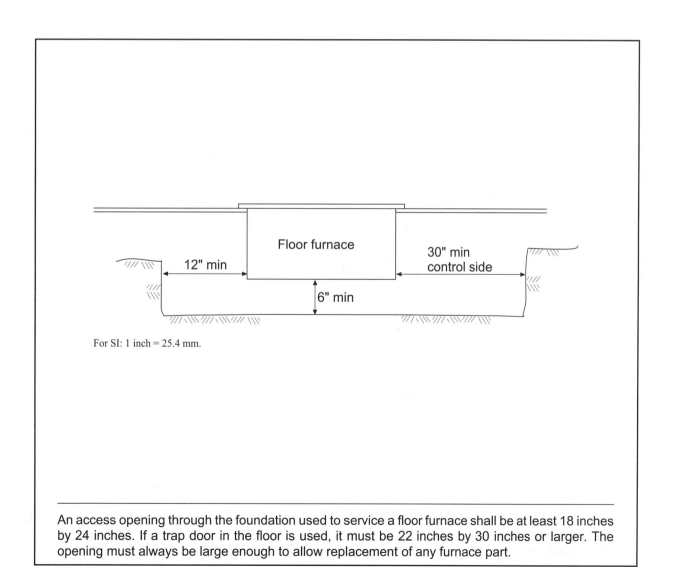

For SI: 1 inch = 25.4 mm.

An access opening through the foundation used to service a floor furnace shall be at least 18 inches by 24 inches. If a trap door in the floor is used, it must be 22 inches by 30 inches or larger. The opening must always be large enough to allow replacement of any furnace part.

Topic: Location

Category: Heating and Cooling Equipment

Reference: IRC M1409.2

Subject: Vented Wall Furnaces

Code Text: *Vented wall furnaces shall be located where they will not cause a fire hazard to walls, floors, combustible furnishings or doors. Vented wall furnaces installed between bathrooms and adjoining rooms shall not circulate air from bathrooms to other parts of the building. Vented wall furnaces shall not be located where a door can swing within 12 inches (305 mm) of the furnace air inlet or outlet measured at right angles to the opening. Doorstops or door closers shall not be installed to obtain this clearance.*

Discussion and Commentary: Provided with the necessary internal clearances, wall furnaces are designed for installation in combustible stud cavities. To allow for necessary air movement and reduce the potential for the ignition of combustible materials, adequate clearances are mandated.

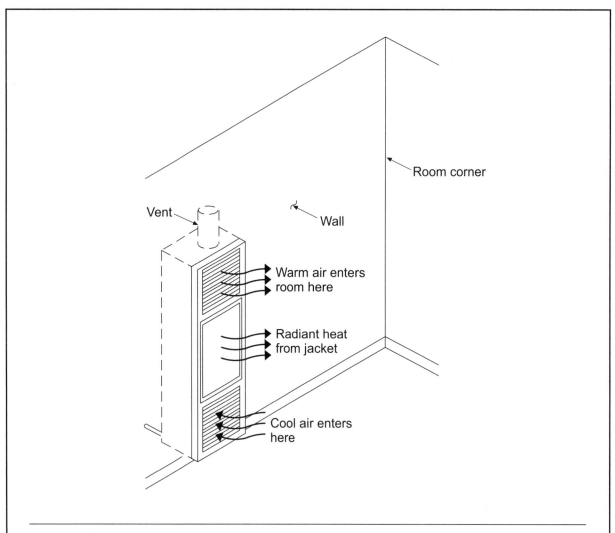

Unlike forced-air furnaces, wall furnaces are not designed for the distribution of hot air through ductwork. Wall furnaces may be of the gravity-type or possibly constructed with small blowers. The attachment of ducts to either type of unit could potentially cause overheating.

Topic: Condensate Disposal

Reference: IRC M1411.3, M1411.3.1

Category: Heating and Cooling Equipment

Subject: Refrigeration Cooling Equipment

Code Text: *Condensate from all cooling coils or evaporators shall be conveyed from the drain pan outlet to an approved place of disposal. Condensate shall not discharge into a street, alley or other areas so as to cause a nuisance. In addition to the requirements of Section M1411.3, a secondary drain or auxiliary drain pan shall be required for each cooling or evaporator coil where damage to any building components will occur as a result of overflow from the equipment drain pan or stoppage in the condensate drain piping. Such piping shall maintain a minimum horizontal slope in the direction of discharge of not less than $^1/_8$ unit vertical in 12 units horizontal (1-percent slope). Drain piping shall be a minimum of $^3/_4$-inch (19 mm) nominal pipe size.*

Discussion and Commentary: The condensate created by cooling coils and evaporators can be a nuisance at best, a hazard at worst. Condensation can damage or ruin the appearance of building finish material, and over time undetected water damage can lead to possible structural failure.

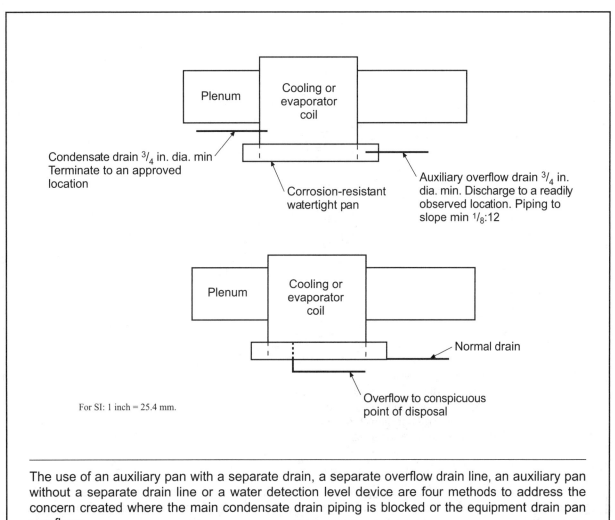

For SI: 1 inch = 25.4 mm.

The use of an auxiliary pan with a separate drain, a separate overflow drain line, an auxiliary pan without a separate drain line or a water detection level device are four methods to address the concern created where the main condensate drain piping is blocked or the equipment drain pan overflows.

Code Text: *Fireplace stoves shall be listed, labeled and installed in accordance with the terms of the listing. Fireplace stoves shall be tested in accordance with UL 737. Hearth extensions for fireplace stoves shall be installed in accordance with the listing of the fireplace stove. The supporting structure for a hearth extension for a fireplace stove shall be at the same level as the supporting structure for the fireplace unit. The hearth extension shall be readily distinguishable from the surrounding floor area.*

Discussion and Commentary: A fireplace stove is a free-standing, chimney-connected solid fuel-burning heater designed to be operated with the fire chamber doors in either the open or closed position. A fireplace stove, including its hearth extension, must be installed in strict accordance with the listing.

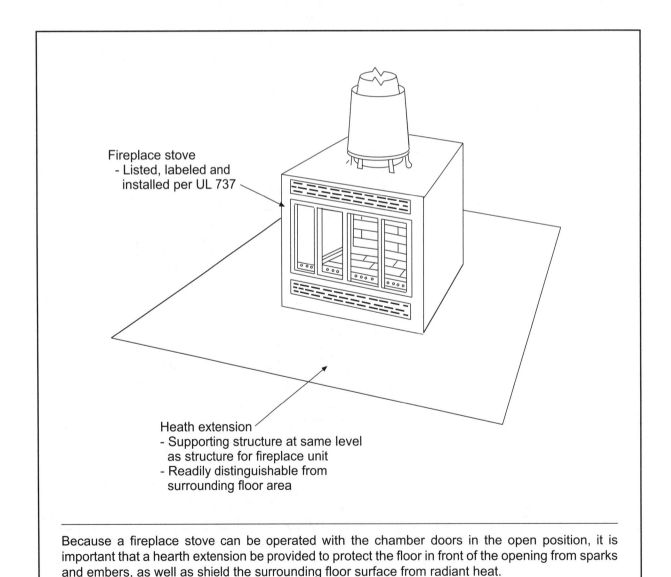

Fireplace stove
- Listed, labeled and installed per UL 737

Heath extension
- Supporting structure at same level as structure for fireplace unit
- Readily distinguishable from surrounding floor area

Because a fireplace stove can be operated with the chamber doors in the open position, it is important that a hearth extension be provided to protect the floor in front of the opening from sparks and embers, as well as shield the surrounding floor surface from radiant heat.

Quiz

Study Session 10
IRC Chapters 13 and 14

1. In order for someone to service an appliance other than a room heater, a minimum working space of _____ inches deep by 30 inches wide shall be provided in front of the control side of the appliance.

    a. 24   b. 30

    c. 36   d. 48

    Reference _____

2. For the purpose of providing working space, a furnace or air handler located within a closet or similar compartment shall be provided with a minimum of _____ clearance along the sides, back and top of the appliance, with the total width of the enclosing space a minimum of _____ wider than the furnace.

    a. 3 inches, 6 inches   b. 3 inches, 12 inches

    c. 6 inches, 12 inches   d. 6 inches, 30 inches

    Reference _____

3. A minimum _____ working space shall be provided along the front combustion chamber side of a central furnace having a firebox open to the atmosphere.

    a. 6-inch   b. 12-inch

    c. 24-inch   d. 30-inch

    Reference _____

4. Where an appliance is installed in a compartment accessed by a passageway, the passageway shall have a minimum unobstructed width of _____, but not less than necessary to allow removal of the largest appliance in the space.

    a. 24 inches   b. 30 inches

    c. 36 inches   d. 42 inches

Reference _____

5. Where an appliance is installed in an attic, the clear access opening shall be a minimum of _____ , but in no case less than necessary to allow removal of the largest appliance.

    a. 20-inches by 30-inches   b. 22-inches by 30-inches

    c. 30-inches by 30-inches   d. 36-inches by 36-inches

Reference _____

6. Appliances suspended from the floor shall have a minimum clearance of _____ from the ground.

    a. 3 inches   b. 4 inches

    c. 6 inches   d. 8 inches

Reference _____

7. Excavations for appliance installations shall extend a minimum depth of _____ below the appliance with a minimum clearance of _____ on all sides except the control side.

    a. 3 inches, 6 inches   b. 6 inches, 6 inches

    c. 6 inches, 12 inches   d. 12 inches, 12 inches

Reference _____

8. Where ventilated air space is used to reduce the required clearance between an appliance and unprotected combustible materials, what is the minimum air space that is typically required?

    a. $\frac{1}{2}$ inch   b. 1 inch

    c. 2 inches   d. 3 inches

Reference _____

9. An appliance requires a minimum clearance without protection of 12 inches above the appliance to combustible material. If $^1/_2$-inch-thick insulation board with a ventilated air space is used as protection, the minimum clearance may be reduced to _____.

    a. 4 inches   b. 6 inches

    c. 9 inches   d. 12 inches

    Reference _____

10. A solid-fuel-burning appliance requires a minimum clearance without protection of 18 inches from the sides and back to combustible material. If 24 gage sheet metal with a ventilated air space is used as protection, the minimum clearance may be reduced to _____ .

    a. 4 inches   b. 6 inches

    c. 9 inches   d. 12 inches

    Reference _____

11. Where an appliance with an ignition source is located in the garage, the source of ignition shall be located a minimum of _____ above the garage floor.

    a. 12 inches   b. 18 inches

    c. 36 inches   d. 48 inches

    Reference _____

12. A combustion air opening for a central furnace shall be unobstructed for a minimum distance of _____ in front of the opening.

    a. 3 inches   b. 6 inches

    c. 12 inches   d. 30 inches

    Reference _____

13. The unobstructed total area of the outside and return air ducts or openings to a heat pump shall be a minimum of _____ square inches per 1,000 Btu/h output rating or as indicated by the conditions of the listing of the heat pump.

    a. 6   b. 10

    c. 12.5   d. 20

    Reference _____

14. In order to permit the free drainage of defrost water, the supports or foundation for the outdoor unit of a heat pump shall be raised a minimum of _____ above the ground.

    a. 2 inches
    b. 3 inches

    c. 4 inches
    d. 6 inches

Reference _____

15. Radiant heating panels installed on wood framing shall be fastened a minimum of _____ inch from an element.

    a. $^1/_4$
    b. $^1/_2$

    c. $^3/_4$
    d. 1

Reference _____

16. Unless listed and labeled for closer installation, a duct heater shall be located a minimum of _____ from a heat pump or air conditioner.

    a. 18 inches
    b. 24 inches

    c. 36 inches
    d. 48 inches

Reference _____

17. The furnace register of a vented floor furnace shall be located a minimum of _____ from doors in any position, draperies and other similar combustible objects.

    a. 3 inches
    b. 6 inches

    c. 12 inches
    d. 60 inches

Reference _____

18. The access opening through the foundation wall to a vented floor furnace shall have a minimum size of _____ .

    a. 18-inches by 24-inches
    b. 20-inches by 30-inches

    c. 22-inches by 30-inches
    d. 30-inches by 30-inches

Reference _____

19. Where the lower 6 inches of a vented floor furnace is sealed to prevent water entry, a minimum _____ clearance shall be provided between the furnace and the ground.

    a. 2-inch   b. 3-inch

    c. 4-inch   d. 6-inch

Reference _____

20. A minimum clearance of _____ shall be provided between the inlet or outlet of a vented wall furnace and a door at any point in its swing, measured at a right angle to the opening.

    a. 6 inches   b. 12 inches

    c. 18 inches   d. 36 inches

Reference _____

21. Unless listed and labeled for such use, cooling coils of refrigeration cooling equipment shall not be located _____ of heat exchangers.

    a. within 12 inches   b. within 36 inches

    c. upstream   d. downstream

Reference _____

22. Condensate drain lines from all cooling coils or evaporators shall be a minimum of _____ inch(es) internal diameter.

    a. $^1/_2$ b. $^3/_4$

    c. 1   d. $1^1/_4$

Reference _____

23. Where an auxiliary drain pan with a separate drain is installed under the cooling coils where condensate will occur, the minimum depth of the pan shall be _____ .

    a. 1 inch   b. $1^1/_2$ inches

    c. 2 inches   d. 3 inches

Reference _____

24. Evaporative coolers shall be installed on a level platform or base located a minimum of _____ above the adjoining ground.

    a. 3 inches   b. 4 inches

    c. 6 inches   d. 12 inches

Reference _____

25. The supporting structure for a hearth extension for a fireplace stove shall be located _____ the supporting structure for the fireplace unit.

    a. at least 6 inches below   b. at least 3 inches below

    c. at the same level as   d. at least 3 inches above

Reference _____

26. Where an appliance is located in an underfloor space, a minimum _____ level service space shall be provided at the front or service side of the appliance.

    a. 30-inch by 30-inch   b. 36-inch by 30-inch

    c. 36-inch by 36-inch   d. 42-inch by 36-inch

Reference _____

27. The required lower strapping for a water heater in Seismic Design Categories D_1 and D_2 shall be located a minimum of _____ inches above the controls.

    a. 3   b. 4

    c. 6   d. 8

Reference _____

28. A private garage containing a hydrogen-generating appliance shall be limited to a maximum of _____ square feet in area.

    a. 720   b. 800

    c. 850   d. 1,000

Reference _____

29. Where piping other than cast-iron or galvanized steel is installed through bored holes of studs within a concealed wall space, steel shield plates are not required where the holes are a minimum of _____ inch(es) from the nearest edge of the stud.

    a. $^3/_4$ b. $1^1/_4$

    c. $1^3/_8$ d. $1^1/_2$

    Reference _____

30. Unless listed for use on combustible floors without floor protection, a floor-mounted vented room heater shall be located so that noncombustible flooring materials extend a minimum of _____ inches beyond the appliance on all sides.

    a. 12   b. 18

    c. 24   d. 36

    Reference _____

31. Where access to an appliance located in an attic is provided by an unobstructed passageway at least 6 feet high and 22 inches wide, the passageway shall have a maximum length of _____.

    a. 10 feet   b. 20 feet

    c. 50 feet   d. an unlimited distance

    Reference _____

32. The two permanent outdoor openings required to be provided in a private garage containing a hydrogen-generating appliance shall have a minimum free area of _____ per 1,000 cubic feet of garage volume.

    a. $^1/_2$ square foot   b. 1 square foot

    c. 2 square feet   d. 5 square feet

    Reference _____

33. The piping and fittings for the refrigerant vapor (suction) lines of cooling equipment shall be insulated with insulation having a minimum thermal resistivity of _____.

    a. R-4   b. R-5

    c. R-8   d. R-11

    Reference _____

34. The register of a vented floor furnace shall be located a minimum of _____ below any projecting combustible materials.

    a. 12 inches   b. 2 feet

    c. 3 feet   d. 5 feet

Reference _____

35. Secondary drain piping from a cooling coil or evaporator shall maintain a minimum horizontal slope in the direction of discharge of _____ unit vertical in 12 units horizontal.

    a. $^1/_{12}$ b. $^1/_8$

    c. $^1/_4$ d. $^1/_2$

Reference _____

Study Session

11

2012 IRC Chapters 15, 16 and 19

Exhaust Systems, Duct Systems and Special Fuel-Burning Equipment

OBJECTIVE: To gain an understanding of the general provisions for the installation of exhaust systems, duct systems and special fuel-burning equipment, such as ranges, ovens and sauna heaters.

REFERENCE: Chapters 15, 16 and 19, 2012 *International Residential Code*

KEY POINTS:
- Where must mechanical exhaust air terminate? What locations are strictly prohibited?
- What are the installation requirements for clothes dryer exhaust systems? What is the maximum length of a flexible transition duct? Where is such a duct permitted?
- What is the maximum permitted length of a clothes dryer exhaust duct? How does the presence of one or more bends affect the allowable length?
- How is a range hood required to be ducted? What are the physical characteristics of the duct? What materials are permitted for duct construction?
- How must a domestic open-top broiler unit be ducted? What is the required clearance between the overhead exhaust hood and combustible construction? The minimum clearance between the cooking surface and the hood?
- What are the minimum exhaust rates for ventilating kitchens? Bathrooms?
- Where are air exhaust openings required to terminate?
- How is a whole-house mechanical ventilation system required to be designed?
- What is the maximum discharge air temperature permitted for equipment connected to an above-ground duct system? The maximum temperature when gypsum products are used? The maximum flame spread?
- Under what conditions can stud cavities be used as return air ducts?
- What are the limitations on the installation of an underground duct system? What is the maximum duct temperature permitted for underground ducts? What is the minimum size of the access opening to an under-floor plenum?
- What is the maximum permitted flame spread for duct insulation coverings and linings? The maximum permitted smoke-developed index?
- How should external duct insulation be identified? At what intervals must the identification be provided?
- What is the maximum permitted length of vibration isolators?

- What means of support are addressed for metal ducts? Nonmetallic ducts?
- Where are underfloor plenums permitted? How are underfloor plenums to be formed? How shall the plenum connect to the furnace? What degree of access to the plenum is required?
- Which sources of outside or return air are prohibited?
- What is the minimum vertical clearance required between the cook-top of a range and any unprotected combustible material? Under what conditions may the clearance be reduced?
- How must sauna heaters be installed and protected from contact? What is the temperature limit for the heater's thermostat?

Code Text: *Exhaust ducts shall terminate on the outside of the building. Exhaust duct terminations shall be in accordance with the dryer manufacturer's installation instructions. If the manufacturer's instructions do not specify a termination location, the exhaust duct shall terminate not less than 3 feet (914 mm) in any direction from openings into buildings. Exhaust duct terminations shall be equipped with a backdraft damper. Screens shall not be installed at the duct termination.*

Discussion and Commentary: Dryer exhaust contains high levels of moisture and highly combustible clothes fibers; therefore, they must terminate at the building's exterior. To avoid any hazard created by the exhaust termination, there must be a clearance of at least 3 feet to any other openings into the building unless a different distance is prescribed by the manufacturer. This would include operable windows, doors and air intakes.

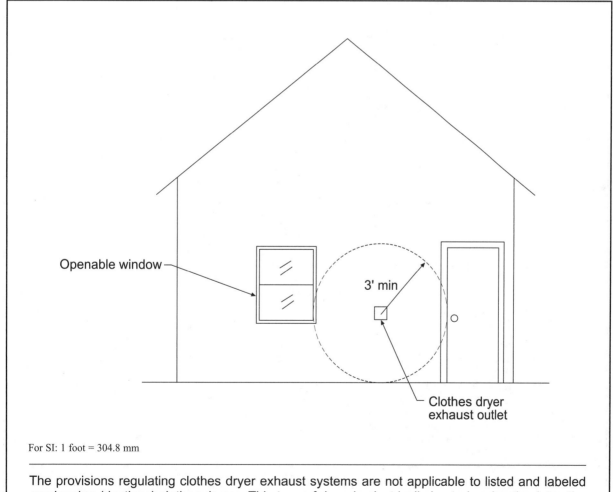

Openable window

3' min

Clothes dryer exhaust outlet

For SI: 1 foot = 304.8 mm

The provisions regulating clothes dryer exhaust systems are not applicable to listed and labeled condensing (ductless) clothes dryers. This type of dryer is electrically heated and recirculates the same air through the clothes drum.

Code Text: *The maximum length of the* clothes dryer *exhaust duct shall be 35 feet (10 668 mm) from the connection to the transition duct from the dryer to the outlet terminal. Where fittings are used, the maximum length of the exhaust duct shall be reduced in accordance with Table M1502.4.4.1.* As an alternate, *the size and maximum length of the exhaust duct shall be determined by the dryer manufacturer's installation instructions. The code official shall be provided with a copy of the installation instructions for the make and model of the dryer at the concealment inspection. In the absence of fitting equivalent length calculations from the manufacturer, Table M1502.4.4.1 shall be used.*

Discussion and Commentary: Clothes dryer exhausts present a different problem than other exhaust systems because the exhaust air is laden with moisture and lint. The air must be vented to the outside and not discharged into an attic or crawl space because wood structural members could be adversely affected. Exhaust outlets must be equipped with a backdraft damper to prevent cold air, rain, snow, rodents and vermin from entering the vent.

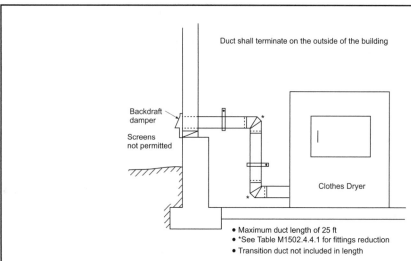

Duct shall terminate on the outside of the building

Backdraft damper

Screens not permitted

Clothes Dryer

- Maximum duct length of 25 ft
- *See Table M1502.4.4.1 for fittings reduction
- Transition duct not included in length

For SI: 1 foot = 304.8 mm, 1 degree = 0.1745 rad.

TABLE M1502.4.4.1
DRYER EXHAUST DUCT FITTING EQUIVALENT LENGTH

DRYER EXHAUST DUCT FITTING TYPE	EQUIVALENT LENGTH
4 inch radius mitered 45 degree elbow	2 feet 6 inches
4 inch radius mitered 90 degree elbow	5 feet
6 inch radius smooth 45 degree elbow	1 foot
6 inch radius smooth 90 degree elbow	1 foot 9 inches
8 inch radius smooth 45 degree elbow	1 foot
8 inch radius smooth 90 degree elbow	1 foot 7 inches
10 inch radius smooth 45 degree elbow	9 inches
10 inch radius smooth 90 degree elbow	1 foot 6 inches

For SI: 1 inch = 25.4 mm, 1 foot = 304.8 mm, 1 degree = 0.0175 rad.

The length restrictions placed on the vent ensure that the dryer exhaust blower will be able to drive sufficient air volume to carry off the moisture. Only when the make and model of the dryer is known, or where an engineered calculation is provided, can the length possibly be increased.

Topic: General Provisions

Category: Exhaust Systems

Reference: IRC M1503.1

Subject: Range Hoods

Code Text: *Range hoods shall discharge to the outdoors through a single-wall duct. The duct serving the hood shall have a smooth interior surface, shall be air tight, shall be equipped with a backdraft damper, and shall be independent of all other exhaust systems. Ducts serving range hoods shall not terminate in an attic or crawl space or areas inside the building. See exception for ductless range hoods.*

Discussion and Commentary: A duct serving a range hood must extend to the exterior of the building. Further clarification indicates that the duct cannot terminate in an attic, crawl space or other areas within the building. Duct construction requires smooth inner walls, prohibiting the use of flexible and semirigid corrugated ducts for range hood exhaust. A backdraft damper is required to prevent the infiltration of outdoor air when the exhaust system is not operating.

As an option to the general requirement for ducting a range hood to the exterior of the building, the use of a ductless range hood is acceptable, provided the hood is listed and labeled for such a condition and either mechanical or natural ventilation is provided within the space where the range is located.

Topic: Domestic Open-Top Broiler Units   **Category:** Exhaust Systems
Reference: IRC M1505.1   **Subject:** Overhead Exhaust Hoods

Code Text: *Domestic open-top broiler units shall have a metal exhaust hood, having a minimum thickness of 0.0157-inch (0.3950 mm) (No. 28 gage), with a $^1/_4$-inch (6 mm) clearance between the hood and the underside of combustible materials or cabinets. A clearance of at least 24 inches (610 mm) shall be maintained between the cooking surface and the combustible material or cabinet.*

Discussion and Commentary: A minimal clearance is needed between the hood and any cabinets or similar items to ensure that the surface material of the hood does not come in direct contact with the combustible material. The vertical clearance protects overhead cabinets from ignition should a fire initiate on the cooktop. It is important that the length of the hood be at least that of the boiler unit to allow for complete coverage.

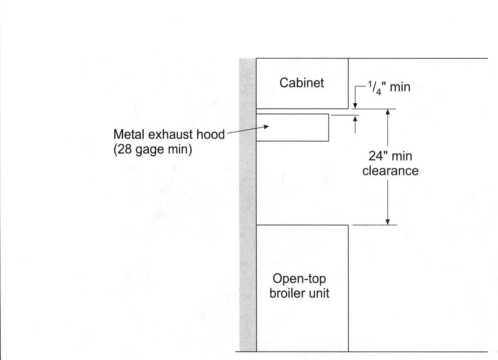

For SI: 1 inch = 25.4 mm.

A boiler unit exhaust hood must discharge to the exterior and be provided with a backdraft damper. A range hood must also exhaust directly to the outdoors unless a listed and labeled ductless range hood is installed in accordance with the manufacturer's instructions.

Topic: Kitchens and Bathrooms

Category: Exhaust Systems

Reference: IRC M1507

Subject: Mechanical Ventilation

Code Text: *Where local exhaust or whole-house mechanical ventilation is provided, the equipment shall be designed in accordance with Section M1507. Exhaust air from bathrooms and toilet rooms shall not be recirculated within a residence or to another dwelling unit and shall be exhausted directly to the outdoors. Exhaust air from bathrooms and toilet rooms shall not discharge into an attic, crawl space or other areas inside the building. Local exhaust systems shall be designed to have the capacity to exhaust the minimum air flow rate determined in accordance with Table M1507.4.*

Discussion and Commentary: The mechanical ventilation rates for bathrooms and water closet compartments are established in Section M1507.4 as an exception to the general requirement in Section R303.3 for natural ventilation. The fundamental provisions require such spaces to be provided with natural ventilation by way of openable windows to the exterior.

TABLE M1507.4
MINIMUM REQUIRED LOCAL EXHAUST RATES FOR
ONE- AND TWO-FAMILY DWELLINGS

AREA TO BE EXHAUSTED	EXHAUST RATES
Kitchens	100 cfm intermittent or 25 cfm continuous
Bathrooms-Toilet Rooms	Mechanical exhaust capacity of 50 cfm intermittent or 20 cfm continuous

For SI: 1 cubic foot per minute = 0.0004719 m^3/s.

Kitchen ventilation can be addressed by two different methods, as is the case for bathroom ventilation. Where continuous kitchen ventilation is provided, a minimum rate of 25 cfm is required. If an intermittent exhaust system is utilized, it must be rated for at least 100 cfm.

Code Text: *The whole-house ventilation system shall consist of one or more supply or exhaust fans, or a combination of such, and associated ducts and controls. Local exhaust or supply fans are permitted to serve as such a system. Outdoor air ducts connected to the return side of an air handler shall be considered to provide supply ventilation. The whole-house mechanical ventilation system shall be provided with controls that enable manual override. The whole-house mechanical ventilation system shall provide outdoor air at a continuous rate of not less than that determined in accordance with Table M1507.3.3(1). See exception permitting intermittent operation.*

Discussion and Commentary: A whole-house mechanical ventilation system is defined as *an exhaust system, supply system, or combination thereof that is designed to mechanically exchange indoor air for outdoor air when operating continuously or through a programmed intermittent schedule to satisfy the whole-house ventilation rate.* Exception 1 to Section R303.1indicates that such a system is required where complying natural ventilation is not provided.

TABLE M1507.3.3(1)
CONTINUOUS WHOLE-HOUSE MECHANICAL VENTILATION SYSTEM AIRFLOW RATE REQUIREMENTS

DWELLING UNIT FLOOR AREA (square feet)	NUMBER OF BEDROOMS				
	0 – 1	2 – 3	4 – 5	6 – 7	> 7
	Airflow in CFM				
< 1,500	30	45	60	75	90
1,501 – 3,000	45	60	75	90	105
3,001 – 4,500	60	75	90	105	120
4,501 – 6,000	75	90	105	120	135
6,001 – 7,500	90	105	120	135	150
> 7,500	105	120	135	150	165

For SI: 1 square foot = 0.0929 m², 1 cubic foot per minute = 0.0004719 m³/s.

Although tight construction is beneficial in reducing energy consumption, closed-house conditions in the heating or cooling season may lead to inadequate fresh air and poor indoor air quality. Whole-house ventilation simply exchanges indoor air for outdoor air at the minimum airflow rates prescribed in Section M1507 based on the area of the dwelling unit and the number of bedrooms.

Code Text: *Equipment connected to duct systems shall be designed to limit discharge air temperature to a maximum of 250°F (121°C). Factory-made air ducts shall be constructed of Class 0 or Class 1 materials as designated in Table M1601.1.1(1). Minimum thicknesses of metal duct material shall be as listed in Table M1601.1.1(2). Use of gypsum products to construct return air ducts or plenums is permitted, provided that the air temperature does not exceed 125°F (52°C). Duct systems shall be constructed of materials having a flame spread index not greater than 200.*

Discussion and Commentary: The temperature limit for discharge air is based on the possible use of nonmetallic ducts. Higher air temperatures could cause their rapid deterioration. The flame spread is strictly limited to 25 for Class 1 ducts and 0 for Class 0 ducts. Where metal ducts are used, they shall be of substantial construction to ensure that their structural integrity is maintained. Ducts fabricated from lighter gage materials will experience an unacceptable amount of distortion in use.

TABLE M1601.1.1(1)
CLASSIFICATION OF FACTORY-MADE AIR DUCTS

DUCT CLASS	MAXIMUM FLAME-SPREAD RATING
0	0
1	25

TABLE M1601.1.1(2)
GAGES OF METAL DUCTS AND PLENUMS USED FOR HEATING OR COOLING

DUCT SIZE	GALVANIZED		ALUMINUM
	Minimum Thickness (inches)	Equivalent Galvanized Gage No.	Minimum Thickness (inches)
Round ducts and enclosed rectangular ducts 14 inches or less 16 and 18 inches 20 inches and over	0.0157 0.0187 0.0236	28 26 24	0.0145 0.018 0.023
Exposed rectangular ducts 14 inches or Over 14ª inches	0.0157 0.0187	28 26	0.0145 0.018

For SI: 1 inch = 25.4 mm.

a. For duct gages and reinforcement requirements at static pressures of ¹/₂ inch, 1 inch and 2 inches w.g., SMACNA *Duct Construction Standard*, Tables 2-1; 2-2 and 2-3 shall apply.

Gypsum products may be used in the fabrication of return air ducts or plenums, provided the air temperature is limited and condensation on the duct surfaces is not a concern. Excessive temperatures and high humidity are both potential causes of damage to the gypsum materials.

Code Text: *Stud wall cavities and the spaces between solid floor joists to be utilized as air plenums shall comply with the following conditions: 7.1) these cavities or spaces shall not be utilized as a plenum for supply air; 7.2) these cavities or spaces shall not be part of a required fire-resistance-rated assembly; 7.3) stud wall cavities shall not convey air from more than one floor level; 7.4) stud wall cavities and joist space plenums shall be isolated from adjacent concealed spaces by tight-fitting fire blocking in accordance with Section R602.8; and 7.5) stud wall cavities in the outside walls of building envelope assemblies shall not be utilized as air plenums.*

Discussion and Commentary: The use of stud spaces and panned joist spaces as return air ducts is acceptable since the air temperature is such not to cause a hazard or deterioration of the construction materials. Isolation of the return air plenum from other unused spaces is necessary so that debris will not be drawn into the furnace.

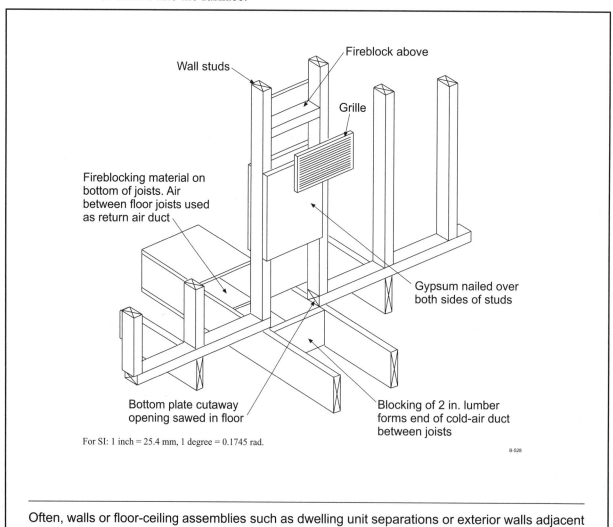

Wall studs

Fireblock above

Grille

Fireblocking material on bottom of joists. Air between floor joists used as return air duct

Gypsum nailed over both sides of studs

Bottom plate cutaway opening sawed in floor

Blocking of 2 in. lumber forms end of cold-air duct between joists

For SI: 1 inch = 25.4 mm, 1 degree = 0.1745 rad.

B-528

Often, walls or floor-ceiling assemblies such as dwelling unit separations or exterior walls adjacent to lot lines must be of fire-resistant-rated construction. Such assemblies must not be used as air plenums.

Code Text: *Underground duct systems shall be constructed of approved concrete, clay, metal or plastic. The maximum duct temperature for plastic ducts shall not be greater than 150°F (66°C). Metal ducts shall be protected from corrosion in an approved manner or shall be completely encased in concrete not less than 2 inches (51 mm) thick. Nonmetallic ducts shall be installed in accordance with the manufacturer's installation instructions.*

Discussion and Commentary: Ducts fabricated from the listed materials should not collapse when installed underground or within concrete. Ducts installed in or beneath concrete should be properly sealed and secured before the concrete pour. Otherwise, leakage of concrete into the ducts could block the flow of air when the system is placed in operation. For the same reason, underground ducts should be secured to avoid damage during backfilling.

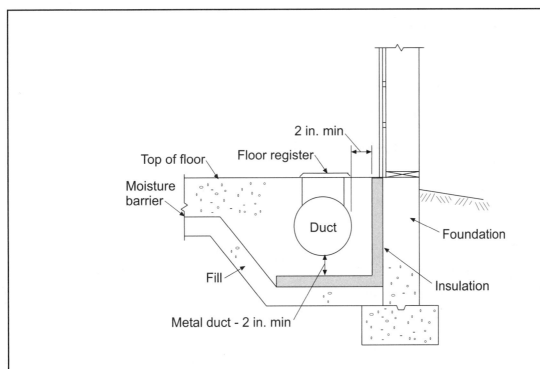

Note:
1. Min of 2 in. concrete encasement around metal duct.
2. Approved material designed for this use and installed in accordance with manufacturer's instructions.

For SI: 1 inch = 25.4 mm.

Where plastic pipe is used in an underground duct system, the duct temperature is limited to avoid the possibility of deterioration or distortion of the plastic materials. To ensure the adequacy of plastic ducts, specific material and structural criteria are referenced in several ASTM standards.

Topic: Duct Insulation Materials

Category: Duct Systems

Reference: IRC M1601.3

Subject: Duct Construction

Code Text: *Duct coverings and linings, including adhesives where used, shall have a flame-spread index not higher than 25, and a smoke-developed index not over 50 when tested in accordance with ASTM E 84 or UL 723. See exception where polyurethane foam is spray-applied to ducts in attics and crawl spaces. External duct insulation and factory-insulated flexible ducts shall be legibly printed or identified at intervals not greater than 36 inches (914 mm) with the name of the manufacturer; the thermal resistance R-value at the specified installed thickness; and the flame spread and smoke-developed indexes of the composite materials.*

Discussion and Commentary: Ducts that pass through nonconditioned areas should be insulated to prevent excessive heat loss or heat gain in cooling installations. Insulation also prevents the cool surfaces of air-conditioning ducts from condensing moisture from the surrounding unconditioned air.

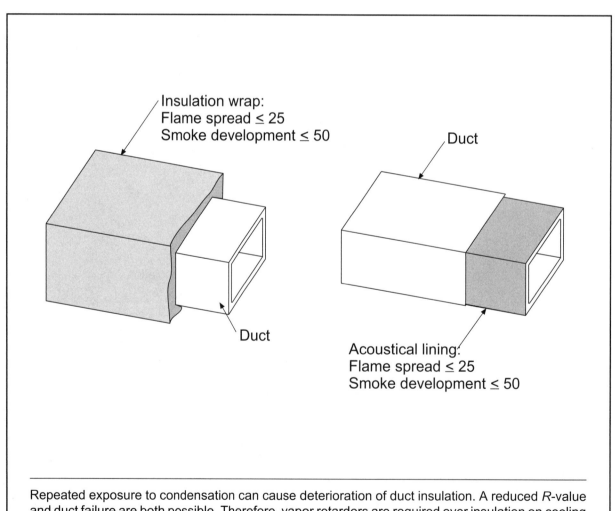

Repeated exposure to condensation can cause deterioration of duct insulation. A reduced *R*-value and duct failure are both possible. Therefore, vapor retarders are required over insulation on cooling supply ducts that pass through nonconditioned areas where condensation may occur.

Code Text: *All joints, longitudinal and transverse seams, and connections in ductwork shall be securely fastened and sealed with welds, gaskets, mastics (adhesives), mastic-plus-embedded-fabric systems or tapes. Duct connections to flanges of air distribution system equipment shall be sealed and mechanically fastened. Crimp joints for round metal ducts shall have a contact lap of not less than 1 inch (25.4 mm) and shall be mechanically fastened by means of not less than three sheet metal screws or rivets equally spaced around the joint. Metal ducts shall be supported by $^1/_2$-inch (13 mm) wide 18-gage metal straps or 12-gage galvanized wire at intervals not exceeding 10 feet (3048 mm) or other approved means.*

Discussion and Commentary: Duct leakage should be minimized to prevent energy loss. Adequate duct support is necessary to maintain alignment and limit deflection.

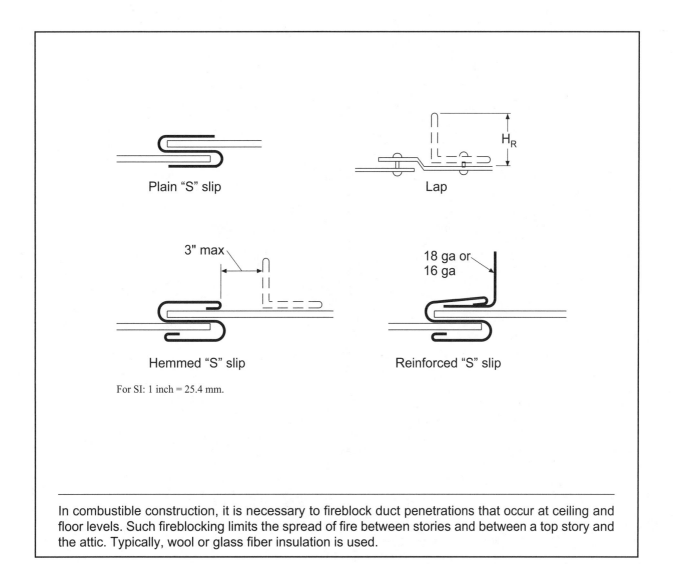

Plain "S" slip

Lap

3" max

Hemmed "S" slip

18 ga or 16 ga

Reinforced "S" slip

For SI: 1 inch = 25.4 mm.

In combustible construction, it is necessary to fireblock duct penetrations that occur at ceiling and floor levels. Such fireblocking limits the spread of fire between stories and between a top story and the attic. Typically, wool or glass fiber insulation is used.

Code Text: *Return air shall be taken from inside the dwelling. Dilution of return air with outdoor air shall be permitted.*

Discussion and Commentary: Recirculation of air contaminated with objectionable odors, fumes or flammable vapors is prohibited because of potential health and safety hazards. Therefore, return air ducts must not serve kitchens, bathrooms, closets, attached garages or other dwelling units. Also, outside air intakes must be located carefully to preclude objectionable or contaminated air from being drawn into the air distribution system. Air intakes above the roof must be located a minimum of 10 feet laterally from a plumbing vent or fuel-burning appliance vent, unless the source of noxious fumes is more than 3 feet higher than the outside air intake. In that case, a lesser lateral separation is acceptable.

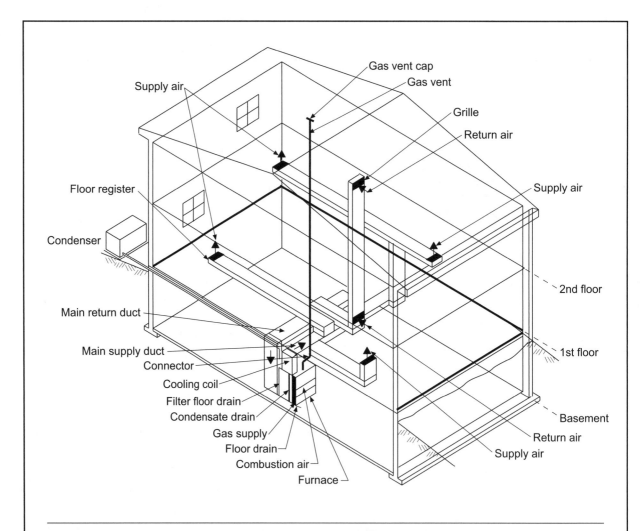

To prevent the intrusion of animals and insects, it is necessary to provide any outdoor air inlet with a mesh screen having openings no greater than $^1/_2$ inch. However, in order to maintain the necessary openings free and clear from obstruction, a minimum dimension of $^1/_4$ inch is mandated.

Code Text: *Freestanding or built-in ranges shall have a vertical clearance above the cooking top of not less than 30 inches (762 mm) to unprotected combustible material. Reduced clearances are permitted in accordance with the listing and labeling of the range hoods or appliances. The installation of a listed and labeled cooking appliance or microwave oven over a listed and labeled cooking appliance shall be in accordance with Section M1504.1.*

Discussion and Commentary: Vertical clearance to combustibles is important because cooking range tops produce considerable heat. Additionally, overheated utensils, particularly those containing grease, may be the source of flame. An adequate clear space above the range also ensures that hoods, shelves and other potential obstructions do not encroach into the working space needed for cooking operations. It is also important that the installation of any household cooking appliance not interfere with combustion air.

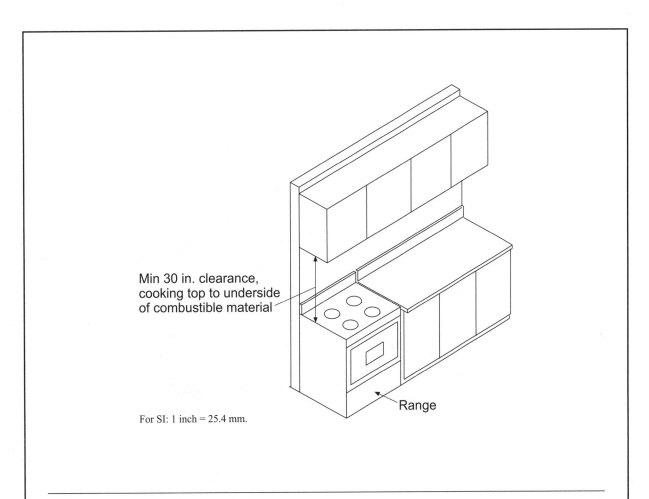

Min 30 in. clearance, cooking top to underside of combustible material

Range

For SI: 1 inch = 25.4 mm.

Although vented or unvented range hoods are often installed over ranges, the code does not require hoods to be installed. Where installed however, the hoods and exhaust methods must be in compliance with the code and the manufacturer's installation instructions.

Code Text: *Sauna heaters shall be protected from accidental contact by persons with a guard of material having a low thermal conductivity, such as wood. The guard shall have no substantial effect on the transfer of heat from the heater to the room. Sauna heaters shall be equipped with a thermostat that will limit room temperature to not greater than 194°F (90°C). Where the thermostat is not an integral part of the heater, the heat-sensing element shall be located within 6 inches (152 mm) of the ceiling.*

Discussion and Commentary: In a small space such as a sauna room, it is quite possible that contact with the heating element will occur unless the element is adequately guarded. The guard must be of a type that allows the heated air to circulate throughout the room, as concentrated heat and a combustible guard could create a fire hazard.

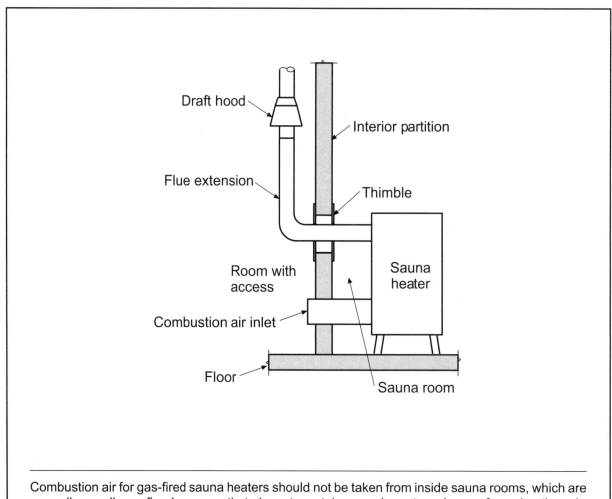

Combustion air for gas-fired sauna heaters should not be taken from inside sauna rooms, which are normally small, confined spaces that do not contain an adequate volume of combustion air. Combustion air and venting shall be in compliance with Chapters 17 and 18.

Quiz

Study Session 11
IRC Chapters 15, 16 and 19

1. Screws that join clothes dryer exhaust ducts may protrude a maximum of
    _____ inch into the inside of the duct.

    a. 0, the use of screws is prohibited

    b. $^1/_8$

    c. $^3/_{16}$

    d. $^1/_4$

    Reference _____

2. Unless modified by the manufacturer's installation instructions, what is the maximum
    permitted length of a clothes dryer exhaust duct when measured from the dryer location
    to the outlet termination?

    a. 18 feet   b. 25 feet

    c. 30 feet   d. 35 feet

    Reference _____

3. Unless modified by the manufacturer's installation instructions or determined by
    engineering calculation, what is the maximum permitted length of a clothes dryer exhaust
    duct that has one 6-inch-radius smooth 90-degree elbow and two 6-inch-radius smooth
    45-degree elbows?

    a. 25 feet, 0 inches   b. 28 feet, 6 inches

    c. 31 feet, 3 inches   d. 35 feet, 0 inches

    Reference _____

4. Which of the following requirements is not applicable to a duct serving a range hood?

    a. It shall have a smooth interior surface.

    b. It shall be airtight.

    c. It shall be equipped with a backdraft damper.

    d. It may terminate in an attic space.

Reference _____

5. When serving a domestic kitchen cooking appliance equipped with a down-draft exhaust system, a schedule 40 PVC exhaust duct shall extend a maximum of _____ above grade outside the building.

    a. 1 inch   b. 6 inches

    c. 12 inches   d. 36 inches

Reference _____

6. A domestic open-top broiler unit shall be provided with a minimum clearance of _____ between the hood and the underside of combustible material or cabinets.

    a. 0 inches; no clearance is required

    b. $^1/_4$ inch

    c. $^1/_2$ inch

    d. 1 inch

Reference _____

7. For a domestic open-top boiler unit, a minimum of _____ shall be maintained between the cooking surface and combustible material or cabinet.

    a. 24 inches   b. 28 inches

    c. 30 inches   d. 32 inches

Reference _____

8. Heating equipment connected to an above-ground duct system shall be designed to limit discharge air temperature to a maximum of _____ .

    a. 125°F   b. 200°F

    c. 225°F   d. 250°F

Reference _____

9. When serving heating, cooling or ventilation equipment, above-ground factory-made duct systems shall be constructed of materials classified as_____ .

    a. Class 0 only   b. Class 1 only

    c. Class 0 or Class 1   d. Class 1 or Class 2

Reference _____

10. A 16-inch exposed rectangular duct that is a portion of an above-ground duct system shall have a minimum equivalent galvanized sheet gage of _____ .

    a. No. 22   b. No. 26

    c. No. 28   d. No. 30

Reference _____

11. Gypsum board may be used in the construction of a return air plenum in an above-ground duct system, provided the maximum air temperature is _____ and the exposed surfaces are not subject to condensation.

    a. 125°F   b. 150°F

    c. 180°F   d. 210°F

Reference _____

12. What is the maximum flame spread index permitted for materials used to construct above-ground duct systems?

    a. 25   b. 75

    c. 200   d. 450

Reference _____

13. Where used in an underground duct system, the maximum permitted duct temperature for plastic ducts shall be _____ .

    a. 125°F   b. 150°F

    c. 180°F   d. 210°F

Reference _____

14. When protected by concrete, metal ducts installed in an underground duct system shall be encased with a minimum thickness of _____ .

    a. 1 inch   b. 2 inches

    c. 3 inches   d. 4 inches

    Reference _____

15. As a general rule, duct coverings and linings shall have a maximum flame spread index of _____ and a maximum smoke-developed index of _____ .

    a. 25, 50   b. 25, 200

    c. 50, 200   d. 50, 450

    Reference _____

16. Factory-insulated flexible factory-made ducts shall be identified at maximum intervals of _____ with the name of the manufacturer, the thermal resistance R-value, the flame spread index and the smoke-developed index.

    a. 3 feet   b. 5 feet

    c. 6 feet   d. 10 feet

    Reference _____

17. Where vibration isolators are installed between mechanical equipment and metal ducts, they shall have a maximum length of _____ .

    a. 6 inches   b. 10 inches

    c. 12 inches   d. 24 inches

    Reference _____

18. Where metal ducts are supported by $^1/_2$-inch wide metal straps, the straps shall be minimum _____ and located a maximum of _____ on center.

    a. 12-gage, 6 feet   b. 18-gage, 6 feet

    c. 18-gage, 10 feet   d. 22-gage, 10 feet

    Reference _____

19. In an existing structure, an under-floor space used as a supply plenum shall be formed by materials having a maximum flame spread rating of _____ .

    a. 25   b. 50

    c. 75   d. 200

    Reference _____

20. Where the under-floor space is permitted to be used as a supply plenum, a duct shall extend from the furnace supply outlet to a minimum of _____ inch(es) below the combustible framing.

    a. 1   b. 3

    c. 6   d. 8

    Reference _____

21. Unless the discharge outlet is a minimum of _____ above the outside air inlet, outside air for a forced-air heating system shall be taken a minimum of _____ from an appliance vent outlet, a vent opening from a plumbing drainage system, or the discharge outlet of an exhaust fan.

    a. 2 feet, 6 feet   b. 2 feet, 10 feet

    c. 3 feet, 6 feet   d. 3 feet, 10 feet

    Reference _____

22. Which one of the following spaces is acceptable as a source for return air for a forced-air heating system?

    a. sleeping room   b. bathroom

    c. kitchen   d. closet

    Reference _____

23. Air exhaust openings shall terminate a minimum of _____ feet from property lines.

    a. 3   b. 5

    c. 8   d. 10

    Reference _____

24. Built-in or freestanding ranges shall have a minimum vertical clearance of _____ above the cooking top to an unprotected combustible construction.

    a. 24 inches   b. 30 inches

    c. 32 inches   d. 36 inches

Reference _____

25. Sauna heaters shall be equipped with a thermostat that will limit room temperature to a maximum of _____ .

    a. 120°F   b. 140°F

    c. 172°F   d. 194°F

Reference _____

26. A mechanically ventilated kitchen area shall be provided with a minimum ventilation rate of _____ cfm where continuous ventilation is utilized.

    a. 20   b. 25

    c. 50   d. 100

Reference _____

27. A bathroom ventilated by an intermittent mechanical means shall have a minimum mechanical exhaust capacity of _____ cfm.

    a. 20   b. 25

    c. 50   d. 100

Reference _____

28. The access through the floor to an under-floor plenum shall provide a minimum opening size of _____ .

    a. 18 inches by 24 inches   b. 20 inches by 24 inches

    c. 22 inches by 30 inches   d. 24 inches by 30 inches

Reference _____

29. Where an under-floor space is used as a supply plenum in an existing structure, the furnace shall be equipped with an approved automatic control that limits the outlet air temperature to a maximum of _____ °F.

    a. 150   b. 200

    c. 225   d. 250

Reference _____

30. Where the thermostat used to limit sauna room temperature is not an integral part of the sauna heater, the heat-sensing element shall be located a maximum of _____ inches below the ceiling.

    a. 3   b. 4

    c. 6   d. 8

Reference _____

31. Unless modified by the manufacturer's installation instructions, a clothes dryer exhaust duct shall terminate outside of the building a minimum of _____ feet in any direction from openings into the building.

    a. 2   b. 3

    c. 5   d. 10

Reference _____

32. A flexible transition duct for a clothes dryer exhaust system is limited to a maximum length of _____ feet.

    a. 3   b. 5

    c. 6   d. 8

Reference _____

33. Where used in an above-ground duct system, factory-made air ducts constructed of Class 0 materials shall have a maximum flame spread rating of _____.

    a. 0   b. 25

    c. 75   d. 200

Reference _____

34. Crimp joints for round metal ducts shall have a minimum contact lap of _____ inch(es).

    a. $^3/_4$ b. 1

    c. $1^1/_2$ d. 2

Reference _____

35. Unless in conformance with the criteria for underground duct systems, ducts shall be installed with a minimum of _____ inches separation from earth.

    a. 3   b. 4

    c. 6   d. 12

Reference _____

Study Session

12

2012 IRC Chapters 17, 18, 20, 21, 22 and 23

Combustion Air, Chimneys and Vents, Boilers/Water Heaters, Hydronic Piping, Special Piping and Storage Systems, and Solar Systems

OBJECTIVE: To provide an understanding of the general provisions for combustion air, chimneys, vents, boilers, water heaters, hydronic piping, special piping and storage systems, and solar systems.

REFERENCE: Chapters 17, 18, 20, 21, 22 and 23, 2012 *International Residential Code*

KEY POINTS:
- What criteria must be used to regulate combustion air for solid-fuel-burning appliances? Oil-fired appliances? Gas-fired appliances?
- What venting method is to be used where two or more listed appliances are connected to a common natural draft venting system?
- What is the minimum thickness for a single-wall metal pipe connector?
- How far above the highest connected appliance outlet must a natural draft appliance vent terminate? How far above the bottom of a wall vent must a natural draft gas vent terminate?
- Where must a chimney connector enter a masonry chimney? How is the size of a chimney flue that is connected to more than one appliance calculated?
- Where are valves required for a boiler? What type of gauges are mandated? What is a boiler low-water cutoff?
- What requirements apply to expansion tanks used with hot water boilers?
- Where is the installation of fuel-fired water heaters prohibited? In what manner can access be provided to a water heater in an under-floor or attic area?
- What are the general provisions related to hydronic piping systems? How must floor heating systems be tested?
- How are special piping and storage systems to be installed?
- What special conditions apply to the installation of a solar energy system?

Topic: Scope   **Category:** Combustion Air

Reference: IRC M1701.1   **Subject:** General Provisions

Code Text: *Solid-fuel-burning appliances shall be provided with combustion air in accordance with the appliance manufacturer's installation instructions. Oil-fired appliances shall be provided with combustion air in accordance with NFPA 31. The methods of providing combustion air in* Chapter 17 *do not apply to fireplaces, fireplace stoves and direct-vent appliances. The requirements for combustion and dilution air for gas-fired appliances shall be in accordance with Chapter 24.*

Discussion and Commentary: Complete combustion of solid and liquid fuel is essential for the proper operation of appliances, for control of harmful emissions and for achieving maximum fuel efficiency. The primary use of combustion air is to supply the oxygen necessary for the complete and efficient burning of fuel. The byproducts of incomplete combustion are poisonous, corrosive and combustible, and can cause serious appliance or equipment malfunctions that pose fire or explosion hazards.

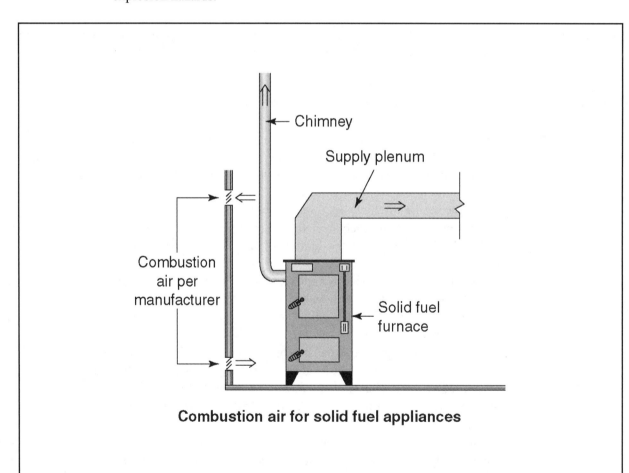

Combustion air for solid fuel appliances

An insufficient combustion air supply can cause an appliance to overheat and discharge combustion byproducts into the building. A combustion air supply is also necessary to prevent oxygen depletion, which threatens the safety of the building occupants.

Code Text: *Manually operated dampers shall not be installed except in connectors or chimneys serving solid-fuel-burning appliances. Automatically operated dampers shall conform to UL 17* (Vent or Chimney Connector Dampers for Oil-fired Appliances) *and be installed in accordance with the terms of their listing and label. The installation shall prevent firing of the burner when the damper is not opened to a safe position.*

Discussion and Commentary: A damper installed in a solid fuel-burning appliance is usually opened manually during the process of starting a fire. If the damper is not opened, smoke will spill back into the room, alerting the user. For gas or liquid fuel-fired appliances, however, the user may not be aware of a partially or completely closed damper, and a hazardous condition could develop.

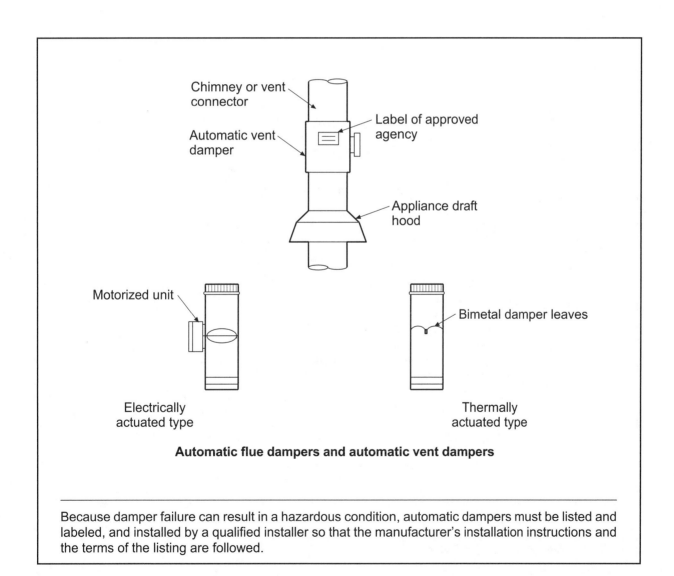

Automatic flue dampers and automatic vent dampers

Because damper failure can result in a hazardous condition, automatic dampers must be listed and labeled, and installed by a qualified installer so that the manufacturer's installation instructions and the terms of the listing are followed.

Code Text: *Vent and chimney connectors shall be installed in accordance with the manufacturer's installation instructions and within the space where the appliance is located. A chimney connector or vent connector shall not pass through any floor or ceiling. The horizontal run of an uninsulated connector to a natural draft chimney shall not exceed 75 percent of the height of the vertical portion of the chimney above the connector. The horizontal run of a listed connector to a natural draft chimney shall not exceed 100 percent of the height of the vertical portion of the chimney above the connector.*

Discussion and Commentary: A chimney connector is simply a pipe that connects a fuel-burning appliance to a chimney. A vent connector serves a similar purpose. The ideal chimney or vent configuration is a vertical system, even though it is not always practical. The connector length is limited because of flow resistance of the connector and heat loss through the connector wall.

TABLE M1803.3.4
CHIMNEY AND VENT CONNECTOR CLEARANCES
TO COMBUSTIBLE MATERIALS[a]

TYPE OF CONNECTOR	MINIMUM CLEARANCE (inches)
Single–wall metal pipe connectors:	
Oil and solid-fuel appliances	18
Oil appliances listed for use with Type L vents	9
Type L vent piping connectors:	
Oil and solid-fuel appliances	9
Oil appliances listed for use with Type L vents	3[b]

For SI: 1 inch = 25.4 mm.

a. These minimum clearances apply to unlisted single-wall chimney and vent connectors. Reduction of required clearances is permitted as in Table M1306.2.

b. When listed Type L vent piping is used, the clearance shall be in accordance with the vent listing.

Connectors must be provided with minimum clearances to combustibles. The clearances established in Table M1803.3.4 may be reduced if the combustible material is protected by one of the methods established in Table M1306.2.

Code Text: *Vents passing through a roof shall extend through flashing and terminate in accordance with the manufacturer's installation requirements. Decorative shrouds shall not be installed at the termination of vents except where such shrouds are listed and labeled for use with the specific venting system and are installed in accordance with the manufacturer's installation instructions.*

Discussion and Commentary: The vent system must be physically protected to prevent damage to the vent and to prevent combustibles from coming into contact with or being placed too close to the vents. This protection is usually provided by enclosing the vent in chases, shafts or cavities during the building construction. Physical protection is not required in the room or space where the vent originates (at the appliance connection) and is not required in such locations as attics that are not occupied or used for storage.

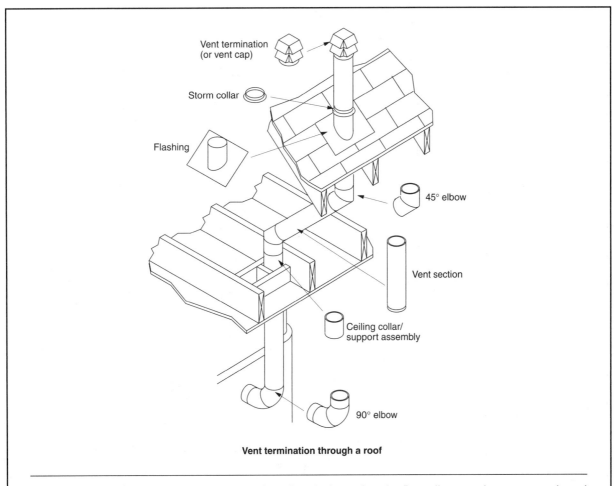

Vent termination through a roof

Direct-vent appliances, often referenced as "sealed combustion" appliances, have an enclosed combustion chamber and a conduit for bringing all combustion air directly from the outdoors. Rather than prescribe specific provisions, the code relies on the manufacturer's installation requirements when addressing vent terminals.

Code Text: *Vents for natural draft appliances shall terminate at least 5 feet (1524 mm) above the highest connected appliance outlet, and natural draft gas vents serving wall furnaces shall terminate at an elevation at least 12 feet (3658 mm) above the bottom of the furnace. Type L venting systems shall terminate with a listed and labeled cap in accordance with the vent manufacturer's installation instructions not less than 2 feet (610 mm) above the roof and not less than 2 feet (610 mm) above any portion of the building within 10 feet (3048 mm).*

Discussion and Commentary: Natural draft appliances must terminate at a high enough point to ensure that the venting system will perform properly under all conditions. Type L vents are used for oil-burning appliances and are designed for higher temperature flue gases than those encountered in gas-fired appliances.

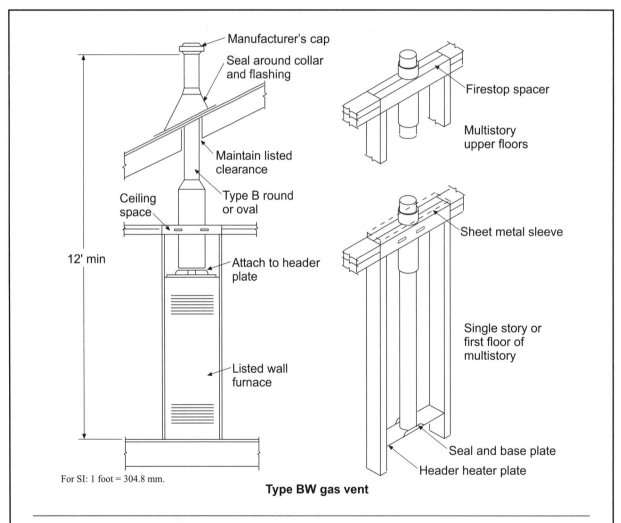

For SI: 1 foot = 304.8 mm.

Type BW gas vent

Unless the vent is an integral part of a listed and labeled appliance, an individual vent for a single appliance must have a cross-sectional area equivalent to or greater than that of the connector to the appliance, with a minimum area of 7 square inches.

Topic: Chimney Connection	Category: Chimneys and Vents
Reference: IRC M1805.2	Subject: Masonry and Factory-Built Chimneys

Code Text: *A chimney connector shall enter a masonry chimney not less than 6 inches (152 mm) above the bottom of the chimney. Where it is not possible to locate the connector entry at least 6 inches (152 mm) above the bottom of the chimney flue, a cleanout shall be provided by installing a capped tee in the connector next to the chimney. A connector entering a masonry chimney shall extend through, but not beyond, the wall and shall be flush with the inner face of the liner. Connectors, or thimbles where used, shall be firmly cemented into the masonry.*

Discussion and Commentary: Chimney connectors must pass through a masonry chimney wall to the inner face of the liner, but not beyond. A connector that extends into the chimney passageway can restrict the flow of flue gases and provide a ledge for creosote, soot and debris to accumulate. The installation of a thimble at the chimney opening is permitted to provide for easy removal of the connector to facilitate cleaning, provided the thimble is permanently secured in place.

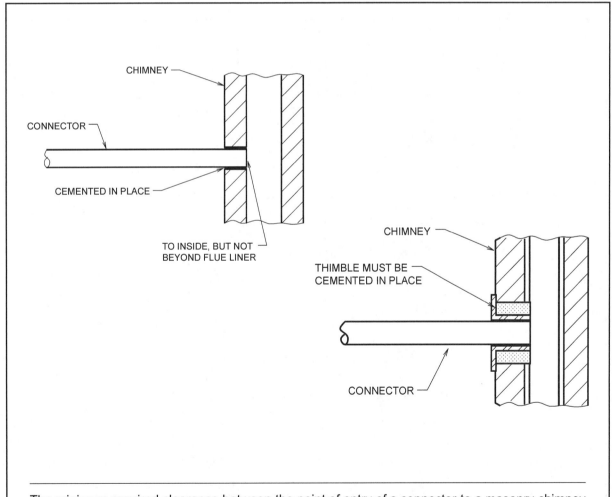

The minimum required clearance between the point of entry of a connector to a masonry chimney and the bottom of the chimney flue is intended to prevent blockage of the connector opening by debris that has collected at the bottom of the flue passage. A cleanout can be provided as an alternative where the 6-inch clearance is not available.

Code Text: *In addition to the requirements of the IRC, the installation of boilers shall conform to the manufacturer's instructions. The manufacturer's rating data, the nameplate and operating instructions of a permanent type shall be attached to the boiler. Boilers shall have all controls set, adjusted and tested by the installer. Boilers shall be equipped with pressure-relief valves with minimum rated capacities for the equipment served. Pressure-relief valves shall be set at the maximum rating of the boiler. Discharge shall be piped to drains by gravity to within 18 inches (457 mm) of the floor or to an open receptor.*

Discussion and   A boiler is defined as "a self-contained appliance from which hot water is circulated for
Commentary: heating purposes and then returned to the appliance. It operates at a maximum water pressure of 160 psig and at a maximum water temperature of 250°F." To ensure a safe and proper installation, boilers must be constructed to the appropriate standards and installed in accordance with the manufacturer's instructions.

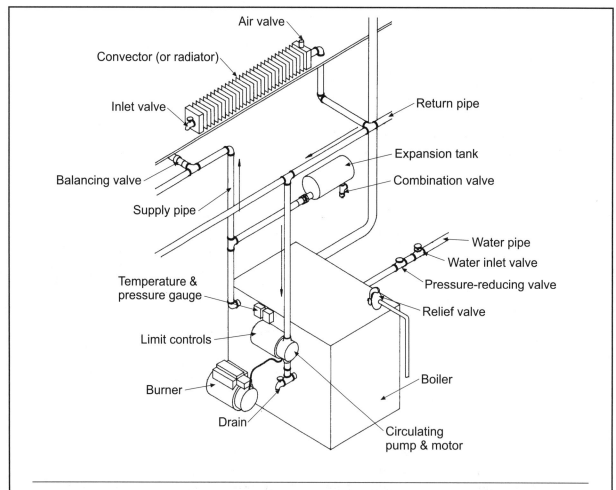

The installer must provide a complete set of boiler operating instructions for use by the owner. It is wise to attach such instructions directly to the boiler for immediate access. In addition, a complete control diagram must be furnished and located with the operating instructions.

Topic: Expansion Tanks   **Category:** Boilers/Water Heaters
Reference: IRC M2003   **Subject:** Boilers

Code Text: *Hot water boilers shall be provided with expansion tanks. Nonpressurized expansion tanks shall be securely fastened to the structure or boiler and supported to carry twice the weight of the tank filled with water. Pressurized expansion tanks shall be consistent with the volume and capacity of the system. Tanks shall be capable of withstanding a hydrostatic test pressure of two and one-half times the allowable working pressure of the system.*

Discussion and Commentary: Inactive hydronic heating systems are filled with water at room temperature. When the boiler is fired, the temperature of the water increases, causing the volume of the water to increase. Without an expansion tank, hydrostatic pressure would rapidly build within the boiler and piping system, activating the pressure-relief valve. An expansion tank provides space for the water to expand as it is heated and keeps the water pressure in the normal operating range while the boiler is operating.

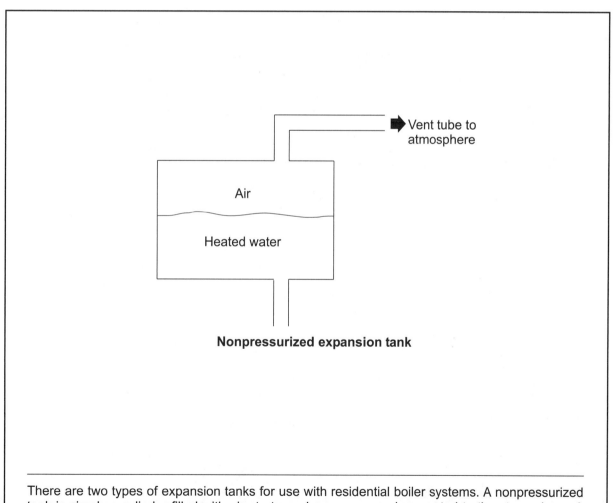

There are two types of expansion tanks for use with residential boiler systems. A nonpressurized tank is simply a cylinder filled with air at atmosphere pressure, i.e., vented to the atmosphere. A pressurized tank is configured as a sealed cylinder divided by a flexible diaphragm.

Code Text: *Fuel-fired water heaters shall not be installed in a room used as a storage closet. Water heaters located in a bedroom or bathroom shall be installed in a sealed enclosure so that combustion air will not be taken from the living space. Installation of direct-vent water heaters within an enclosure is not required. Access to water heaters that are located in an attic or under floor crawl space is permitted to be through a closet located in a sleeping room or bathroom where ventilation of those spaces is in accordance with* the IRC.

Discussion and Commentary: The prohibitions regarding the locations of fuel-fired water heaters is intended to reduce the potential threats of depleted oxygen levels; elevated levels of carbon dioxide, nitrous oxides, carbon monoxide and other combustion gases; ignition of combustibles; and elevated levels of flammable gases. In small rooms, such as closets, bedrooms and bathrooms, the doors are often closed, which could allow combustion gases to build up to life-threatening levels.

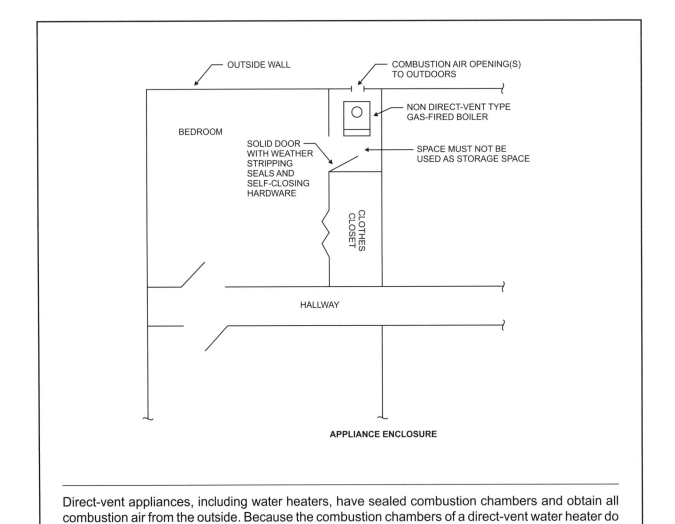

Direct-vent appliances, including water heaters, have sealed combustion chambers and obtain all combustion air from the outside. Because the combustion chambers of a direct-vent water heater do not communicate with the room atmosphere, a requirement for enclosure is not necessary.

Code Text: *Hydronic piping shall conform to Table M2101.1. Approved piping, valves, fittings and connections shall be installed in accordance with the manufacturer's installation instructions. The potable water system shall be protected from backflow in accordance with the provisions listed in Section P2902. Piping shall be installed so that piping, connections and equipment shall not be subjected to excessive strains or stresses. Provisions shall be made to compensate for expansion, contraction, shrinkage and structural settlement. Hydronic piping shall be tested hydrostatically at a pressure of not less than 100 pounds per square inch (690kPa) for a duration of not less than 15 minutes.*

Discussion and Commentary: A variety of piping materials is acceptable for use in hydronic piping systems. Table M2101.9 establishes the maximum horizontal and vertical hanger spacing intervals for various piping materials than can be used in hydronic piping systems.

TABLE M2101.1
HYDRONIC PIPING MATERIALS

MATERIAL	USE CODE[a]	STANDARD[b]	JOINTS	NOTES
Brass pipe	1	ASTM B 43	Brazed, welded, threaded, mechanical and flanged fittings	
Brass tubing	1	ASTM B 135	Brazed, soldered and mechanical fittings	
Chlorinated poly (vinyl chloride) (CPVC) pipe and tubing	1, 2, 3	ASTM D 2846	Solvent cement joints, compression joints and threaded adapters	
Copper pipe	1	ASTM B 42, B 302	Brazed, soldered and mechanical fittings threaded, welded and flanged	
Copper tubing (type K, L or M)	1, 2	ASTM B 75, B 88, B 251, B 306	Brazed, soldered and flared mechanical fittings	Joints embedded in concrete
Cross-linked polyethylene (PEX)	1, 2, 3	ASTM F 876, F 877	(See PEX fittings)	Install in accordance with manufacturer's instructions
Cross-linked polyethylene/aluminum/cross-linked polyethylene-(PEX-AL-PEX) pressure pipe	1, 2	ASTM F 1281 or CAN/CSA B137.10	Mechanical, crimp/insert	Install in accordance with manufacturer's instructions
PEX fittings		ASTM F 877 ASTM F 1807 ASTM F 1960 ASTM F 2098 ASTM F 2159 ASTM F 2735	Copper-crimp/insert fittings, cold expansion fittings, stainless steel clamp, insert fittings	Install in accordance with manufacturer's instructions
Polybutylene (PB) pipe and tubing	1, 2, 3	ASTM D 3309	Heat-fusion, crimp/insert and compression	Joints in concrete shall be heat-fused
Polyethylene (PE) pipe, tubing and fittings (for ground source heat pump loop systems)	1, 2, 4	ASTM D 2513 ASTM D 3350 ASTM D 2513 ASTM D 3035 ASTM D 2447 ASTM D 2683 ASTM F 1055 ASTM D 2837 ASTM D 3350 ASTM D 1693	Heat-fusion	
Polyethylene/aluminum/polyethylene (PE-AL-PE) pressure pipe	1, 2, 3	ASTM F 1282 CSA B 137.9	Mechanical, crimp/insert	
Polypropylene (PP)	1, 2, 3	ISO 15874 ASTM F 2389	Heat-fusion joints, mechanical fittings, threaded adapters, compression joints	
Raised temperature polyethylene (PE-RT)	1, 2, 3	ASTM F 2623 ASTM F 2769	Copper crimp/insert fitting stainless steel clamp, insert fittings	
Soldering fluxes	1	ASTM B 813	Copper tube joints	
Steel pipe	1, 2	ASTM A 53, A 106	Brazed, welded, threaded, flanged and mechanical fittings	Joints in concrete shall be welded. Galvanized pipe shall not be welded or brazed.
Steel tubing	1	ASTM A 254	Mechanical fittings, welded	

For SI: °C = [(°F)-32]/1.8.
a. Use code: 3. Temperatures below 180°F only.
1. Above ground.   4. Low temperature (below 130°F) applications only.
2. Embedded in radiant systems.   b. Standards as listed in Chapter 44.

Pipe openings through concrete or masonry must be sleeved to compensate for expansion of the piping due to hot water. As the flow of water causes the pipe to move and expand when the boiler cycles, sufficient clearance is necessary to avoid stress damage to the piping.

Code Text: *Radiant floor heating systems shall have a thermal barrier in accordance with Sections M2103.2.1 through M2103.2.4. See exception for engineered systems. Radiant piping used in slab-on grade applications shall have insulating materials having a minimum R-value of 5 installed beneath the piping. In suspended floor applications, insulation shall be installed in the joist bay cavity serving the heating space above and shall consist of materials having a minimum R-value of 11. Insulating materials used in thermal barriers shall be installed so that the manufacturer's R-value mark is readily observable upon inspection.*

Discussion and Commentary: Radiant floor heating systems cannot operate as intended without the installation of an appropriate thermal barrier. In slab-on-grade conditions, the ground will require a substantial charging of energy in order to hit a point of equilibrium where the thermal energy starts coming upwards instead of going downwards

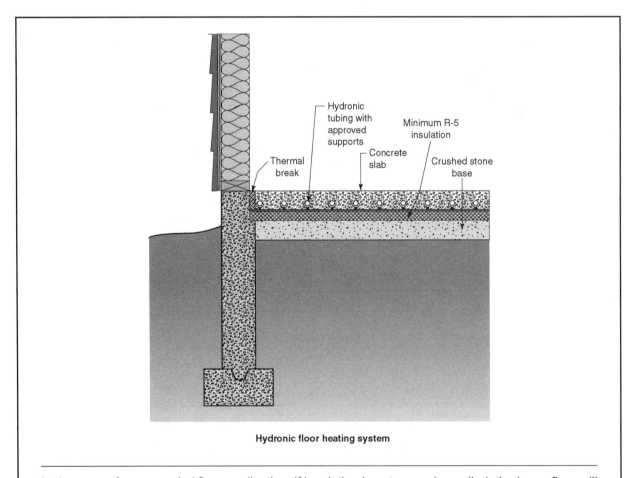

Hydronic floor heating system

In the case of a suspended floor application, if insulation is not properly applied, the lower floor will have a tendency to overheat and the floor that is trying to be heated will be underheated. If the system is installed without the complying insulation, it will be very difficult to balance the heat distribution.

Code Text: *Supply tanks shall be listed and labeled and shall conform to UL 58 for underground tanks and UL 80 for inside tanks. The maximum amount of fuel oil stored above ground or inside of a building shall be 660 gallons (2498 L). See exception for storage of fuel oil used for space or water heating. Supply tanks (located within a building) larger than 10 gallons (38 L) shall be placed not less than 5 feet (1524 mm) from any fire or flame either within or external to any fuel-burning appliance. Tanks installed outside above ground shall be a minimum of 5 feet (1524 mm) from an adjoining property line. Such tanks shall be suitably protected from the weather and from physical damage.*

Discussion and Commentary: The cross connection of two supply tanks on the same level is permitted if the total capacity of the system does not exceed 660 gallons. This procedure, which uses gravity flow from one tank to the other, provides an option for installation purposes.

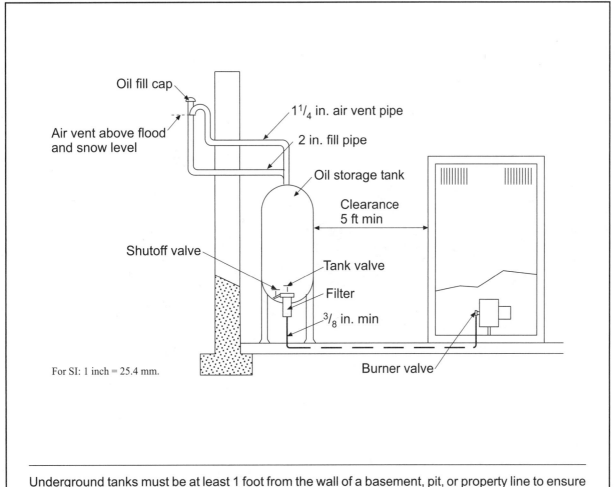

Oil fill cap

Air vent above flood and snow level

$1^1/_4$ in. air vent pipe

2 in. fill pipe

Oil storage tank

Clearance 5 ft min

Shutoff valve

Tank valve

Filter

$^3/_8$ in. min

For SI: 1 inch = 25.4 mm.

Burner valve

Underground tanks must be at least 1 foot from the wall of a basement, pit, or property line to ensure that settlement of surrounding materials does not place a stress on the tank or piping. Tanks must also be covered with at least 12 inches of earth for physical protection.

Code Text: *Solar energy collectors, controls, dampers, fans, blowers and pumps shall be accessible for inspection, maintenance, repair and replacement. Where mounted on or above the roof coverings, the collectors and supporting structure shall be constructed of noncombustible materials or fire-retardant-treated wood equivalent to that required for the roof construction. Roof and wall penetrations shall be flashed and sealed in accordance with Chapter 9 of the IRC to prevent entry of water, rodents and insects .*

Discussion and Commentary: Developed due to an emphasis placed on conservation of our natural resources, solar energy systems allow for an alternative method in providing space heating or cooling, domestic water heating, and pool heating. The effectiveness of these systems depends on the size of the solar array, the energy requirements of the dwelling and the availability of sunlight.

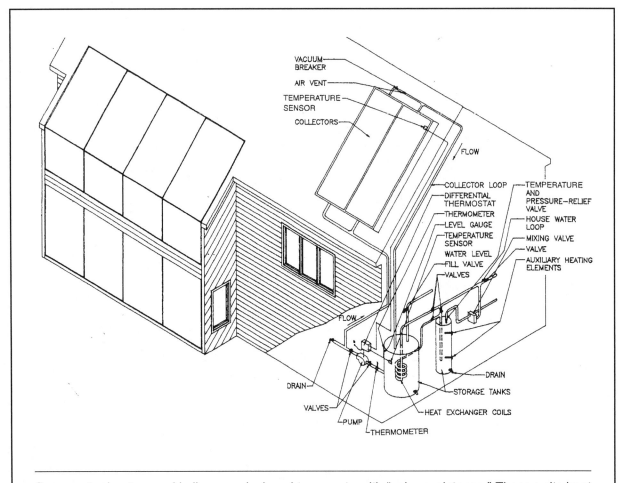

Some water heaters and boilers are designed to operate with "solar assistance." These units heat water by ordinary methods and have additional connections for solar preheating. Preheated water reduces the amount of energy that is consumed with ordinary heating methods.

Quiz

Study Session 12
IRC Chapters 17, 18, 20, 21, 22 and 23

1. Which of the following types of appliances is required to be provided with combustion air in accordance with NFPA 31?

    a. solid-fuel burning

    b. oil-fired

    c. direct-vent

    d. gas-fired

    Reference _____

2. Manually-operated dampers shall not be installed in vents except in connectors or chimneys serving _____ appliances.

    a. solid-fuel-burning

    b. oil-fired

    c. direct vent

    d. gas-fired

    Reference _____

3. Where a single-wall metal pipe vent connector serving an oil-fired appliance listed for a Type L vent passes through a wall, the pipe shall be guarded by a ventilated metal thimble with a minimum diameter of _____ inches larger than the vent connector.

    a. 3

    b. 4

    c. 6

    d. 8

    Reference _____

4. The bottom of the vent terminal for a mechanical draft system shall be located a minimum of _____ inches above finished ground level.

    a. 3   b. 4

    c. 6   d. 12

Reference _____

5. Unless an integral part of a listed and labeled appliance, an individual Type L vent for a single appliance shall have a minimum cross-sectional area of _____ inches.

    a. 5   b. 6

    c. 7   d. 8

Reference _____

6. A chimney connector shall enter a masonry chimney a minimum of _____ inches above the bottom of the chimney.

    a. 6   b. 8

    c. 9   d. 12

Reference _____

7. Nonpressurized expansion tanks shall be supported to carry _____ times the weight of the tank filled with water.

    a. $1^1/_2$ b. 2

    c. $2^1/_2$ c. 3

Reference _____

8. Where hydronic piping is installed as part of a radiant floor heating system for a slab-on-grade application, insulating materials having a minimum *R*-value of _____ shall be installed beneath the piping.

    a. 4   b. 5

    c. 8   d. 11

Reference _____

9. In suspended floor applications containing the hydronic piping for a radiant floor heating system, insulation with a minimum *R*-value of _____ shall be installed in the joist bay cavity.

    a. 7   b. 9

    c. 11   d. 13

    Reference _____

10. A ground source heat pump loop system shall be pressure tested with water at _____ psi for _____ minutes with no observed leaks prior to the backfilling of the connection trenches.

    a. 60, 10   b. 80, 15

    c. 80, 30   d. 100, 30

    Reference _____

11. Oil supply tanks used within a building with a capacity of more than 10 gallons shall be placed a minimum of _____ feet from any fire or flame within a fuel-burning appliance.

    a. 5   b. 10

    c. 12   d. 15

    Reference _____

12. The horizontal run of an uninsulated connector to a natural draft chimney shall be a maximum of _____ percent of the height of the vertical portion of the chimney above the connector.

    a. 50   b. 75

    c. 100   d. 125

    Reference _____

13. For other than direct vent appliances, the vent termination for a mechanical draft system shall be mounted a minimum of _____ feet horizontally from an oil tank or gas meter.

    a. 3   b. 4

    c. 15   d. 10

    Reference _____

14. Vent piping for an oil tank shall terminate a minimum of _____ feet, measured vertically or horizontally, from any building opening.

    a. 2   b. 3

    c. 5   d. 10

Reference _____

15. In low temperature hydronic piping, solder joints in a metal pipe shall occur a minimum of _____ inches from any transition from the metal pipe to PE-AL-PE pressure pipe.

    a. 12   b. 18

    c. 24   d. 36

Reference _____

16. Unless specified otherwise in the manufacturer's installation instructions, the cross-sectional area of a chimney flue connected to a solid-fuel-burning appliance shall be a maximum of _____ times the area of the flue collar.

    a. $1^1/_2$ b. 2

    c. 3   d. 4

Reference _____

17. Vent and chimney connectors shall have a minimum slope of _____ rise per foot of run.

    a. $^1/_8$ inch   b. $^1/_4$ inch

    c. $^3/_8$ inch   d. $^1/_2$ inch

Reference _____

18. The horizontal run of a listed vent connector to a natural draft chimney shall be a maximum of _____ of the height of the vertical portion of the chimney above the connector.

    a. 50 percent   b. 75 percent

    c. 100 percent   d. 125 percent

Reference _____

19. Unless modified by Table M1306.2, what is the minimum required clearance between an unlisted Type L vent piping connector serving a solid fuel appliance and any combustible materials?

    a. 3 inches   b. 6 inches

    c. 9 inches   d. 18 inches

Reference _____

20. A natural draft gas vent serving a wall furnace shall terminate a minimum height of _____ above the bottom of the furnace.

    a. 5 feet   b. 8 feet

    c. 10 feet   d. 12 feet

Reference _____

21. Chimney flues connected to more than one appliance shall be sized based on the area of the largest connector plus _____ of the areas of additional chimney connectors.

    a. 50 percent   b. 75 percent

    c. 100 percent   d. 125 percent

Reference _____

22. The pressurized expansion tank for a boiler shall be capable of withstanding a minimum hydrostatic test pressure _____ the allowable working pressure of the system.

    a. equal to   b. twice

    c. two and one-half times   d. three times

Reference _____

23. Hydronic piping to be embedded in concrete shall be tested by applying a minimum hydrostatic pressure of _____ for a minimum time period of _____ .

    a. 50 psi, 15 minutes   b. 50 psi, 30 minutes

    c. 100 psi, 15 minutes   d. 100 psi, 30 minutes

Reference _____

24. Fill piping for oil tanks shall terminate outside of a building a minimum distance of _____ from any building opening at the same or lower level.

    a. 12 inches   b. 2 feet

    c. 5 feet   d. 10 feet

Reference _____

25. Solar energy systems shall be equipped with means to limit the maximum water temperature of the system fluid entering or exchanging heat with any pressurized vessel inside the dwelling to a maximum of _____ .

    a. 120°F   b. 150°F

    c. 180°F   d. 210°F

Reference _____

26. Fuel-fired water heaters shall not be installed in a _____ .

    a. bathroom   b. bedroom

    c. kitchen   d. storage closet

Reference _____

27. PEX tubing used for hydronic piping shall be supported at maximum intervals of _____ when installed in a horizontal position.

    a. 2 feet, 8 inches   b. 4 feet

    c. 6 feet, 6 inches   d. 8 feet

Reference _____

28. Hydronic piping shall be tested hydrostatically at a minimum pressure of _____ psi for a minimum of _____ minutes.

    a. 5, 15   b. 10, 30

    c. 60, 30   d. 100, 15

Reference _____

29. Unless intended for the storage of fuel oil used for space or water heating, fuel oil storage tank installed inside of a building shall have a maximum capacity of _____ gallons.

    a. 330   b. 660

    c. 1,000   d. 2,000

Reference _____

30. Pressure at the fuel oil supply inlet to an appliance shall be a maximum of _____ psi.

    a. 3   b. 5

    c. 10   d. 15

Reference _____

31. A nonpressurized expansion tank for a hot water boiler shall have a minimum capacity of _____ gallons where the system volume of the forced hot-water system is 50 gallons.

    a. 3.0   b. 6.0

    c. 7.5   d. 12.0

Reference _____

32. Vents for natural draft appliances shall terminate a minimum of _____ feet above the highest connected appliance outlet.

    a. 3   b. 5

    c. 10   d. 12

Reference _____

33. Discharge piping beyond the pressure relief valve of a boiler shall terminate a maximum of _____ inch(es) above the floor or to an open receptor.

    a. 1   b. 6

    c. 12   d. 18

Reference _____

34. Fuel oil tanks installed outside above ground shall be located a minimum of _____ feet from an adjoining property line.

    a. 5   b. 10

    c. 15   d. 20

Reference _____

35. Vent piping for a fuel oil tank shall have a minimum pipe size of _____ inch/es.

    a. $^3/_4$ b. 1

    c. $1^1/_4$ d. $1^1/_2$

Reference _____

2012 IRC Chapter 24
Fuel Gas

OBJECTIVE: To provide an understanding of the general provisions for fuel gas piping systems, fuel gas utilization equipment and related accessories, venting systems and combustion air configurations.

REFERENCE: Chapter 24, 2012 *International Residential Code*

KEY POINTS:
- In what locations are the installation of fuel-fired appliances prohibited? What concerns are applicable to outdoor installations?
- How shall combustion, ventilation and dilution air be provided when all indoor air is used? Outdoor combustion air? Combination indoor and outdoor?
- How can a mechanical air supply system provide for the necessary combustion air?
- Where must equipment and appliances having an ignition source be located in a private garage?
- How may the required clearances between combustible materials and chimneys; vents; fuel gas appliances, devices, and equipment; and kitchen exhaust equipment be reduced?
- What identifying mark is to be used on exposed gas piping? What type of pipe material is exempt from the labeling requirement?
- How is the maximum gas demand determined? How is the gas piping to be sized?
- What materials are acceptable for use as gas piping? When are materials allowed to be used again?
- When is a protective coating mandated for gas piping? How are field threading operations regulated? What methods are acceptable for pipe joints?
- Where is gas piping prohibited? What are the limitations for gas piping in concealed locations? How must such piping be protected where installed through a foundation or basement wall?
- What methods of protection against physical damage are identified for gas piping? How should pipe be protected against corrosion? What is the minimum required burial depth?
- How shall changes in direction be made for metallic pipe? For plastic pipe?

- What are the general conditions for the inspection and testing of fuel gas piping?
- Where must gas shutoff valves be located? Where must appliance shutoff valves be installed? Equipment shutoff valves for a fireplace?
- What is the maximum permitted length for an appliance fuel connector serving a range or domestic clothes dryer? Serving other appliances?
- At what maximum intervals must fuel gas piping be supported?
- What methods are available for terminating a gas vent? Where are decorative shrouds permitted?
- Where must chimneys serving gas-fired appliances terminate? At what minimum height must gas vents terminate above a roof? Where must a Type B or Type L gas vent terminate? A Type B-W vent? A single-wall metal pipe?
- How are vented wall furnaces to be located? Floor furnaces? Unit heaters?
- What are prohibited sources for outside or return air for a forced-air heating system?

Code Text: Chapter 24 *covers those fuel gas piping systems, fuel-gas appliances and related accessories, venting systems and combustion air configurations most commonly encountered in the construction of one- and two-family dwellings and structures regulated by* the IRC. *Coverage of piping systems shall extend from the point of delivery to the outlet of the appliance shutoff valves (see definition of "Point of delivery"). Piping systems requirements shall include design, materials, components, fabrication, assembly, installation, testing, inspection, operation and maintenance. Requirements for gas appliances and related accessories shall include installation, combustion and ventilation air and venting and connections to piping systems.*

Discussion and Commentary: Chapter 24 is extracted from the 2012 edition of the *International Fuel Gas Code* (IFGC) and has been modified where necessary to conform to the scope of application of the IRC. The section numbers appearing in parentheses are the section numbers of the corresponding text in the IFGC.

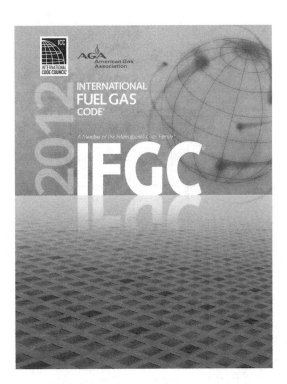

The omission from Chapter 24 of any material or method of installation established in the IFGC does not mean that the use of such material or method is prohibited. Fuel-gas piping systems, appliances, related accessories, venting systems and combustion air configurations not specifically covered in IRC Chapter 24 shall comply with the applicable provisions of the IFGC.

Code Text: *Where exhaust fans, clothes dryers and kitchen ventilation systems interfere with the operation of appliances, makeup air shall be provided.*

Discussion and Commentary: The introduction of makeup air is critical to the proper operation of all fuel-burning appliances located in areas subject to the effects of exhaust systems. Too little makeup air will cause excessive negative pressures to develop, not only reducing the exhaust airflow but also interfering with appliance venting and potentially causing combustion byproducts to discharge into the building.

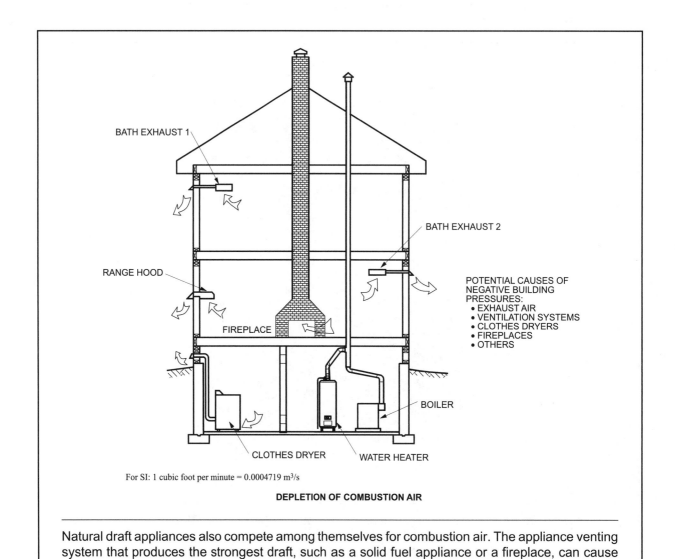

BATH EXHAUST 1

BATH EXHAUST 2

RANGE HOOD

POTENTIAL CAUSES OF
NEGATIVE BUILDING
PRESSURES:
• EXHAUST AIR
• VENTILATION SYSTEMS
• CLOTHES DRYERS
• FIREPLACES
• OTHERS

FIREPLACE

BOILER

CLOTHES DRYER

WATER HEATER

For SI: 1 cubic foot per minute = 0.0004719 m³/s

DEPLETION OF COMBUSTION AIR

Natural draft appliances also compete among themselves for combustion air. The appliance venting system that produces the strongest draft, such as a solid fuel appliance or a fireplace, can cause combustion air shortages for the appliance venting systems that produce a weaker draft.

Code Text: Where all combustion air is taken from the indoors, *each opening used to connect indoor spaces shall have a minimum free area of 1 square inch per 1,000 Btu/h (2,200 mm²/kw) of the total input rating of all appliances in the space, but not less than 100 square inches (0.06m²). One opening shall commence within 12 inches (305 mm) of the top and one opening shall commence with 12 inches (305 mm) of the bottom of the enclosure. The minimum dimension of air openings shall be not less than 3 inches (76 mm).*

Discussion and Commentary: Where adjacent spaces are considered together for the purpose of increasing the available volume of combustion air, the openings must be permanently open. The traditional design of high and low openings creates a thermosiphon air flow with the bottom opening acting as the inlet and the top opening acting as the outlet.

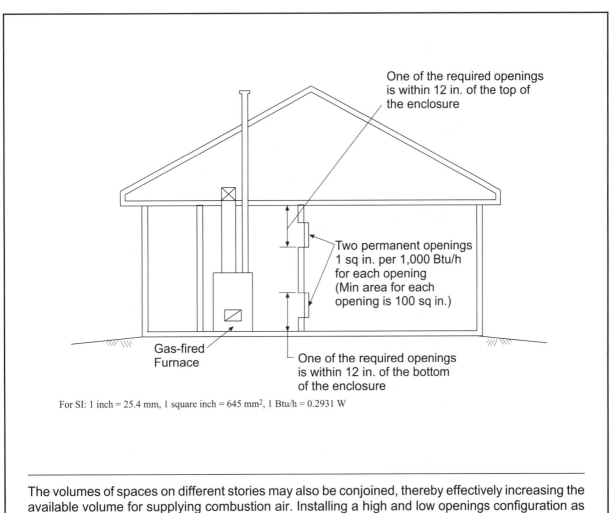

One of the required openings is within 12 in. of the top of the enclosure

Two permanent openings 1 sq in. per 1,000 Btu/h for each opening (Min area for each opening is 100 sq in.)

Gas-fired Furnace

One of the required openings is within 12 in. of the bottom of the enclosure

For SI: 1 inch = 25.4 mm, 1 square inch = 645 mm², 1 Btu/h = 0.2931 W

The volumes of spaces on different stories may also be conjoined, thereby effectively increasing the available volume for supplying combustion air. Installing a high and low openings configuration as required for a single-story condition would serve no purpose when vertically connecting stories; therefore, a single opening is permitted.

Code Text: *The required size of openings for combustion, ventilation and dilution air shall be based on the net free area of each opening. Screens shall have a mesh size not smaller than $^3/_4$ inch (6.4 mm). Nonmotorized louvers and grilles shall be fixed in the open position. Motorized louvers shall be interlocked with the appliance so that they are proven to be in the full open position prior to main burner ignition and during main burner operation.*

Discussion and Commentary: Combustion air openings are sized based upon the available free, unobstructed area for the passage of air into the space where the fuel-burning appliances are located. Louvers or grilles placed over these openings reduce the available area on account of the solid portions of the louvers or grilles. The reduction in opening area must be considered because only the unobstructed area can be credited toward the required opening size.

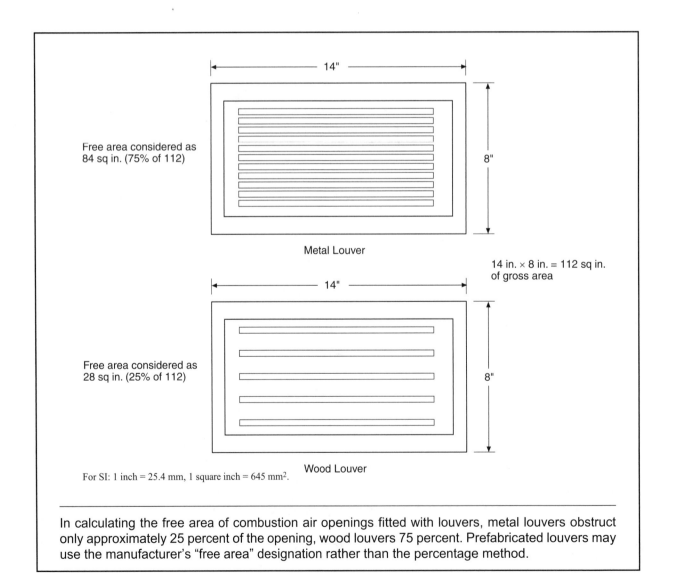

Free area considered as 84 sq in. (75% of 112)

Metal Louver

14 in. × 8 in. = 112 sq in. of gross area

Free area considered as 28 sq in. (25% of 112)

Wood Louver

For SI: 1 inch = 25.4 mm, 1 square inch = 645 mm².

In calculating the free area of combustion air openings fitted with louvers, metal louvers obstruct only approximately 25 percent of the opening, wood louvers 75 percent. Prefabricated louvers may use the manufacturer's "free area" designation rather than the percentage method.

Topic: Protection Against Damage

Category: Fuel Gas

Reference: IRC G2415.7

Subject: Piping System Installation

Code Text: *In concealed locations, where piping other than black or galvanized steel is installed through holes or notches in wood studs, joists, rafters or similar members less than 1¹/₂ inches (38 mm) from the nearest edge of the member, the pipe shall be protected by shield plates. Protective steel shield plates having a minimum thickness of 0.0575-inch (1.463 mm) (No. 16 gage) shall cover the area of the pipe where the member is notched or bored, and shall extend a minimum of 4 inches (102 mm) above sole plates, below top plates and to each side of a stud, joist or rafter.*

Discussion and Commentary: The requirement for shield plates is intended to minimize the possibility that nails or screws will be driven into the gas pipe or tube. Because nails and screws sometimes miss the stud, rafter, joist, or sole or top plates, the shield plates must extend beyond the stud itself.

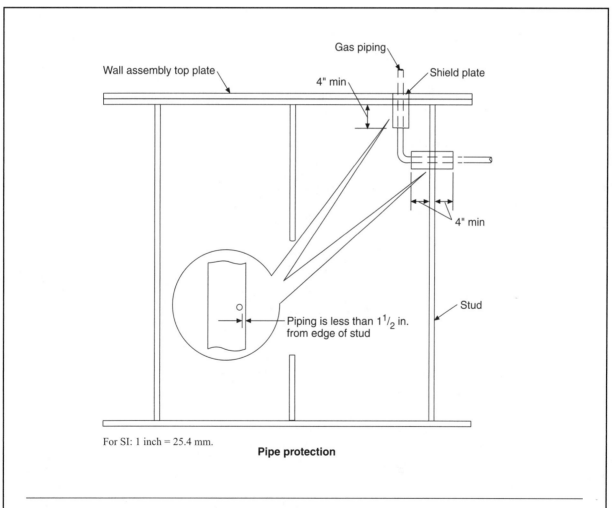

For SI: 1 inch = 25.4 mm.

Pipe protection

Piping must not be installed in any solid concrete or masonry floor construction. The potential for pipe damage due to slab settlement, cracking or the corrosive action of the floor material makes it imperative that the gas piping be installed in channels or casings, or similarly protected.

Topic: Design and Installation

Category: Fuel Gas

Reference: IRC G2418.2, G2424.1

Subject: Piping Support

Code Text: *Piping shall be supported with metal pipe hooks, metal pipe straps, metal bands, metal brackets, metal hangers or building structural components suitable for the size of piping, of adequate strength and quality, and located at intervals so as to prevent or damp out excessive vibration. Piping shall be anchored to prevent undue strains on connected equipment and shall not be supported by other piping. Supports, hangers, and anchors shall be installed so as not to interfere with the free expansion and contraction of the piping between anchors. Piping shall be supported at intervals not exceeding the spacing specified in Table G2424.1.*

Discussion and Commentary: Suspended gas piping must be supported properly by hangers and straps to prevent straining or bending, which could cause leaks and create a hazardous condition. The supporting material should be compatible with the piping material to avoid electrolytic action.

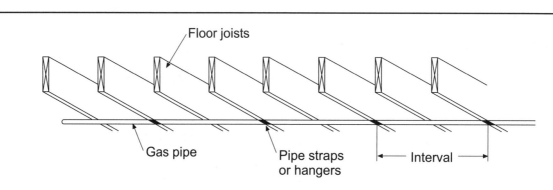

Gas pipe supports

TABLE G2424.1
SUPPORT OF PIPING

STEEL PIPE, NOMINAL SIZE OF PIPE (inches)	SPACING OF SUPPORTS (feet)	NOMINAL SIZE OF TUBING SMOOTH-WALL (inch O.D.)	SPACING OF SUPPORTS (feet)
$1/2$	6	$1/2$	4
$3/4$ or 1	8	$5/8$ or $3/4$	6
$1 1/4$ or larger (horizontal)	10	$7/8$ or 1 (horizontal)	8
$1 1/4$ or larger (vertical)	Every floor level	1 or Larger (vertical)	Every floor level

For SI: 1 inch = 25.4 mm, 1 foot = 304.8 mm.

Buried piping should be laid on a solid bed. This could be 4 to 6 inches of sand, gravel or tamped earth. The backfill should not contain any large rocks or frozen chunks of earth that might impose loads that could damage the pipe.

Topic: Sediment Trap   **Category:** Fuel Gas
Reference: IRC G2419.4   **Subject:** Drips and Sloped Piping

Code Text: *Where a sediment trap is not incorporated as a part of the appliance, a sediment trap shall be installed downstream of the appliance shut-off valve as close to the inlet of the appliance as practical. The sediment trap shall be either a tee fitting having a capped nipple of any length installed vertically in the bottom-most opening of the tee as illustrated in Figure G2419.4 or other device approved as an effective sediment trap. Illuminating appliances, ranges, clothes dryers, decorative vented appliances for installation in vented fireplaces, gas fireplaces, and outdoor grills need not be so equipped.*

Discussion and Commentary: Sediment traps cause the gas flow to change direction 90 degrees at the sediment collection point, thus causing solid or liquid contaminants to drop out of the gas flow. Sediment traps prevent debris from entering the gas controls where such contaminants could cause hazardous appliance malfunction.

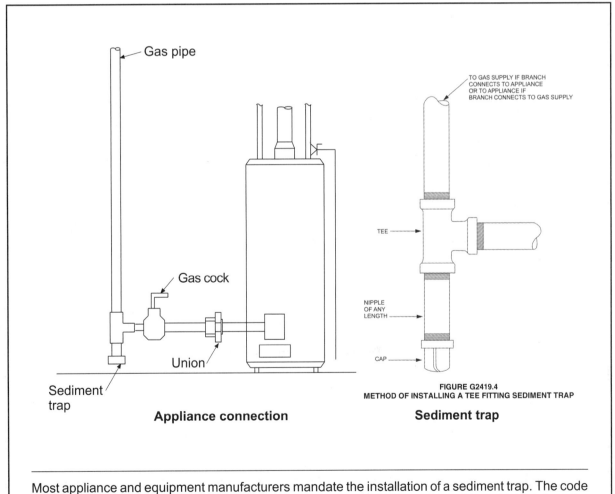

FIGURE G2419.4
METHOD OF INSTALLING A TEE FITTING SEDIMENT TRAP

Appliance connection   **Sediment trap**

Most appliance and equipment manufacturers mandate the installation of a sediment trap. The code requires compliance with manufacturers' instructions and listings, so it is consistent that the installation of a trap is a basic code provision.

Code Text: *Each appliance shall be provided with a shutoff valve in accordance with Section G2430.5.1 (located within the same room), G2420.5.2 (vented decorative appliances and room heaters) or G2520.5.3 (located at manifold). The shutoff valve shall be located in the same room as the appliance. The shutoff valve shall be within 6 feet (1829 mm) of the appliance, and shall be installed upstream of the union, connector or quick disconnect device it serves. Such shutoff valves shall be provided with access.*

Discussion and Commentary: The fuel gas connection to every gas appliance must be provided with an individual gas shutoff valve to permit maintenance, repair, replacement or temporary disconnection. The shutoff valve must be adjacent to the appliance, conspicuously located, and within reach to permit it to be easily located and operated in the event of an emergency. For example, a gas cock serving an appliance on the first floor could not be located in the basement.

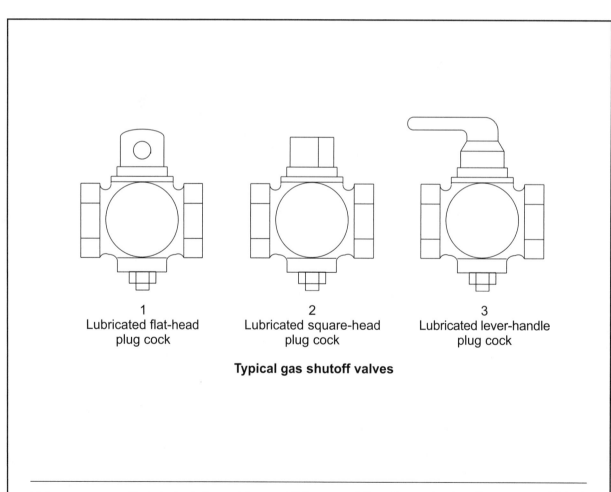

1
Lubricated flat-head
plug cock

2
Lubricated square-head
plug cock

3
Lubricated lever-handle
plug cock

Typical gas shutoff valves

Valves are more likely to leak than piping and fittings, and therefore valve locations are restricted. Locating fuel gas valves in concealed spaces unduly restricts access to the valve in the event of an emergency or servicing the equipment to which the valve is attached.

Code Text: *Appliance fuel connectors shall be installed in accordance with the manufacturer's instructions and Sections G2433.1.2.1 through G2422.1.2.4. Connectors shall not exceed 6 feet (1829 mm) in overall length.* See exception where rigid metallic piping is used. *Only one connector shall be used for each appliance. A union fitting shall be provided for appliances connected by rigid metallic pipe. Such unions shall be accessible and located within 6 feet (1829 mm) of the appliance.*

Discussion and Commentary: Flexible (semirigid) appliance fuel connectors are primarily for use with cooking ranges and clothes dryers where the gas connection is located behind the appliance and some degree of flexibility is necessary to facilitate the hook-up. Longer lengths for range and dryer installations allow the connector to be coiled or otherwise arranged to allow limited movement of the appliance without stressing the connector. Rigid connectors are not practical for such movable appliances.

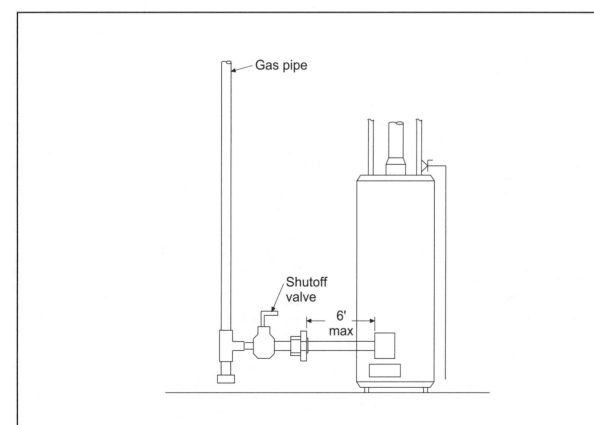

For SI: 1 foot = 304.8 mm.

Flexible connectors, usually constructed of brass or stainless steel, are labeled with tags of metal rings placed over the tubing. The manufacturer's installation instructions must be followed to protect the connector from damage and to prevent leakage.

Code Text: *The type of venting system to be used shall be in accordance with Table G2427.4. Plastic piping used for venting appliances listed for use with such venting materials shall be approved. Special gas vents shall be listed and installed in accordance with the special gas vent manufacturer's installation instructions.*

Discussion and Commentary: Various types of vents are listed in Table G2427.4 with the corresponding types of appliances that can be served by the vent.

TABLE G2427.4
TYPE OF VENTING SYSTEM TO BE USED

APPLIANCES	TYPE OF VENTING SYSTEM
Listed Category I appliances Listed appliances equipped with draft hood Appliances listed for use with Type B gas vent	Type B gas vent (Section G2427.6) Chimney (Section G2427.5) Single-wall metal pipe (Section G2427.7) Listed chimney lining system for gas venting (Section G2427.5.2) Special gas vent listed for these appliances (Section G2427.4.2)
Listed vented wall furnaces	Type B-W gas vent (Sections G2427.6, G2436)
Category II appliances	As specified or furnished by manufacturers of listed appliances (Sections G2427.4.1, G2427.4.2)
Category III appliances	As specified or furnished by manufacturers of listed appliances (Sections G2427.4.1, G2427.4.2)
Category IV appliances	As specified or furnished by manufacturers of listed appliances (Sections G2427.4.1, G2427.4.2)
Unlisted appliances	Chimney (Section G2427.5)
Decorative appliances in vented fireplaces	Chimney
Direct-vent appliances	See Section G2427.2.1
Appliances with integral vent	See Section G2427.2.2

The type and size of the vent must be as dictated by the manufacturer's installation instructions for the appliance. The design and installation instructions prevented by the vent manufacturer must also be consulted when designing a vent system for any particular application.

Topic: Factory-Built Chimneys

Category: Fuel Gas

Reference: IRC G2427.5.1

Subject: Venting of Equipment

Code Text: *Factory-built chimneys shall be installed in accordance with the manufacturers' instructions. Factory-built chimneys used to vent appliances that operate at positive vent pressure shall be listed for such application.*

Discussion and Commentary: All prefabricated chimney systems must bear the label of an approved agency. The label states information such as the type of appliance the chimney was tested for use with, a reference to the manufacturer's installation instructions and the minimum required clearances to combustibles. The manufacturer's instructions contain sizing criteria and the requirements for every aspect of a factory-built chimney installation, which include component assembly, clearances to combustibles, support, terminations, connections, protection from damage and fireblocking.

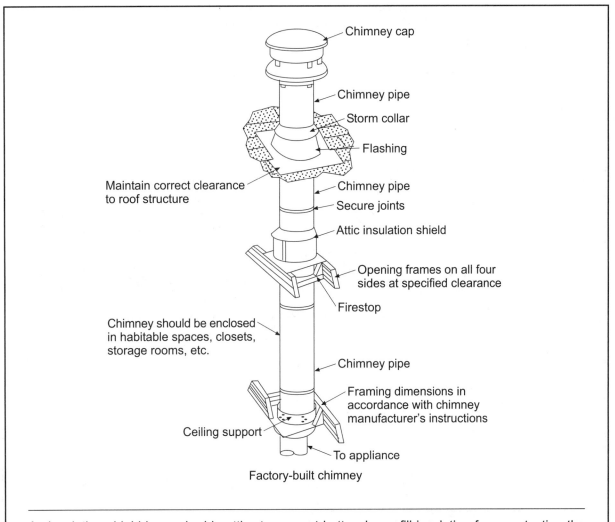

Factory-built chimney

An insulation shield is required in attics to prevent batt or loose-fill insulation from contacting the chimney. The attic shield maintains the airspace clearance between the chimney and the insulation material to reduce the chimney temperatures transmitted from the chimney surfaces.

Topic: Gas Vent Termination

Category: Fuel Gas

Reference: IRC G2427.6.3

Subject: Venting of Equipment

Code Text: *Gas vents that are 12 inches (305 mm) or less in size and located not less than 8 feet (2438 mm) from a vertical wall or similar obstruction shall terminate above the roof in accordance with Figure G2427.6.3. Gas vents that are over 12 inches (305 mm) in size or are located less than 8 feet (2438 mm) from a vertical wall or similar obstruction shall terminate not less than 2 feet (610 mm) above the highest point where they pass through the roof and not less than 2 feet (610 mm) above any portion of a building within 10 feet (3048 mm) horizontally.* See other requirements for direct-vent systems, appliances with integral vents, and appliances utilizing mechanical draft systems.

Discussion and Commentary: Figure G2427.6.3 indicates the need for a greater vent height above the roof as the roof approaches being a vertical surface. The greater the roof pitch, the greater effect of the wind striking the roof surface. However, the height should be limited as much as possible because of several concerns. Vent piping exposed to the outdoors encourages condensation in cold weather, could require guy wires or braces depending on height and is considered aesthetically unattractive.

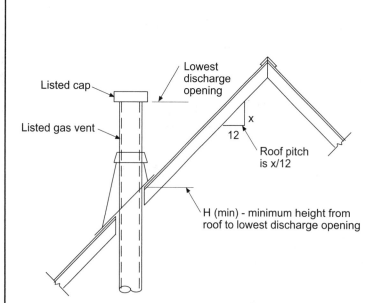

Roof pitch	H (min) ft	m
Flat to 6/12	1.0	0.30
6/12 to 7/12	1.25	0.38
Over 7/12 to 8/12	1.5	0.46
Over 8/12 to 9/12	2.0	0.61
Over 9/12 to 10/12	2.5	0.76
Over 10/12 to 11/12	3.25	0.99
Over 11/12 to 12/12	4.0	1.22
Over 12/12 to 14/12	5.0	1.52
Over 14/12 to 16/12	6.0	1.83
Over 16/12 to 18/12	7.0	2.13
Over 18/12 to 20/12	7.5	2.27
Over 20/12 to 21/12	8.0	2.44

Gas vent termination

It is a common misapplication for code users to apply chimney termination height requirements to vents, thereby causing vents to extend above roofs much higher than required in many cases. For roof pitches up to a 6/12 pitch, only a 1-foot extension is necessary.

Topic: Common Connector/Manifold   **Category:** Fuel Gas
Reference: IRC G2427.10.3.4   **Subject:** Venting of Equipment

Code Text: *Where two or more gas appliances are vented through a common vent connector or vent manifold, the common vent connector or vent manifold shall be located at the highest level consistent with available headroom and the required clearance to combustible materials and shall be sized in accordance with Section G2428 or other approved engineering methods.*

Discussion and Commentary: A manifold is defined as a vent or connector that is a lateral (horizontal) extension of the lower end of a common vent. Manifolds must be installed as high as possible to provide for the most appliance connector rise, which is always beneficial for venting performance. Connector rise takes advantage of the energy of hot gases discharging directly from the appliance and develops flow velocity and draft in addition to that produced by the common vent.

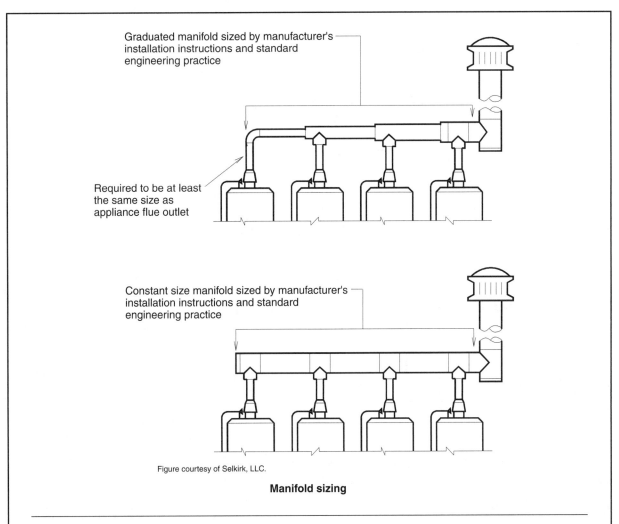

Figure courtesy of Selkirk, LLC.

Manifold sizing

An option is available where there are only two draft-hood-equipped appliances. This alternative method is intended for use only with the alternative sizing method of Section G2427.6.9.1 and is not compatible with Section G2428.

Code Text: *An automatically operated vent damper shall be of a listed type.*

Discussion and Commentary: The installation of an automatic damper must be in strict accordance with the manufacturer's installation instructions. A malfunctioning or improperly installed flue damper could cause the appliance to malfunction and discharge the products of combustion directly into the building interior. Because automatic damper failure could result in a hazardous condition, automatic dampers are required to be listed and labeled. Automatic flue dampers are energy-saving devices designed to close off or restrict an appliance flue passageway when the appliance is not operating and is in its "off" cycle. Such devices save energy by trapping residual heat in a heat exchanger after the burners shut off and by preventing the escape of conditioned room air up the vent. Thus, the appliance efficiency can be boosted, and building air infiltration can be reduced.

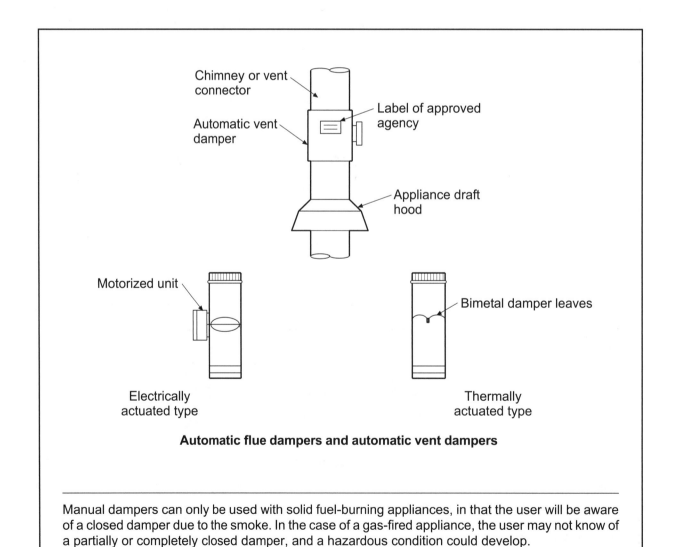

Chimney or vent connector

Label of approved agency

Automatic vent damper

Appliance draft hood

Motorized unit

Bimetal damper leaves

Electrically actuated type

Thermally actuated type

Automatic flue dampers and automatic vent dampers

Manual dampers can only be used with solid fuel-burning appliances, in that the user will be aware of a closed damper due to the smoke. In the case of a gas-fired appliance, the user may not know of a partially or completely closed damper, and a hazardous condition could develop.

Code Text: *Decorative appliances for installation in approved solid fuel-burning fireplaces, with the exception of those tested in accordance with ANSI Z21.84, shall utilize a direct ignition device, an ignitor or a pilot flame to ignite the fuel at the main burner, and shall be equipped with a flame safeguard device. The flame safeguard device shall automatically shut off the fuel supply to a main burner or group of burners when the means of ignition of such burners becomes inoperative.*

Discussion and Commentary: The typical flame safeguard device used with gas log set appliances is a combination manual control valve, pressure regulator, pilot feed and magnetic pilot safety mechanism with a thermocouple generator. If the pilot flame is extinguished, the drop in thermocouple output voltage will cause the control valve to "lock out" in the closed position, thereby preventing the flow of gas to the main burner and the pilot burner.

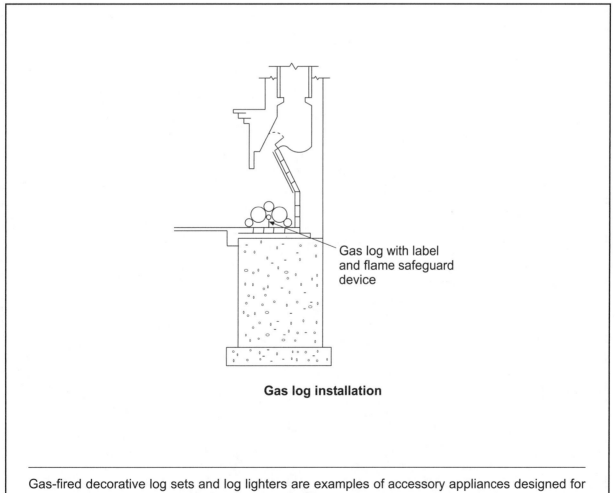

Gas log with label and flame safeguard device

Gas log installation

Gas-fired decorative log sets and log lighters are examples of accessory appliances designed for installation in solid fuel-burning fireplaces. Gas log sets provide some radiant heat; however, their primary function is to create an aesthetically pleasant simulation of a wood log fire.

Code Text: *Suspended-type unit heaters shall be supported by elements that are designed and constructed to accommodate the weight and dynamic loads. Hangers and brackets shall be of noncombustible material. Suspended-type unit heaters shall be installed with clearances to combustible materials of not less than 18 inches (457 mm) at the sides, 12 inches (305 mm) at the bottom and 6 inches (152 mm) above the top where the unit heater has an internal draft hood or 1 inch (25 mm) above the top of the sloping side of the vertical draft hood.*

Discussion and Commentary: As with all suspended fuel-fired appliances, a support failure could result in a fire, explosion or injury to building occupants. The supports themselves must be properly designed. Equally important are the structural members to which the supports are attached, such as rafters, beams, joists and purlins. All brackets, pipes, rods, angle iron, structural members and fasteners must be designed for the dead and dynamic loads.

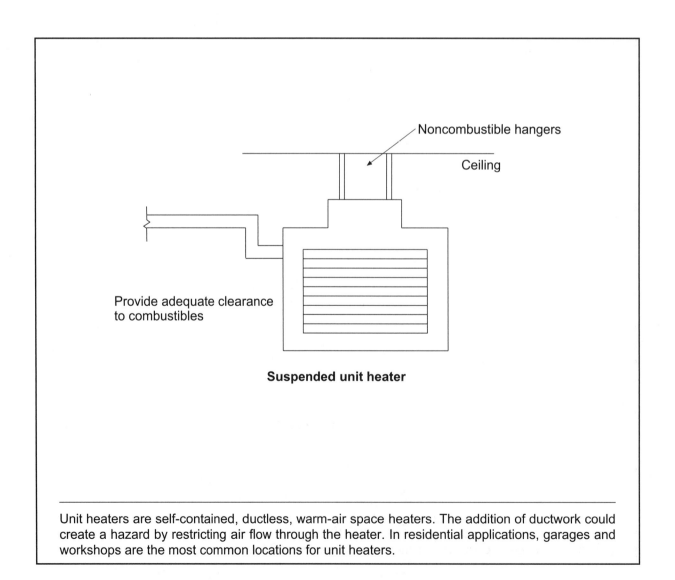

Suspended unit heater

Unit heaters are self-contained, ductless, warm-air space heaters. The addition of ductwork could create a hazard by restricting air flow through the heater. In residential applications, garages and workshops are the most common locations for unit heaters.

Code Text: *Unvented room heaters shall be equipped with an oxygen-depletion-sensitive safety shutoff system. The system shall shut off the gas supply to the main and pilot burners when the oxygen in the surrounding atmosphere is depleted to the percent concentration specified by the manufacturer, but not lower than 18 percent. The system shall not incorporate field adjustment means capable of changing the set point at which the system acts to shut off the gas supply to the room heater.*

Discussion and Commentary: The oxygen depletion sensor is basically a pilot burner that is extremely sensitive to the oxygen content in the combustion air. If the oxygen content (approximately 80 percent) drops to a predetermined level, the pilot flame will destabilize and extinguish or become incapable of sufficiently heating the thermocouple or thermopile generator, resulting in main gas control valve shutdown and lockout.

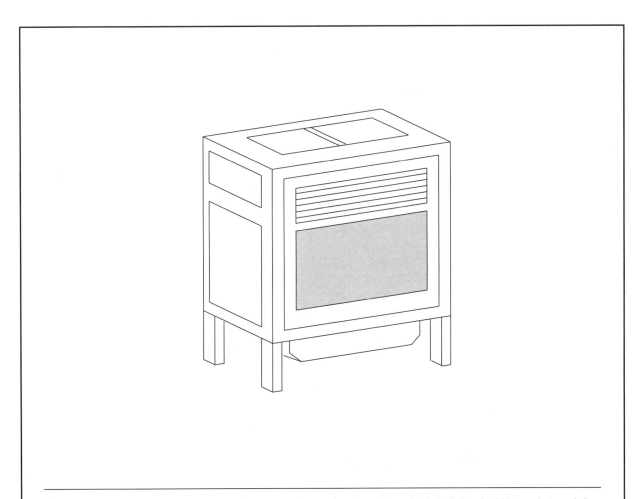

Unvented room heaters are limited to a maximum input rating of 40,000 Btu/h. When determining the maximum unvented room heater rating allowed, the aggregate input rating of all unvented appliances in a room is limited to 20 Btu/h per cubic foot of volume of the room.

Quiz

Study Session 13
IRC Chapter 24

1. In general, fuel-fired appliances are permitted to be located in which one of the following spaces?

    a. sleeping rooms

    b. storage closets

    c. bathrooms

    d. garages

    Reference _____

2. Where located in a hazardous location, fuel-gas equipment and appliances having an ignition source shall be elevated such that the ignition source is located a minimum of _____ above the floor.

    a. 6 inches

    b. 12 inches

    c. 18 inches

    d. 30 inches

    Reference _____

3. Unless protected from motor vehicle damage, fuel-gas appliances located in private garages shall be installed with a minimum clearance of _____ above the floor.

    a. 60 inches

    b. 72 inches

    c. 80 inches

    d. 84 inches

    Reference _____

4. Assume that a fuel-gas appliance requires a minimum clearance without protection of 36 inches above the appliance to combustible material. If $^1/_2$-inch thick insulation board over 1-inch glass fiber batts is used as protection, the minimum clearance may be reduced to _____ .

   a.   9 inches
   b.   12 inches
   c.   18 inches
   d.   24 inches

   Reference _____

5. Assume that a fuel-gas appliance requires a minimum clearance without protection of 9 inches from the sides and back to combustible material. If 0.024 sheet metal with a ventilated air space is used as protection, the minimum clearance may be reduced to _____ .

   a.   2 inches
   b.   3 inches
   c.   4 inches
   d.   6 inches

   Reference _____

6. For other than steel pipe, exposed gas piping shall be identified by a yellow label marked "Gas" spaced at maximum intervals of _____ unless in the same room as the appliance served.

   a.   5 feet
   b.   6 feet
   c.   8 feet
   d.   10 feet

   Reference _____

7. Unless special conditions are met, the maximum design operating pressure for gas piping systems located inside buildings shall be _____ .

   a.   5 psig
   b.   10 psig
   c.   15 psig
   d.   20 psig

   Reference _____

8. LP-gas systems designed to operate below _____ shall be designed to either accommodate liquid LP-gas or prevent LP-gas vapor from condensing into a liquid.

   a.   −5°F
   b.   0°F
   c.   20°F
   d.   32°F

   Reference _____

9. Copper tubing to be used with gases shall comply with standard _____ of ASTM B 88 or ASTM B 280.

    a. Type K only   b. Type L only

    c. Type K or Type L   d. Type L or Type M

Reference _____

10. For the field threading of 1-inch metallic pipe, a length of _____ (approximate) shall be provided for the threaded portion.

    a. $^3/_4$ inch   b. $^7/_8$ inch

    c. 1 inch   d. $1^1/_8$ inches

Reference _____

11. In which of the following locations is the installation of gas piping not specifically prohibited?

    a. in a supply air duct   b. through a clothes chute

    c. in an elevator shaft   d. adjacent to an exterior stairway

Reference _____

12. For gas piping, other than black or galvanized steel, installed in a concealed location, a shield plate need not be provided for protection against physical damage where a bored hole is located a minimum of _____ from the edge of the wood member.

    a. $^5/_8$ inch   b. 1 inch

    c. $1^1/_4$ inches   d. $1^1/_2$ inches

Reference _____

13. In general, underground gas piping shall be installed a minimum depth of _____ below grade.

    a. 6 inches   b. 12 inches

    c. 18 inches   d. 24 inches

Reference _____

14. The unthreaded portion of gas piping outlets shall extend a minimum of _____ through finished ceilings and walls.

    a. 1 inch   b. 2 inches

    c. 3 inches   d. 4 inches

    Reference _____

15. The tracer wire required adjacent to underground nonmetallic gas piping shall be a minimum of _____ wire size.

    a. 14 AWG   b. 16 AWG

    c. 18 AWG   d. 20 AWG

    Reference _____

16. The inside radius of a bend in a metallic gas pipe shall be a minimum of _____ the outside diameter of the pipe.

    a. six times   b. eight times

    c. twelve times   d twenty-five times

    Reference _____

17. Gas piping shall be tested at a minimum pressure _____ the proposed maximum working pressure, but not less than 3 psig.

    a. equal to   b. of one and one-half times

    c. of twice   d. of two and one-half times

    Reference _____

18. Where outdoor combustion air is provided for gas-fired appliances through openings to the outdoors, the minimum dimension of such air openings shall be _____ inches.

    a. 3   b. 4

    c. 10   d. 12

    Reference _____

19. Gas piping for other than dry gas conditions shall be sloped a minimum of _____ to prevent traps.

   a. $^1/_8$ inch in 10 feet   b. $^1/_8$ inch in 15 feet

   c. $^1/_4$ inch in 10 feet   d. $^1/_4$ inch in 15 feet

   Reference _____

20. Unless located at the manifold, a gas shutoff valve for an appliance shall be provided within a maximum distance of _____ between the valve and the appliance.

   a. 3 feet   b. 4 feet

   c. 6 feet   d. 10 feet

   Reference _____

21. In general, appliance fuel connectors shall have a maximum overall length of _____ .

   a. 3 feet   b. 5 feet

   c. 6 feet   d. 10 feet

   Reference _____

22. The radius of the inner curve of plastic gas pipe bends shall be a minimum of _____ times the inside diameter of the pipe.

   a. 12   b. 16

   c. 20   d. 25

   Reference _____

23. Gas appliance connectors, where connected to a masonry chimney, shall connect at a point a minimum of _____ above the lowest portion of the interior of the chimney flue.

   a. 12 inches   b. 18 inches

   c. 30 inches   d. 36 inches

   Reference _____

24. A sauna room provided with a gas-fired sauna heater shall be provided with a minimum _____ ventilation opening located near the top of the door into the sauna room.

   a. 4-inch by 8-inch   b. 4-inch by 12-inch

   c. 8-inch by 8-inch   d. 8-inch by 12-inch

   Reference _____

25. Unvented gas room heaters shall be limited to a maximum input rating of _____ .

   a. 35,000 Btu/h   b. 40,000 Btu/h

   c. 50,000 Btu/h   d. 65,000 Btu/h

   Reference _____

26. Gas-fired equipment and appliances, where suspended above grade level, shall be provided with a minimum clearance of _____ inches from adjoining grade.

   a. 2   b. 4

   c. 6   d. 8

   Reference _____

27. Fuel gas piping installed across a roof surface shall be elevated above the roof a minimum of _____ inches.

   a. $1^1/_2$ b. 2

   c. $3^1/_2$ d. 6

   Reference _____

28. Which of the following test mediums shall not be used in the testing of a fuel gas piping system?

   a. air   b. carbon dioxide

   c. nitrogen   d. oxygen

   Reference _____

29. One-inch O.D. smooth-wall tubing used as fuel gas piping shall be supported horizontally at maximum intervals of _____ feet.

    a. 4   b 6

    c. 8   d. 10

Reference _____

30. Where a post-mounted illuminating gas appliance is installed on a steel pipe post having a height of 30 inches, what is the minimum size Schedule 40 steel pipe required?

    a. $^3/_4$ inch   b. 1 inch

    c. $1^1/_2$ inch   d. $2^1/_2$ inch

Reference _____

31. In the sizing of a fuel gas piping system, a barbeque without a manufacturer's input rating shall be rated for _____ input btu/h.

    a. 25,000   b. 35,000

    c. 40,000   d. 50,000

Reference _____

32. The unthreaded portion of a gas piping outlet passing up through an outdoor patio shall extend a minimum of _____ inch(es) above the patio.

    a. 1   b. 2

    c. 4   d. 6

Reference _____

33. Where approved and not susceptible to physical damage, an individual gas line to an outdoor barbeque grill shall be installed a minimum of _____ inches below finished grade.

    a. 6   b. 8

    c. 12   d. 18

Reference _____

34. The union fitting required for gas-fired appliances connected by rigid metallic pipe shall be accessible and located a maximum of _____ feet from the appliance.

     a. 3   b. 5

     c. 6   d. 10

Reference _____

35. Suspended-type gas-fired unit heaters shall have a minimum clearance to combustible materials of _____ inches on the sides.

     a. 6   b. 12

     c. 15   d. 18

Reference _____

Study Session

14

2012 IRC Chapters 25, 26 and 27
General Plumbing Requirements and Plumbing Fixtures

OBJECTIVE: To provide an understanding of the general requirements for the installation of plumbing systems, including inspections and tests; piping installation, protection and support; and plumbing fixtures.

REFERENCE: Chapters 25, 26 and 27, 2012 *International Residential Code*

KEY POINTS:
- By what method must the building sewer system be tested and inspected? The DWV system? The water-supply system?
- Where installed through framing members in concealed locations, how shall piping be protected? How should it be protected from breakage where passing through or under walls?
- How shall a soil pipe, waste pipe or building drain be installed where passing under a footing or through a foundation wall?
- What are the installation limitations for piping installed in areas subject to freezing temperatures? To what minimum depth must water service pipe be installed? How shall trenches be located in relationship to footings or bearing walls?
- How must piping be installed when located in a trench? How is backfilling to be accomplished? How is trenching installed parallel to footings to be located?
- What are the general provisions for the support of piping? What is the maximum horizontal spacing between supports for various types of pipe material? Between vertical supports?
- How are exterior openings for pipes to be waterproofed?
- How must piping be listed and identified? When is third-party testing required? Third-party certification?
- What is the minimum required diameter for fixture tail pieces serving sinks, dishwashers, laundry tubs, bathtubs and similar fixtures? For bidets, lavatories and similar fixtures?
- What type of fixture connection must be provided with an access panel or utility space? What size of access opening must be provided?
- What are the general requirements for the installation of plumbing fixtures regarding support? Water tightness?

- What horizontal side clearances are mandated for water closets and bidets? What minimum clearance is required at the front of water closets, bidets and lavatories?

- What is the minimum required extension for a standpipe? The maximum allowable extension? Under what conditions may a laundry tray waste connect into a standpipe for an automatic clothes washer drain?

- What is the minimum permitted size for a shower compartment? What alternative size is permitted? Which direction must hinged shower doors swing? What is the minimum required opening size? How are shower receptors to be constructed and installed?

- How are water closets regulated in regard to flushing devices, water supplies and flush valves?

- What is the minimum waste outlet size for a lavatory? A bathtub? A sink? A laundry tub? A food-waste grinder? A floor drain?

- How may a sink and dishwasher discharge through a single trap? A sink, dishwasher and food grinder?

- What is temperature limit for water supplied to a shower? A bathtub? A whirlpool bathtub? A bidet?

Code Text: *New plumbing work and parts of existing systems affected by new work or alterations shall be inspected by the building official to ensure compliance with the requirements of the IRC. DWV systems shall be tested on completion of the rough piping installation by water or for piping systems other than plastic, by air with no evidence of leakage. After the plumbing fixtures have been set and their traps filled with water, their connections shall be tested and proved gas tight and/or water tight.*

Discussion and Commentary: All plumbing and drainage work must be tested in an appropriate manner to verify that it is leak free. The building sewer must be tested with a minimum 10-foot head of water. Both rough plumbing and finished plumbing are tested with a variety of methods available. Water supply piping and any backflow prevention assemblies must also be inspected and tested.

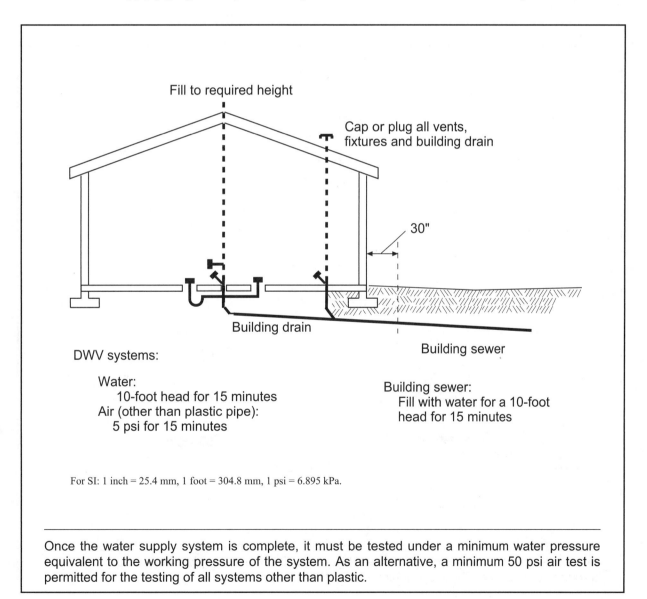

Fill to required height

Cap or plug all vents, fixtures and building drain

30"

Building drain

Building sewer

DWV systems:

  Water:
    10-foot head for 15 minutes
  Air (other than plastic pipe):
    5 psi for 15 minutes

Building sewer:
  Fill with water for a 10-foot
  head for 15 minutes

For SI: 1 inch = 25.4 mm, 1 foot = 304.8 mm, 1 psi = 6.895 kPa.

Once the water supply system is complete, it must be tested under a minimum water pressure equivalent to the working pressure of the system. As an alternative, a minimum 50 psi air test is permitted for the testing of all systems other than plastic.

Code Text: *Pipes passing through concrete or cinder walls and floors, cold-formed steel framing or other corrosive material shall be protected against external corrosion by a protective sheathing or wrapping or other means that will withstand any reaction from lime and acid of concrete, cinder or other corrosive material. Sheathing or wrapping shall allow for movement including expansion and contraction of piping. Minimum wall thickness of material shall be 0.025 inch (0.64 mm).*

Discussion and Pipes made from brass, copper, cast iron and steel are subject to corrosion when exposed to
Commentary: the lime and acid of concrete and cinder or other corrosive material such as soil; therefore, a coating or wrapping is necessary. Typical protective coatings include coal tar wrapper with paper, epoxy or plastic coatings.

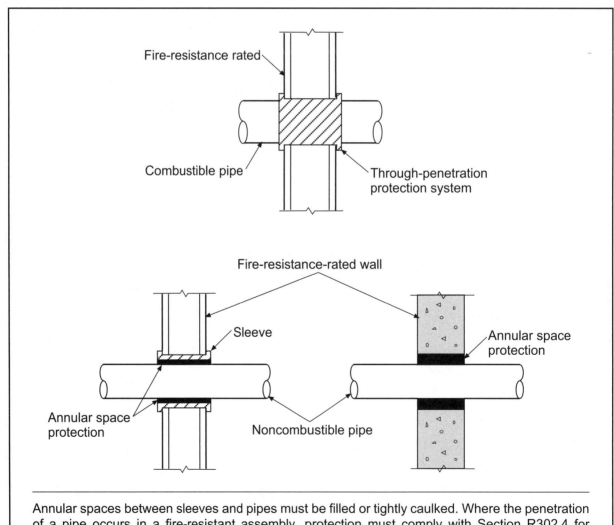

Annular spaces between sleeves and pipes must be filled or tightly caulked. Where the penetration of a pipe occurs in a fire-resistant assembly, protection must comply with Section R302.4 for through-penetrations. This may require a listed firestop system.

Topic: Pipes Through Foundation Walls
Reference: IRC P2603.4

Category: General Plumbing Requirements
Subject: Structural and Piping Protection

Code Text: *A pipe that passes through a foundation wall shall be provided with a relieving arch, or a pipe sleeve shall be built into the foundation wall. The sleeve shall be two pipe sizes greater than the pipe passing through the wall.*

Discussion and Commentary: Piping installed within a foundation wall must be structurally protected from any transferred loading from the foundation wall. This protection may be provided through a relieving arch or a pipe sleeve. When a sleeve is used, it should be sized such that it is two pipe sizes larger than the penetration pipe. This space will allow for any differential movement of the pipe. Providing structural protection to the piping system ensures that the piping will not be subjected to the undue stresses that may cause the piping to rupture and leak.

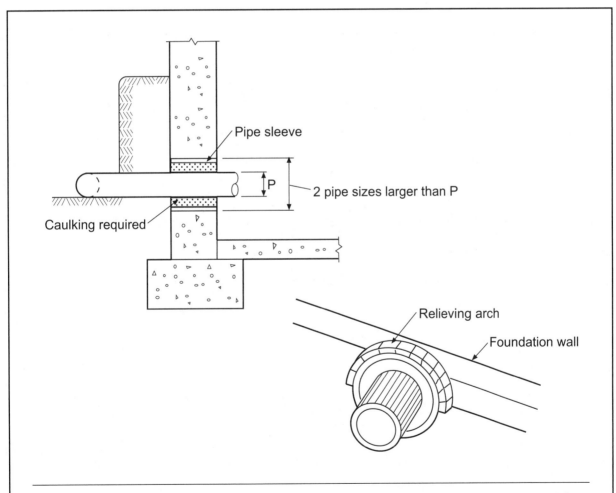

Piping located in concealed spaces within the interior of the building must also be protected from damage if installed through holes or notches in studs, joists and similar framing members. Where the hole or notch is less than $1^{1}/_{2}$ inches from the nearest member edge, steel protective shield plates are required.

Code Text: *In localities having a winter design temperature of 32°F (0°C) or lower . . . a water, soil or waste pipe shall not be installed outside of a building, in exterior walls, in attics or crawl spaces, or in any other place subjected to freezing temperature unless adequate provision is made to protect it from freezing by insulation or heat or both.*

Discussion and Commentary: Occupied spaces utilize the building's comfort heating system to assist in protecting pipes from freezing. Where the temperature of the air surrounding the insulation remains low for a significant period of time, such as in a vacant building, insulation alone will not provide adequate protection without the addition of heat. In such conditions, the water in the pipe will freeze regardless of the amount of insulation used, and heat must be provided—often in the form of electric resistance heat tape or cable inside the insulation next to the piping.

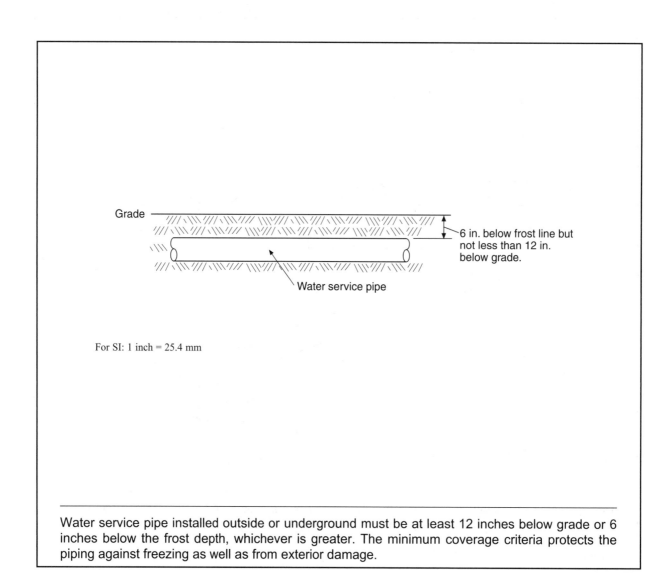

Water service pipe installed outside or underground must be at least 12 inches below grade or 6 inches below the frost depth, whichever is greater. The minimum coverage criteria protects the piping against freezing as well as from exterior damage.

Code Text: *Where trenches are excavated such that the bottom of the trench forms the bed for the pipe, solid and continuous load-bearing support shall be provided between joints. Where over-excavated, the trench shall be backfilled to the proper grade with compacted earth, sand, fine gravel or similar granular material. Piping shall not be supported on rocks or blocks at any point. Rocky or unstable soil shall be over excavated by two or more pipe diameters and brought to the proper grade with suitable compacted granular material.*

Discussion and Commentary: Pipe is protected from damage by proper backfilling techniques. Backfilling is accomplished in 6-inch layers, with each layer being tamped into place. The material used for backfilling must be evenly distributed to avoid any pipe movement. Large rocks or frozen chunks of dirt must not be located within 12 inches of the pipe.

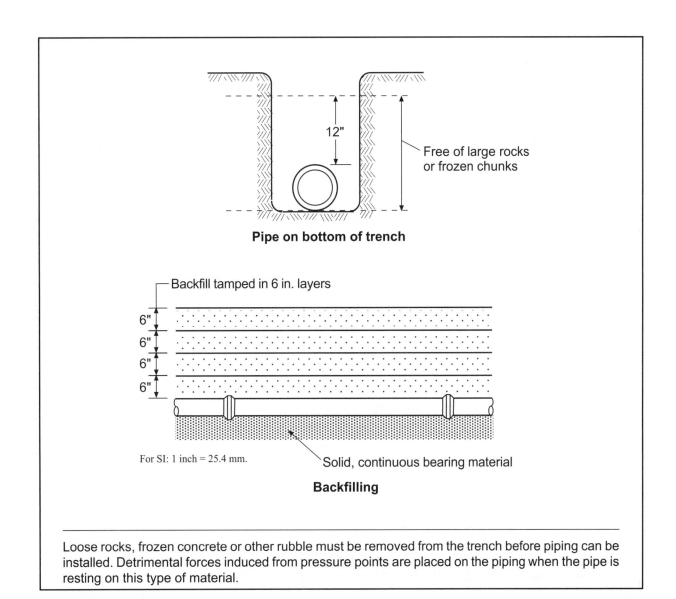

12"

Free of large rocks or frozen chunks

Pipe on bottom of trench

Backfill tamped in 6 in. layers

6"
6"
6"
6"

For SI: 1 inch = 25.4 mm.

Solid, continuous bearing material

Backfilling

Loose rocks, frozen concrete or other rubble must be removed from the trench before piping can be installed. Detrimental forces induced from pressure points are placed on the piping when the pipe is resting on this type of material.

Topic: Protection of Footings

Reference: IRC P2604.4

Category: General Plumbing Requirements

Subject: Trenching and Backfilling

Code Text: *Trenching installed parallel to footings shall not extend below the 45-degree (0.79 mm) bearing plane of the bottom edge of a wall or footing.*

Discussion and Commentary: A footing requires a minimum load-bearing area to distribute the weight of the building. That plane extends downward at approximately a 45-degree angle from the base of the footing. The trench containing the piping must not be located below that load-bearing plane, to avoid undermining the building footing. The key to the provision is the term *parallel*. Where extending a water or sewer pipe beneath a footing, such installation is usually perpendicular to the foundation and footing. To avoid undermining the load-bearing capacity of the soil, extend the trench beyond the load-bearing plane.

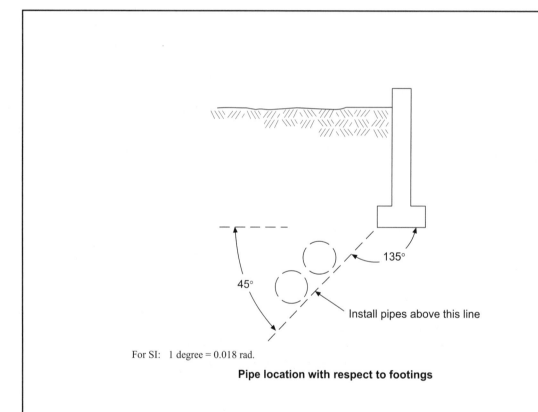

135°

45°

Install pipes above this line

For SI: 1 degree = 0.018 rad.

Pipe location with respect to footings

It is permissible to install water-service pipe in the same trench with a building sewer under the provisions of Section P2905.4.2, provided the sewer is constructed of materials listed for underground use within a building. Otherwise, various methods of separation are described.

Topic: General Provisions

Category: General Plumbing Requirements

Reference: IRC P2605.1

Subject: Support

Code Text: *Piping shall be supported so as to ensure alignment and prevent sagging, and allow movement associated with the expansion and contraction of the piping system. Piping in the ground shall be laid on a firm bed for its entire length, except where support is otherwise provided. Hangers and strapping shall be of approved material that will not promote galvanic action. Rigid support sway bracing shall be provided at changes in direction greater than 45 degrees (0.79 rad) for pipe sizes 4 inches (102 mm) and larger. Piping shall be supported at distances not to exceed those indicated in Table P2605.1.*

Discussion and Commentary: For larger pipes, hangers alone may not be sufficient to resist the forces created by water movement within the piping. Therefore, rigid bracing is required to restrict or eliminate lateral movement of both horizontal and vertical piping.

TABLE P2605.1
PIPING SUPPORT

PIPING MATERIAL	MAXIMUM HORIZONTAL SPACING (feet)	MAXIMUM VERTICAL SPACING
ABS pipe	4	10[b]
Aluminum tubing	10	15
Brass pipe	10	10
Cast-iron pipe	5[a]	15
Copper or copper alloy pipe	12	10
Copper or copper alloy tubing (1$^1/_4$ inches in diameter and smaller)	6	10
Copper or copper alloy tubing (1$^1/_2$ inches in diameter and larger)	10	10
Cross-linked polyethylene (PEX) pipe	2.67 (32 inches)	10[b]
Cross-linked polyethylene/aluminum/cross-linked polyethylene (PEX-AL-PEX) pipe	2.67 (32 inches)	4[b]
CPVC pipe or tubing (1 inch in diameter and smaller)	3	10[b]
CPVC pipe or tubing (1$^1/_4$ inches in diameter and larger)	4	10[b]
Lead pipe	Continuous	4
PB pipe or tubing	2.67 (32 inches)	4
Polyethylene/aluminum/polyethylene (PE-AL-PE) pipe	2.67 (32 inches)	4[b]
Polyethylene of raised temperature (PE-RT) pipe	2.67 (32 inches)	10[b]
Polypropylene (PP) pipe or tubing (1 inch and smaller)	2.67 (32 inches)	10[b]
Polypropylene (PP) pipe or tubing (1$^1/_4$ inches and larger)	4	10[b]
PVC pipe	4	10[b]
Stainless steel drainage systems	10	10[b]
Steel pipe	12	15

For SI: 1 inch = 25.4 mm, 1 foot = 304.8 mm.

a. The maximum horizontal spacing of cast-iron pipe hangers shall be increased to 10 feet where 10-foot lengths of pipe are installed.

b. Midstory guide for sizes 2 inches and smaller.

Depending on the type of pipe material, the maximum vertical and horizontal spacing between supports can vary. Whereas supports must occur at frequent intervals for horizontal piping due to the potential for sagging, vertical piping typically requires support only at each story height.

Code Text: *Fixture tail pieces shall be not less than $1^1/_2$ inches (38 mm) in diameter for sinks, dishwashers, laundry tubs, bathtubs and similar fixtures, and not less than $1^1/_4$ inches (32 mm) in diameter for bidets, lavatories and similar fixtures.*

Discussion and Commentary: Tail pieces are short lengths of pipe attached directly to a fixture by means of a flange for connection to other piping or traps. Tail pieces, as well as traps, continuous wastes, waste fittings and overflow fittings, must be of No. 20 gage minimum thickness when constructed of seamless drawn brass. Plastic tubular fittings are permitted when in conformance with ASTM F 409. The minimum size of $1^1/_2$ inches for sinks, dishwashers, laundry tubs and bathtubs is consistent with the minimum sizes for fixture drains, traps and trap arms.

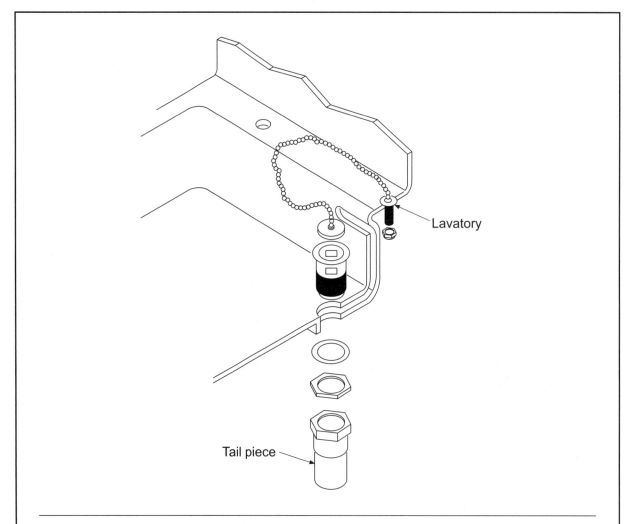

Lavatory

Tail piece

Fixture fittings, faucets and plumbing fixtures require smooth impervious surfaces with no surface defects that could reduce the efficiency, effectiveness or sanitary requirements of the code. Table P2701.1 lists the appropriate material standards for various plumbing items.

Topic: General Provisions
Reference: IRC P2705.1

Category: Plumbing Fixtures
Subject: Installation

Code Text: *Floor-outlet or floor-mounted fixtures shall be secured to the drainage connection and to the floor, when so designed, by screws, bolts, washers, nuts and similar fasteners of copper, brass or other corrosion-resistant material. Wall-hung fixtures shall be rigidly supported so that strain is not transmitted to the plumbing system. Where fixtures come in contact with walls, the contact area shall be water tight. Plumbing fixtures shall be usable. The location of piping, fixtures or equipment shall not interfere with the operation of windows or doors.*

Discussion and Commentary: The installation of fixtures shall be accomplished based on structural, sanitation and operational considerations.

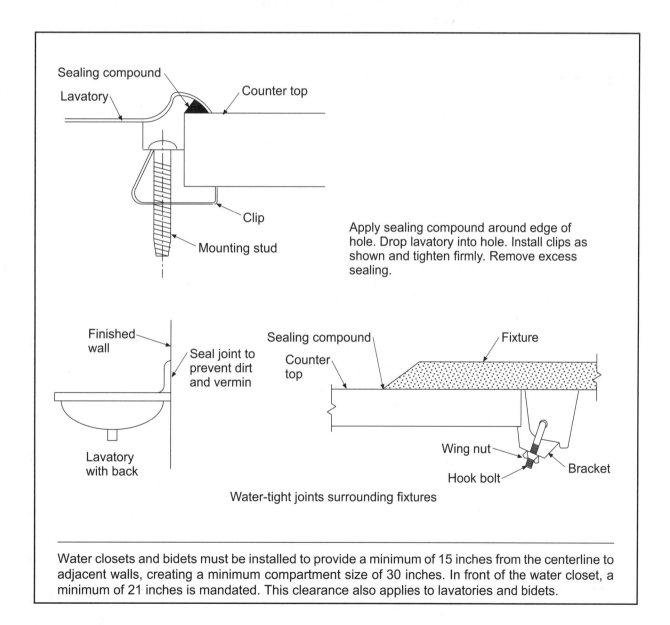

Apply sealing compound around edge of hole. Drop lavatory into hole. Install clips as shown and tighten firmly. Remove excess sealing.

Water-tight joints surrounding fixtures

Water closets and bidets must be installed to provide a minimum of 15 inches from the centerline to adjacent walls, creating a minimum compartment size of 30 inches. In front of the water closet, a minimum of 21 inches is mandated. This clearance also applies to lavatories and bidets.

Code Text: *Standpipes shall extend not less than 18 inches (457 mm) but not greater than 42 inches (1067 mm) above the trap weir. Access shall be provided to all standpipe traps and drains for rodding. A laundry tray waste line is permitted to connect into a standpipe for the automatic clothes washer drain. The standpipes shall extend not less than 30 inches (762 mm) above the trap weir and shall extend above the flood level rim of the laundry tray. The outlet of the laundry tray shall not be greater than 30 inches (762 mm) horizontal distance from the standpipe trap.*

Discussion and Commentary: A standpipe can serve as an indirect waste receptor. The established minimum receptor height aims to provide a minimal retention capacity and head pressure to prevent overflow. This height limitation is especially necessary when an indirect waste pipe receives a high rate of discharge, such as the pumped discharge from a clothes washer.

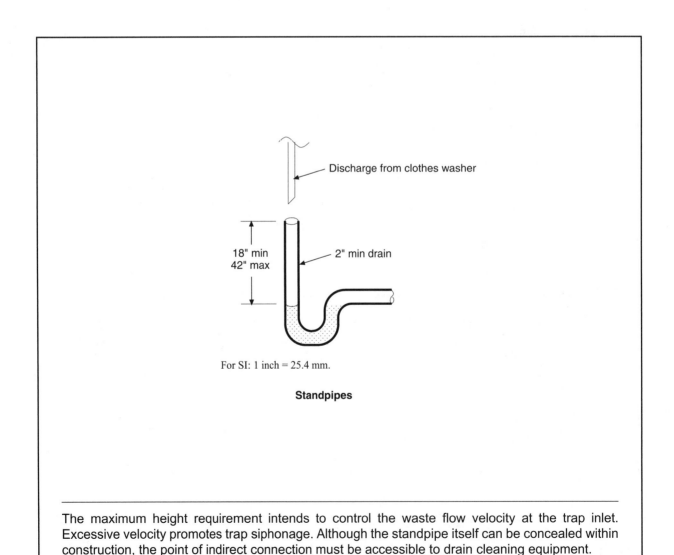

For SI: 1 inch = 25.4 mm.

Standpipes

The maximum height requirement intends to control the waste flow velocity at the trap inlet. Excessive velocity promotes trap siphonage. Although the standpipe itself can be concealed within construction, the point of indirect connection must be accessible to drain cleaning equipment.

Topic: General Provisions	**Category:** Plumbing Fixtures
Reference: IRC P2708.1	**Subject:** Showers

Code Text: *Shower compartments shall have not less than 900 square inches (0.6 m²) of interior cross-sectional area. Shower compartments shall not be less than 30 inches (752 mm) in minimum dimension measured from the finished interior dimension of the shower compartment, exclusive of fixture valves, shower heads, soap dishes, and safety grab bars or rails.* See exceptions for showers 1) with fold-down seats, and 2) having compartments at least 25 inches wide and 1,300 square inches in cross-sectional area. *The minimum required area and dimension shall be measured from the finished interior dimension at a height equal to the top of the threshold at a point tangent to its centerline and shall be continued to a height not less than 70 inches (1778 mm) above the shower drain outlet. Hinged shower doors shall open outward.*

Discussion and Commentary: The dimensional criteria for shower compartments facilitate safety and comfort based on typical body movement while showering. Fold-down seats can encroach into the required dimensions, provided the 900 square inches are available when the seats are folded.

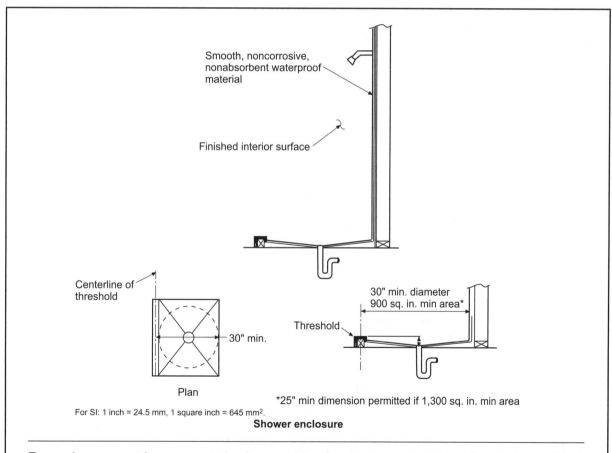

Shower enclosure

For SI: 1 inch = 24.5 mm, 1 square inch = 645 mm².

Every shower must have a control valve capable of protecting an individual from being scalded. These devices are also required to protect against rapid temperature fluctuation by automatically maintaining the discharge temperature to plus or minus 3°F of the selected temperature. The high limit stop must be set to limit water temperature to a maximum of 120°F.

Code Text: *The shower compartment access and egress opening shall have a clear and unobstructed finished width of not less than 22 inches (559 mm).*

Discussion and Commentary: It is important that an adequate width be provided to and from a shower compartment. The requirement for 22 inches of clear width provides for functional access by the shower user, including the service and maintenance of the compartment. In addition, the established width provides access for emergency response and/or rescue, should the need arise.

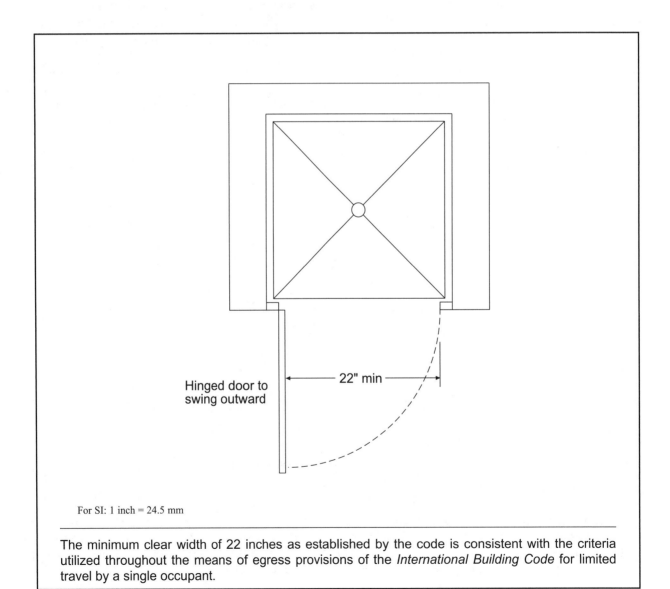

Hinged door to swing outward

22" min

For SI: 1 inch = 24.5 mm

The minimum clear width of 22 inches as established by the code is consistent with the criteria utilized throughout the means of egress provisions of the *International Building Code* for limited travel by a single occupant.

Topic: Construction	**Category:** Plumbing Fixtures
Reference: IRC P2709.1	**Subject:** Shower Receptors

Code Text: *Where a shower receptor has a finished curb threshold, it shall be not less than 1 inch (25 mm) below the sides and back of the receptor. The curb shall be not less than 2 inches (51 mm) and not more than 9 inches (229 mm) deep when measured from the top of the curb to the top of the drain. The finished floor shall slope uniformly toward the drain not less than $^1/_4$ unit vertical in 12 units horizontal (2-percent slope) nor more than $^1/_2$ unit vertical per 12 units horizontal (4-percent slope), and floor drains shall be flanged to provide a water-tight joint in the floor.*

Discussion and Commentary: When a shower is built in place, a shower pan is required under the shower's finished floor. Various materials are used for shower linings, including sheet lead, sheet copper, PVC sheet, chlorinated polyethylene sheet and hot mopping. The shower pan is designed to collect any water that leaks through the shower floor.

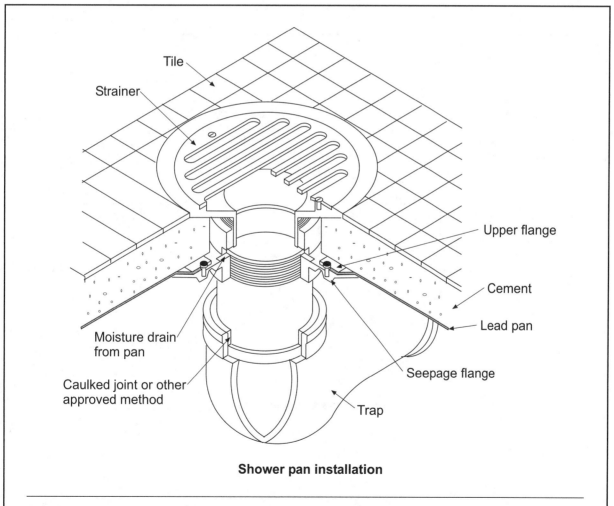

Shower pan installation

The shower pan must connect to the shower drain with a flashing flange with weep holes to allow the leaking water to enter the drainage system. A clamping ring or similar device is used to clamp down the waterproof membrane and make a watertight joint.

Study Session 14   361

Topic: Sink and Dishwasher

Category: Plumbing Fixtures

Reference: IRC P2717.2, P2717.3

Subject: Dishwashing Machines

Code Text: *A sink and dishwasher are permitted to discharge through a single $1^1/_2$-inch (38 mm) trap. The discharge pipe from the dishwasher shall be increased to not less than $^3/_4$ inch (19 mm) in diameter and shall be connected with a wye fitting to the sink tailpiece. The dishwasher waste line shall rise and be securely fastened to the underside of the counter before connecting to the sink tailpiece. The combined discharge from a sink, dishwasher and waste grinder is permitted to discharge through a single $1^1/_2$ -inch (38 mm) trap.*

Discussion and Commentary: Dishwashing machines may connect directly to the drainage system. An indirect connection by air gap or air break is not required. The direct connection is acceptable because the potable water supply is adequately protected against backflow.

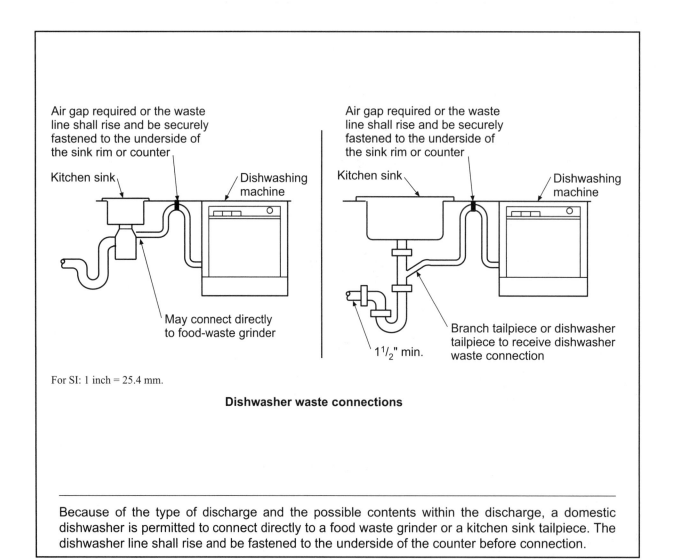

Air gap required or the waste line shall rise and be securely fastened to the underside of the sink rim or counter

Kitchen sink

Dishwashing machine

May connect directly to food-waste grinder

Air gap required or the waste line shall rise and be securely fastened to the underside of the sink rim or counter

Kitchen sink

Dishwashing machine

Branch tailpiece or dishwasher tailpiece to receive dishwasher waste connection

$1^1/_2$" min.

For SI: 1 inch = 25.4 mm.

Dishwasher waste connections

Because of the type of discharge and the possible contents within the discharge, a domestic dishwasher is permitted to connect directly to a food waste grinder or a kitchen sink tailpiece. The dishwasher line shall rise and be fastened to the underside of the counter before connection.

Code Text: *Access shall be provided to circulation pumps in accordance with the fixture or pump manufacturer's installation instructions. Where the manufacturer's instructions do not specify the location and minimum size of field fabricated access openings, an opening of not less than 12-inches by 12-inches (304 mm by 304 mm) shall be installed for access to the circulation pump. A door or panel shall be permitted to close the opening.*

Discussion and Commentary: In order to allow for adequate access to the circulation pump of a whirlpool bathtub, an access opening must be provided. Although a minimum size of 12 inches by 12 inches is established by the code, the opening must also be large enough to permit the removal and replacement of the pump.

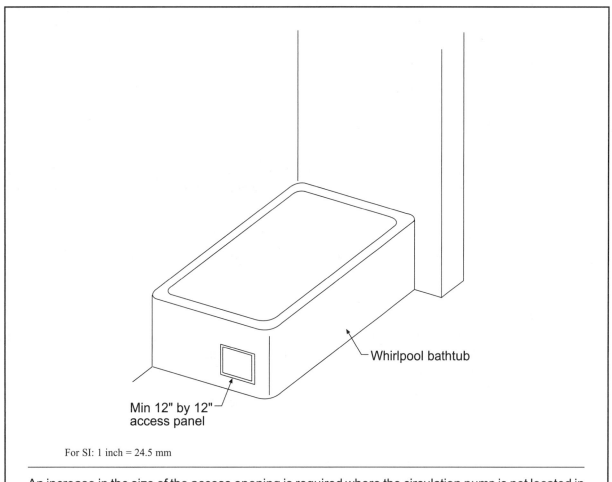

Whirlpool bathtub

Min 12" by 12" access panel

For SI: 1 inch = 24.5 mm

An increase in the size of the access opening is required where the circulation pump is not located in close proximity to the opening. A minimum 18-inch by 18-inch access openings is mandated where the circulation pump is located more than 2 feet from the access opening.

Quiz

Study Session 14
IRC Chapters 25, 26 and 27

1. Testing of the building sewer shall be accomplished with a minimum _____ head of water with the ability to maintain such pressure for a minimum of _____.

    a. 8-foot, 10 minutes

    b. 8-foot, 15 minutes

    c. 10-foot, 10 minutes

    d. 10-foot, 15 minutes

    Reference _____

2. When tested by air, the DWV system shall be tested at _____ or 10 inches of mercury column with the ability to maintain such pressure for a minimum of _____ .

    a. 5 psi, 10 minutes

    b. 5 psi, 15 minutes

    c. 10 psi, 10 minutes

    d. 10 psi, 15 minutes

    Reference _____

3. The water-supply system shall be proved tight under a water pressure not less than the working pressure of the system, or by a minimum _____ air test for other than plastic piping.

    a. 5 psi

    b. 10 psi

    c. 20 psi

    d. 50 psi

    Reference _____

4. Where copper piping is to pass through holes bored in wood studs, the pipe shall be protected by shield plates except for those holes that are a minimum of _____ from the nearest edge.

    a. $^5/_8$ inch   b. $^3/_4$ inch

    c. $1^3/_8$ inches   d. $1^1/_2$ inches

Reference _____

5. Shield plates required to protect piping installed through notches or bored holes in framing members shall extend a minimum of _____ above sole plates and below top plates.

    a. $^1/_2$ inch   b. 1 inch

    c. 2 inches   d. 3 inches

Reference _____

6. Unless provided with a relieving arch, a building drain passing through a foundation wall shall pass through a pipe sleeve sized _____ greater than the pipe passing through.

    a. 1 inch   b. 2 inches

    c. one pipe size   d. two pipe sizes

Reference _____

7. Where buried, water service pipe shall be installed a minimum of _____ deep.

    a. 6 inches   b. 12 inches

    c. 24 inches   d. 30 inches

Reference _____

8. Where a frost line is established at 30 inches, water service pipe shall be installed a minimum of _____ below grade.

    a. 30 inches   b. 32 inches

    c. 36 inches   d. 42 inches

Reference _____

9. Where unstable soil conditions exist in an area to be trenched for a piping installation, the trench shall be over excavated by a minimum of _____ .

    a. 4 inches
    b. 6 inches
    c. one pipe diameter
    d. two pipe diameters

    Reference _____

10. The backfilling of a trench where piping is installed shall be done in maximum _____ layers when using loose earth.

    a. 4-inch
    b. 6-inch
    c. 8-inch
    d. 12-inch

    Reference _____

11. One-half-inch copper water pipe installed horizontally shall be supported at maximum intervals of _____ .

    a. 3 feet
    b. 4 feet
    c. 6 feet
    d. 12 feet

    Reference _____

12. PB pipe shall be supported at maximum intervals of _____ when installed horizontally.

    a. 32 inches
    b. 36 inches
    c. 60 inches
    d. 72 inches

    Reference _____

13. A fixture tail piece for a laundry tub shall be a minimum of _____ in diameter.

    a. 1 inch
    b. $1^1/_4$ inches
    c. $1^1/_2$ inches
    d. 2 inches

    Reference _____

14. Unless provided with an approved arrangement for access purposes, a fixture with a concealed slip-joint connection shall be provided with an access panel or utility space a minimum of _____ in its smallest dimension.

   a. 12 inches   b. 18 inches

   c. 30 inches   d. 36 inches

   Reference _____

15. Standpipes shall extend a minimum of _____ and a maximum of _____ above the trap weir.

   a. 18 inches, 30 inches   b. 18 inches, 42 inches

   c. 24 inches, 30 inches   d. 24 inches, 42 inches

   Reference _____

16. The outlet of a laundry tray shall be a maximum horizontal distance of _____ from the standpipe trap.

   a. 18 inches   b. 30 inches

   c. 42 inches   d. 60 inches

   Reference _____

17. In general, shower compartments shall have a minimum interior cross-sectional area of _____ square inches.

   a. 900   b. 1,080

   c. 1,296   d. 1,440

   Reference _____

18. A water closet shall be set a minimum of _____ inches from its center to any side wall, partition or vanity.

   a. 15   b. 16

   c. 18   d. 20

   Reference _____

19. A lavatory shall have a waste outlet a minimum of _____ in diameter.

    a. 1 inch   b. $1^1/_4$ inches

    c. $1^1/_2$ inches   d. 2 inches

Reference _____

20. Unless an approved alternative is provided, flush valve seats in tanks for flushing water closets shall be a minimum of _____ above the flood-level rim of the bowl to which it is connected.

    a. $^1/_2$ inch   b. 1 inch

    c. $1^1/_2$ inches   d. 2 inches

Reference _____

21. The waste outlet and overflow outlet for a bathtub shall be connected to waste tubing or piping a minimum of _____ in diameter.

    a. 1 inch   b. $1^1/_4$ inches

    c. $1^1/_2$ inches   d. 2 inches

Reference _____

22. Sinks shall have a waste outlet a minimum of _____ in diameter.

    a. $1^1/_4$ inches   b. $1^1/_2$ inches

    c. 2 inches   d. $3^1/_2$ inches

Reference _____

23. Food waste grinders shall be connected to a drain having a minimum diameter of _____ .

    a. $1^1/_4$ inches   b. $1^1/_2$ inches

    c. 2 inches   d. $3^1/_2$ inches

Reference _____

24. Floor drains shall have waste outlets a minimum of _____ in diameter.

    a. $1^1/_4$ inches   b. $1^1/_2$ inches

    c. 2 inches   d. $3^1/_2$ inches

Reference _____

25. What is the minimum size drain required for a macerating toilet system?

    a. $^3/_4$ inch   b. $1^1/_4$ inches

    c. $1^1/_2$ inches   d. 2 inches

    Reference _____

26. Where a smoke test is utilized to verify gas tightness of the DWV system, a pressure equivalent to a 1-inch water column shall be applied and maintained for a minimum time period of _____ minutes.

    a. 10   b. 15

    c. 30   d. 60

    Reference _____

27. Where an air test of the rough plumbing installation requires a gauge pressure of 5 psi, the testing gauge shall have increments of _____ psi.

    a. 0.10   b. 0.50

    c. 1.00   d. 2.00

    Reference _____

28. Trenches used for the installation of piping that are excavated parallel to footings shall extend a maximum of _____ below the bearing plane of the bottom edge of the footing.

    a. 30 degrees   b. 45 degrees

    c. 12 inches   d. 30 inches

    Reference _____

29. Each compartment of a laundry tub shall be provided with a waste outlet a minimum of _____ in diameter.

    a. 1 inch   b. $1^1/_4$ inch

    c. $1^1/_2$ inch   d. 2 inches

    Reference _____

30. There shall be a minimum of _____ inches clearance in front of a water closet to any wall, fixture or door.

    a. 18   b. 21

    c. 24   d. 27

Reference _____

31. The access and egress opening for a shower compartment shall have a minimum and unobstructed clear width of _____ inches.

    a. 22   b. 24

    c. 30   d. 32

Reference _____

32. A water-temperature-limiting device shall be provided to limit the hot water supplied to a bathtub or whirlpool bathtub to a maximum of _____ °F.

    a. 110   b. 115

    c. 120   d. 125

Reference _____

33. Where a sink and a dishwasher discharge through a single $1^1/_2$-inch trap, the discharge pipe from the dishwasher shall be a minimum of _____ inch in diameter.

    a. $^3/_4$ b. 1

    c. $1^1/_4$ d. $1^1/_2$

Reference _____

34. Where the manufacturer's instructions do not specify the minimum required size of a field-fabricated access opening to the circulation pump of a whirlpool bathtub, a minimum _____ opening shall be installed.

    a. 10-inch by 10-inch   b. 12-inch by 12-inch

    c. 10-inch by 16-inch   d. 12-inch by 18-inch

Reference _____

35. The discharge water temperature from a bidet fitting shall be limited to a maximum of _____ by a water-temperature-limiting device.

    a. 98°F   b. 105°F

    c. 110°F   d. 112°F

Reference _____

Study Session

15

2012 IRC Chapters 28, 29 and 30
Water Heaters, Water Supply and Distribution, and Sanitary Drainage

OBJECTIVE: To provide an understanding of the general provisions for the installation of water heaters, the supply and distribution of water, and the materials, design, construction and installation of sanitary drainage systems.

REFERENCE: Chapters 28, 29 and 30, 2012 *International Residential Code*

KEY POINTS:
- When is a water heater required to be installed in a pan? What type of pan is mandated? How big must it be? How shall it be drained?
- What special condition applies where a water heater having an ignition source is installed in a garage?
- What special conditions are applicable where a combination water heater-space heating system is utilized?
- What type of relief valve or valves is required for a water heater? How is a pressure-relief valve regulated? A temperature-relief valve?
- How is the water heater discharge pipe to be installed? What is the minimum required size of the discharge pipe? Where is the termination point to be located?
- How shall backflow preventers be applied?
- What special conditions exist where both potable and nonpotable water distribution systems are installed? How shall such systems be identified?
- How is the potable water supply system to be protected? What methods of backflow protection are acceptable?
- What is the minimum permitted air gap for various fixtures? Where shall the critical level of an atmospheric-type vacuum breaker be set?
- How must the potable water supply to boilers be protected? To heat exchangers? To lawn irrigation systems? To automatic fire sprinkler systems? To solar systems?
- What are the maximum water consumption flow rates and quantities for various plumbing fixtures and fittings?
- What is the minimum permitted static pressure at the building entrance for water service? The maximum static pressure? How should thermal expansion be addressed?

- What is the minimum size for a water service pipe? How should the water distribution system be sized?

- How should water service pipe be installed? What special requirements apply where the building sewer piping is installed in the same trench? What special concerns are addressed in the installation of water distribution pipe?

- What criteria are to be used in the design and installation of residential fire sprinkler systems? Where are sprinklers required to be installed within the dwelling unit?

- What temperature rating is required for residential sprinklers? In what locations must sprinklers have an intermediate temperature rating?

- How must residential sprinkler piping be supported? What piping materials are acceptable for use? How is the system design flow rate to be determined?

- What types of joints and connections are prohibited in the DWV system? How are approved joints to be made?

- How is the load on the DWV-system piping determined? What type of fittings are to be used?

- Where shall cleanouts be located? What should be done at changes in direction? What clearances are needed in front of cleanouts? What is the minimum size of a cleanout?

- What are the conditions of installation for sewage ejectors or sewage pumps?

Topic: Required Pan
Reference: IRC P2801.5

Category: Water Heaters
Subject: General Provisions

Code Text: *Where a storage tank-type water heater or a hot water storage tank is installed in a location where water leakage from the tank will cause damage, the tank shall be installed in a galvanized steel pan having a material thickness of not less than 0.0236 inch (0.6010 mm)(No. 24 gage), or other pans approved for such use. The pan shall be not less than $1^1/_2$ inches (38 mm) deep and shall be of sufficient size and shape to receive all dripping and condensate from the tank or water heater. The pan shall be drained by an indirect waste pipe having a minimum diameter of $^3/_4$ inch (19 mm).*

Discussion and Commentary: The pan catches water from a leaking tank, a leaking connection to the tank or condensate from the tank, and it may also be used as the indirect waste receptor for the relief valve discharge.

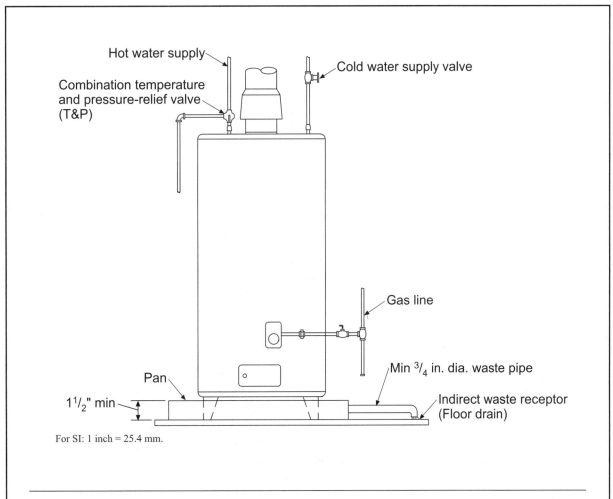

For SI: 1 inch = 25.4 mm.

The pan must not connect directly to the drainage system. An indirect waste pipe is necessary to prevent backflow from the drainage system into the pan. Because the relief valve discharge piping typically terminates in the pan, the waste pipe must be as large as the discharge pipe.

Code Text: *Water heaters having an ignition source shall be elevated such that the source of ignition is not less than 18 inches (457 mm) above the garage floor.* See exception for appliances that are listed as flammable vapor ignition resistant.

Discussion and Commentary: Unless properly elevated, a fuel-burning water heater installed in a garage poses a serious fire hazard. A garage is commonly used as a storage area for gasoline and common household items such as paint, paint thinner and solvents. Fumes from these flammable liquids tend to collect near the floor and produce vapors denser than air. The 18-inch requirement is intended to prevent an energy or heat source from becoming a source of ignition. The ignition source could be a pilot flame, burner, burner igniter or electrical component capable of producing a spark or an arc. The 18-inch elevation must be increased where required by the manufacturer's installation instructions.

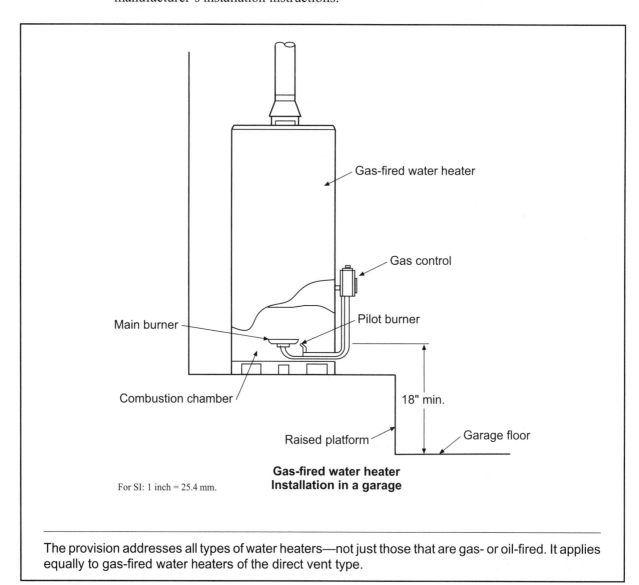

For SI: 1 inch = 25.4 mm.

**Gas-fired water heater
Installation in a garage**

The provision addresses all types of water heaters—not just those that are gas- or oil-fired. It applies equally to gas-fired water heaters of the direct vent type.

Topic: Relief Valves

Reference: IRC P2803.1, P2803.3, P2803.4

Category: Water Heaters

Subject: Relief Valves

Code Text: *Appliances and equipment used for heating water or storing hot water shall be protected by 1) a separate pressure-relief valve and a separate temperature-relief valve; or 2) a combination pressure- and temperature-relief valve. Pressure-relief valves . . . shall be set to open at not less than 25 psi (172 kPa) above the system pressure but not over 150 psi (1034 kPa). The relief-valve setting shall not exceed the tank's rated working pressure. Temperature-relief valves . . . shall be installed such that the temperature-sensing element monitors the water within the top 6 inches (152 mm) of the tank. The valve shall be set to open at a temperature of not greater than 210°F (99°C).*

Discussion and Commentary: A combination temperature- and pressure-relief valve or separate temperature-relief and pressure-relief valves protect the water heater against possible explosion.

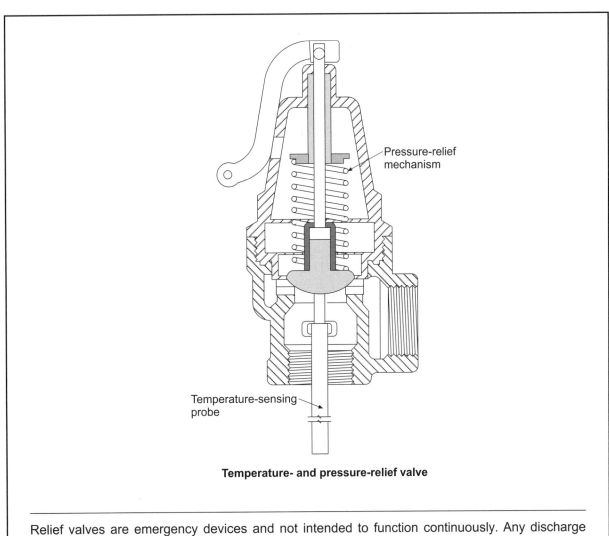

Pressure-relief mechanism

Temperature-sensing probe

Temperature- and pressure-relief valve

Relief valves are emergency devices and not intended to function continuously. Any discharge should not go unnoticed, because when a relief valve discharges, it indicates that something is wrong with the system. The discharge point should always be readily observable.

Code Text: *A potable water supply system shall be designed and installed in such a manner as to prevent contamination from nonpotable liquids, solids or gases being introduced into the potable water supply. Connections shall not be made to a potable water supply in a manner that could contaminate the water supply or provide a cross-connection between the supply and source of contamination except where approved methods are installed to protect the potable water supply. Cross-connections between an individual water supply and a potable public water supply shall be prohibited.*

Discussion and Commentary: Arguably, the most important aspect of the plumbing provisions of the IRC is the protection of potable water systems. History is filled with local and widespread occurrences of sickness and disease caused by not safeguarding the water supply. It is imperative that the potable water supply be maintained in a safe-for-drinking condition at all times and at all outlets.

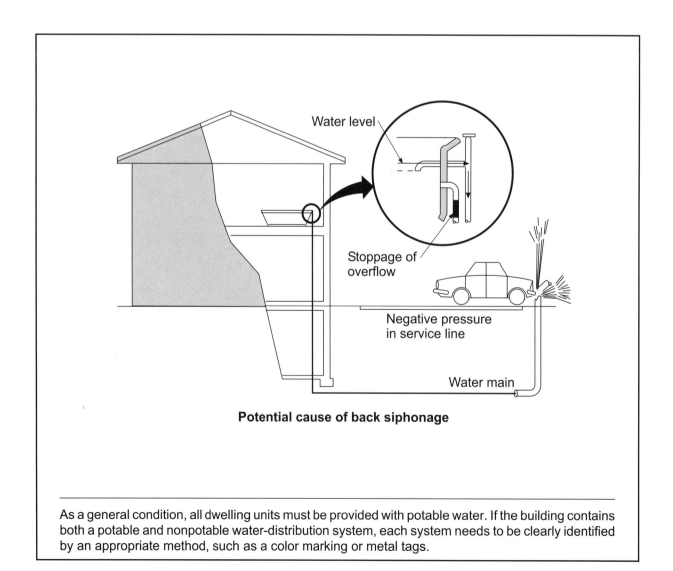

Water level

Stoppage of overflow

Negative pressure in service line

Water main

Potential cause of back siphonage

As a general condition, all dwelling units must be provided with potable water. If the building contains both a potable and nonpotable water-distribution system, each system needs to be clearly identified by an appropriate method, such as a color marking or metal tags.

Code Text: *The minimum air gap shall be measured vertically from the lowest end of a water supply outlet to the flood level rim of the fixture or receptor into which such potable water outlets discharge. The minimum required air gap shall be twice the diameter of the effective opening of the outlet, but in no case less than the values specified in Table P2902.3.1. An air gap is required at the discharge point of a relief valve or piping.*

Discussion and Commentary: An air gap is not a device and has no moving parts. It is, therefore, the most effective and dependable method of preventing backflow. The potable water opening or outlet is terminated at an elevation above the level of the source of contamination. If the appropriate minimum air gap is provided, the potable water could only be contaminated if the entire area or room flooded to a depth that would submerge the potable water opening or outlet.

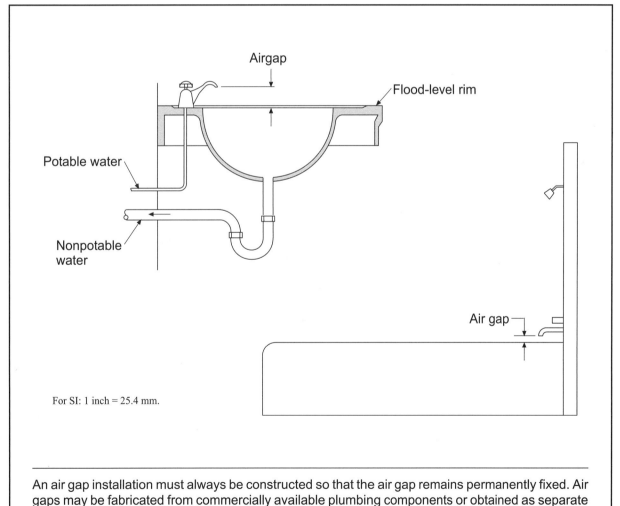

For SI: 1 inch = 25.4 mm.

An air gap installation must always be constructed so that the air gap remains permanently fixed. Air gaps may be fabricated from commercially available plumbing components or obtained as separate units and integrated into plumbing and piping systems.

Code Text: *The water service and water distribution systems shall be designed and pipe sizes shall be selected such that under conditions of peak demand, the capacities at the point of outlet discharge shall not be less than shown in Table P2903.1. The maximum water consumption flow rates and quantities for all plumbing fixtures and fixture fittings shall be in accordance with Table P2903.2.*

Discussion and Commentary: The water distribution system must be capable of supplying a sufficient volume of potable water at pressures adequate to enable all plumbing fixtures, devices and appurtenances to function properly and without undue noise. Each fixture has a given flow rate and minimum water supply pressure needed to properly operate the fixture.

TABLE P2903.1
REQUIRED CAPACITIES AT
POINT OF OUTLET DISCHARGE

FIXTURE AT POINT OF OUTLET	FLOW RATE (gpm)	FLOW PRESSURE (psi)
Bathtub, pressure-balanced or thermostatic mixing valve	4	20
Bidet, thermostatic mixing	2	20
Dishwasher	2.75	8
Laundry tub	4	8
Lavatory	2	8
Shower, pressure-balancing or thermostatic mixing valve	3	20
Shower, temperature controlled	3	20
Sillcock, hose bibb	5	8
Sink	2.5	8
Water closet, flushometer tank	1.6	20
Water closet, tank, close coupled	3	20
Water closet, tank, one-piece	6	20

For SI: 1 gallon per minute = 3.785 L/m,
1 pound per square inch = 6.895 kPa.

TABLE P2903.2
MAXIMUM FLOW RATES AND CONSUMPTION FOR
PLUMBING FIXTURES AND FIXTURE FITTINGS[b]

PLUMBING FIXTURE OR FIXTURE FITTING	PLUMBING FIXTURE OR FIXTURE FITTING
Lavatory faucet	2.2 gpm at 60 psi
Shower head[a]	2.5 gpm at 80 psi
Sink faucet	2.2 gpm at 60 psi
Water closet	1.6 gallons per flushing cycle

For SI: 1 gallon per minute = 3.785 L/m,
1 pound per square inch = 6.895 kPa.
a. A handheld shower spray is also a shower head.
b. Consumption tolerances shall be determined from referenced standards.

The fixtures listed in Table P2903.2 function in a cyclical manner or consume water continuously for the duration of the fixture use. Water is conserved by limiting the flow rate for manually controlled fixtures and by limiting the volume per cycle usage for cyclically operating fixtures.

Code Text: *For water service systems sizes up to and including 2 inches (51 mm), a device for controlling pressure shall be installed where, because of thermal expansion, the pressure on the downstream side of a pressure-reducing valve exceeds the pressure-reducing valve setting. Where a backflow prevention device, check valve or other device is installed on a water supply system using storage water heating equipment such that thermal expansion causes an increase in pressure, a device for controlling pressure shall be installed.*

Discussion and Commentary: In the connection of a water distribution system to water-heating appliances there is the potential for the migration of heated water into the water distribution piping. In a typical water distribution system, the water will expand into the water service and into the public main if the water is not withdrawn from the system at an outlet or connection. If the expansion of water is not accommodated in the system, dangerously high pressures can develop that can cause damage to piping, components and the water heater.

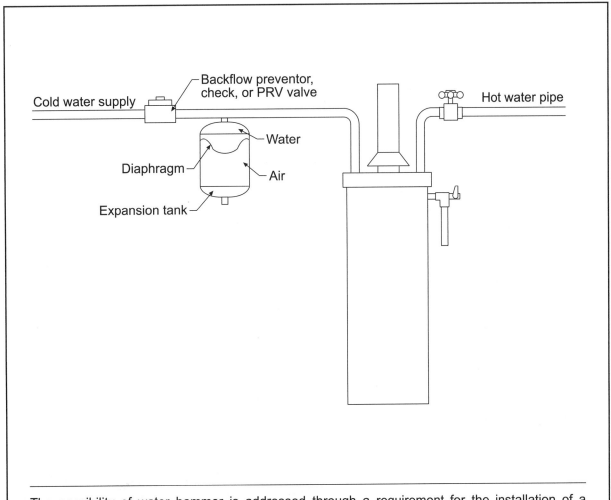

The possibility of water hammer is addressed through a requirement for the installation of a water-hammer arrester where quick-closing valves are used.

Code Text: *Supply loads in the building water-distribution system shall be determined by total load on the pipe being sized, in terms of water-supply fixture units (w.s.f.u.), as shown in Table P2903.6, and gallon per minute (gpm) flow rates. For fixtures not listed, choose a w.s.f.u. value of a fixture with similar flow characteristics.*

Discussion and Commentary: A prescriptive method for estimating the water demand for a building has evolved through the years and has been accepted as a standard method of sizing water systems. The method is based upon weighing fixtures in accordance with their water supply load-producing effects on the water distribution system. This approach couples the flow characteristics of the various fixtures with probability curves of simultaneous flushing of the fixtures.

TABLE P2903.6
WATER-SUPPLY FIXTURE-UNIT VALUES FOR VARIOUS PLUMBING FIXTURES AND FIXTURE GROUPS

TYPE OF FIXTURES OR GROUP OF FIXTURES	WATER-SUPPLY FIXTURE-UNIT VALUE (w.s.f.u.)		
	Hot	Cold	Combined
Bathtub (with/without overhead shower head)	1.0	1.0	1.4
Clothes washer	1.0	1.0	1.4
Dishwasher	1.4	—	1.4
Full-bath group with bathtub (with/without shower head) or shower stall	1.5	2.7	3.6
Half-bath group (water closet and lavatory)	0.5	2.5	2.6
Hose bibb (sillcock)[a]	—	2.5	2.5
Kitchen group (dishwasher and sink with/without garbage grinder)	1.9	1.0	2.5
Kitchen sink	1.0	1.0	1.4
Laundry group (clothes washer standpipe and laundry tub)	1.8	1.8	2.5
Laundry tub	1.0	1.0	1.4
Lavatory	0.5	0.5	0.7
Shower stall	1.0	1.0	1.4
Water closet (tank type)	—	2.2	2.2

For SI: 1 gallon per minute = 3.785 L/m.

a. The fixture unit value 2.5 assumes a flow demand of 2.5 gpm, such as for an individual lawn sprinkler device. If a hose bibb/sill cock will be required to furnish a greater flow, the equivalent fixture-unit value may be obtained from this table or Table P2903.6(1).

The water service pipe must be sized in accordance with Section P2903.7, but in no case may it be less in size than $^3/_4$ inch. Using a prescriptive approach, the size of the water distribution system including the service pipe, meter and main distribution pipe can be determined.

Code Text: *Each dwelling unit shall be provided with an accessible main shutoff valve near the entrance of the water service. The valve shall be of a full-open type having nominal restriction to flow, with provision for drainage such as a bleed orifice or installation of a separate drain valve. A readily accessible full-open valve shall be installed in the cold-water supply pipe to each water heater at or near the water heater. Valves serving individual fixtures, appliances, risers and branches shall be provided with access. An individual shutoff valve shall be required on the fixture supply pipe to each plumbing fixture other than bathtubs and showers.*

Discussion and Commentary: All gate valves and *rated* globe valves are fullway. Gate valves are easily identified by observing the stem when opening the valve. All gate valve stems have stationary stems, unlike globe valves, which have rising stems. Nonrated globe valves are not fullway and are referred to as compression stops.

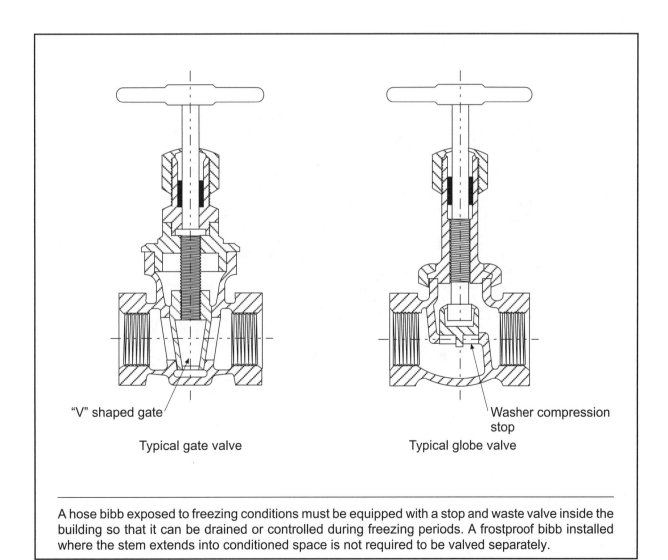

"V" shaped gate

Typical gate valve

Washer compression stop

Typical globe valve

A hose bibb exposed to freezing conditions must be equipped with a stop and waste valve inside the building so that it can be drained or controlled during freezing periods. A frostproof bibb installed where the stem extends into conditioned space is not required to be valved separately.

Code Text: *The design and installation of residential fire sprinkler systems shall be in accordance with NFPA 13D or Section P2904, which shall be considered equivalent to NFPA 13D. Section P2904 shall apply to stand-alone and multipurpose wet-pipe sprinkler systems that do not include the use of antifreeze. A multipurpose fire sprinkler system shall supply domestic water to both fire sprinklers and plumbing fixtures. A stand-alone sprinkler system shall be separate and independent from the water distribution system. A backflow flow preventer shall not be required to separate a stand-alone sprinkler system from the water distribution system.*

Discussion and Commentary: A simple, prescriptive approach to the design of dwelling fire sprinkler systems is provided as an alternate to NFPA 13R. Sprinkler protection, as required by Section R313.1, is intended to allow for the detection and control of fires in residential occupancies and is expected to prevent total fire involvement (flashover) in the room of fire origin.

Section P2904.1 requires the use of new sprinklers listed for residential applications. They are equipped with a thermal element using either a fusible link or fangible bulb filled with conductive liquid that is designed to operate approximately five times faster when compared with a standard spray sprinkler required by NFPA 13.

Topic: Required Locations	**Category:** Water Supply and Distribution
Reference: IRC P2904.1.1	**Subject:** Fire Sprinkler Systems

Code Text: *Sprinklers shall be installed to protect all areas of a dwelling unit.* See exceptions for 1) attics, crawl spaces and other normally unoccupied concealed spaces, 2) gypsum-board-covered closets and pantries not more than 24 square feet in area with the smallest dimension not greater than 3 feet, 3) bathrooms not more than 55 square feet in area, and 4) garages, carports, exterior porches, unheated entry areas such as mud rooms adjacent to an exterior door, and similar spaces.

Discussion and Commentary: Sprinkler protection is not required in those areas of a dwelling unit where statistical evidence has shown through fire incident loss data to not significantly contribute to injuries or death. Sprinklers may be omitted in small closets that are enclosed with gypsum board walls and ceilings, as well as bathrooms no more than 55 square feet in area. Sprinklers are also not required in typically unconditioned areas not intended for living purposes, such as garages, porches, attics and crawl spaces.

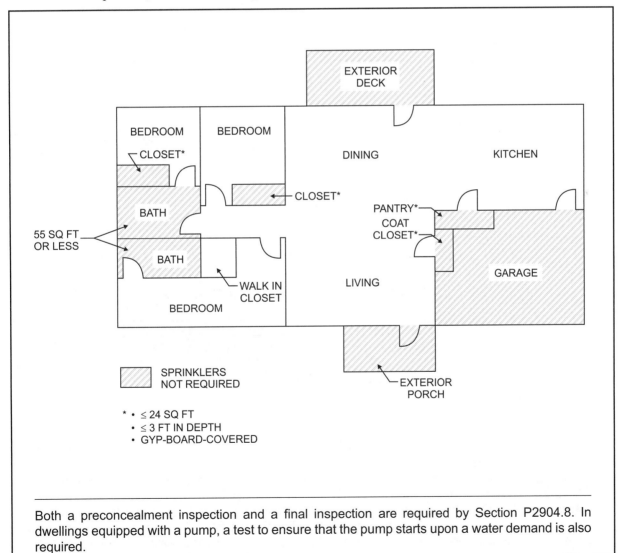

Both a preconcealment inspection and a final inspection are required by Section P2904.8. In dwellings equipped with a pump, a test to ensure that the pump starts upon a water demand is also required.

Code Text: *Except as provided for in Section P2904.2.2* (intermediate temperature sprinklers), *sprinklers shall have a temperature rating of not less than 135°F (57°C) and not more than 170°F (77°C). Sprinklers shall have an intermediate temperature rating not less than 175°F (79°C) and not more than 225°F (107°C) where installed in the following locations: 1) directly under skylights, where the sprinkler is exposed to direct sunlight; 2) in attics; 3) in concealed spaces located directly beneath a roof; and 4) within the distance to a heat source as specified in Table P2904.2.2 Piping shall be protected from freezing as required by Section P2603.6. Where sprinklers are required in areas that are subject to freezing, dry-sidewall or dry-pendent sprinklers extending from a nonfreezing area into a freezing area shall be installed.*

Discussion and Commentary: In addition to establishing the required temperature rating of sprinklers in Section P2904.2, sprinkler coverage requirements are set forth. The coverage area shall be based on the sprinkler listing and the sprinkler manufacturer's installation instructions, but in no case shall the area of coverage for a single sprinkler exceed 400 square feet.

TABLE P2904.2.2
LOCATIONS WHERE INTERMEDIATE TEMPERATURE SPRINKLERS ARE REQUIRED

HEAT SOURCE	RANGE OF DISTANCE FROM HEAT SOURCE WITHIN WHICH INTERMEDIATE TEMPERATURE SPRINKLERS ARE REQUIRED[a, b] (inches)
Fireplace, side of open or recessed fireplace	12 to 36
Fireplace, front of recessed fireplace	36 to 60
Coal and wood burning stove	12 to 42
Kitchen range top	9 to 18
Oven	9 to 18
Vent connector or chimney connector	9 to 18
Heating duct, not insulated	9 to 18
Hot water pipe, not insulated	6 to 12
Side of ceiling or wall warm air register	12 to 24
Front of wall mounted warm air register	18 to 36
Water heater, furnace or boiler	3 to 6
Luminaire up to 250 watts	3 to 6
Luminaire 250 watts up to 499 watts	6 to 12

For SI: 1 inch = 25.4 mm.

a. Sprinklers shall not be located at distances less than the minimum table distance unless the sprinkler listing allows a lesser distance.
b. Distances shall be measured in a straight line from the nearest edge of the heat source to the nearest edge of the sprinkler.

Intermediate temperature sprinklers are mandated where sprinklers are located in close proximity to the various heat sources identified in Table P2904.2.2. The minimum separation between a sprinkler and the heat source set forth in the table must be maintained unless the sprinkler listing allows for a lesser distance. An intermediate temperature sprinkler is not required where the sprinkler/heat source separation distance exceeds the top end of the range established in the table.

Code Text: *Water-service pipe is permitted to be located in the same trench with a building sewer provided such sewer is constructed of materials listed for underground use with a building in Section P3002.1. If the building sewer is not constructed of materials listed in Section P3002.1, the water-service pipe shall be separated from the building sewer by not less than 5 feet (1524 mm), measured horizontally, of undisturbed or compacted earth or place on a solid ledge not less than 12 inches (305 mm) above and to one side of the highest point in the sewer line.* See exception where water service piping crosses sewer piping.

Discussion and Commentary: The separation of water service from the building sewer is intended to reduce the possibility of contaminating the potable water supply. When there is a leak in the building sewer, the soil becomes contaminated around the water pipe. This would put the potable water supply at risk should the water service pipe have a subsequent failure.

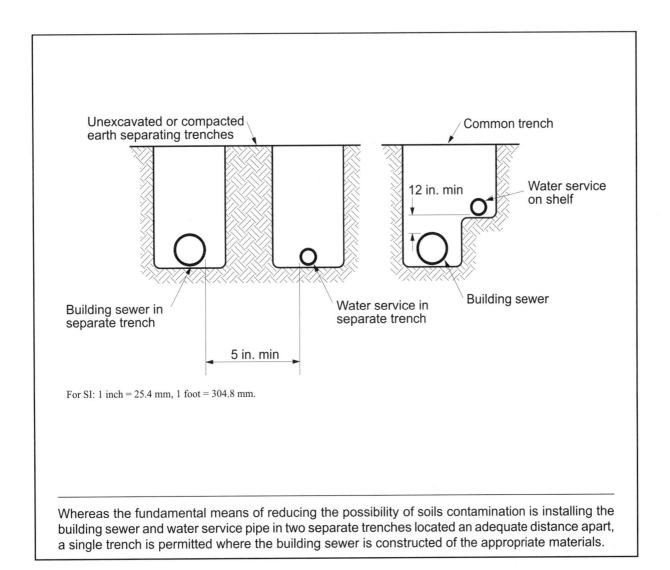

Unexcavated or compacted earth separating trenches

Common trench

12 in. min

Water service on shelf

Building sewer in separate trench

Water service in separate trench

Building sewer

5 in. min

For SI: 1 inch = 25.4 mm, 1 foot = 304.8 mm.

Whereas the fundamental means of reducing the possibility of soils contamination is installing the building sewer and water service pipe in two separate trenches located an adequate distance apart, a single trench is permitted where the building sewer is constructed of the appropriate materials.

Code Text: *Pipe fittings shall be approved for installation with the piping material installed and shall comply with the applicable standards listed in Table P3002.3. Drainage fittings shall have a smooth interior waterway of the same diameter as the piping served. Threaded drainage pipe fittings shall be of the recessed drainage type, black or galvanized. Drainage fittings shall be designed to maintain one-fourth unit vertical in 12 units horizontal (2-percent slope) grade. Section P3002.3.1 shall not be applicable to tubular waste fittings used to convey vertical flow upstream of the trap seal liquid level of a fixture trap.*

Discussion and Commentary: Sanitary, DWV, or water-pipe fittings are acceptable in the dry section of the vent systems. Drainage fittings should have no ledges, shoulders or reductions that can retard or obstruct drainage flow in the piping.

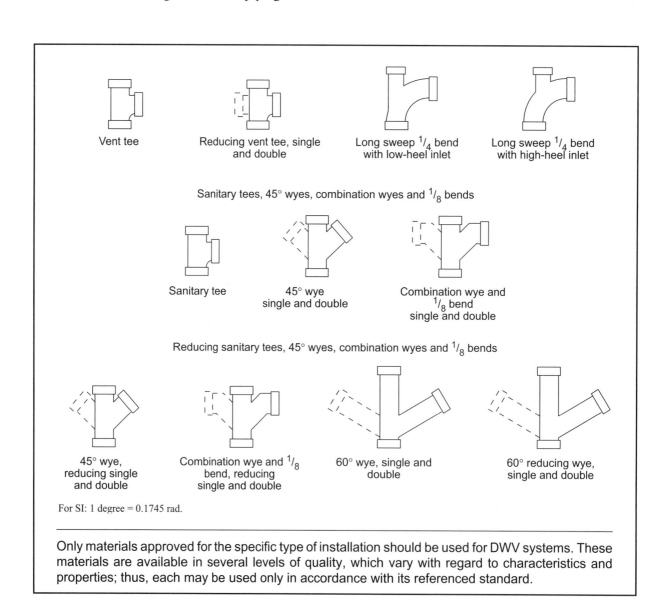

Vent tee

Reducing vent tee, single and double

Long sweep $^1/_4$ bend with low-heel inlet

Long sweep $^1/_4$ bend with high-heel inlet

Sanitary tees, 45° wyes, combination wyes and $^1/_8$ bends

Sanitary tee

45° wye single and double

Combination wye and $^1/_8$ bend single and double

Reducing sanitary tees, 45° wyes, combination wyes and $^1/_8$ bends

45° wye, reducing single and double

Combination wye and $^1/_8$ bend, reducing single and double

60° wye, single and double

60° reducing wye, single and double

For SI: 1 degree = 0.1745 rad.

Only materials approved for the specific type of installation should be used for DWV systems. These materials are available in several levels of quality, which vary with regard to characteristics and properties; thus, each may be used only in accordance with its referenced standard.

Topic: Cleanouts	**Category:** Sanitary Drainage
Reference: IRC P3005.2.2, P3005.2.4	**Subject:** Drainage System

Code Text: Drainage pipe *cleanouts shall be installed not more than 100 feet (30 480 mm) apart in horizontal drainage lines measured from the upstream entrance of the cleanout. Cleanouts shall be installed at each fitting with a change of direction more than 45 degrees (0.79 rad) in the building sewer, building drain and horizontal waste or soil lines. Where more than one change of direction occurs in a run of piping, only one cleanout shall be required in each 40 feet (12 192 mm) of developed length of the drainage piping.*

Discussion and Commentary: Because a cleanout is designed to provide access into the drainage system, the cleanout itself must be accessible, regardless if the piping it serves is concealed or in a location not readily accessible. The cleanout is required to extend up to or above the finished grade either inside or outside the building. A cleanout must be accessible for use wherever it is located in the drainage system and must not be covered by any permanent, nonaccessible construction.

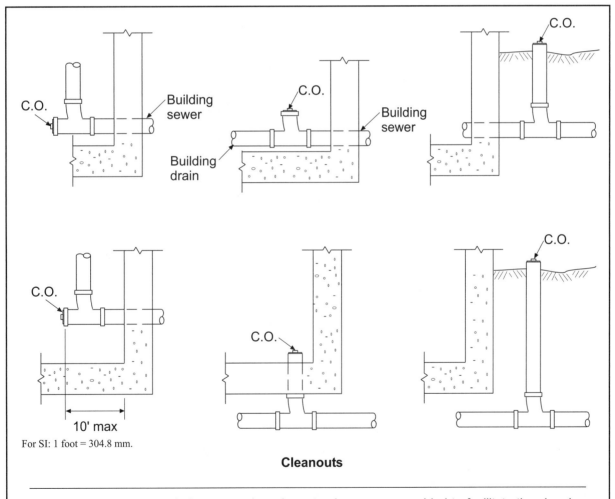

For SI: 1 foot = 304.8 mm.

Cleanouts

For a cleanout to be used, there must be adequate clear space provided to facilitate the cleaning operation. For pipes 3 inches or larger, at least 18 inches must be provided in front of the cleanout, measured perpendicular to the opening. Only 12 inches are needed on smaller pipes.

Code Text: *An offset in a vertical drain, with a change of direction of 45 degrees (0.79 rad) or less from the vertical, shall be sized as a straight vertical drain. A stack with an offset of more than 45 degrees (0.79 rad) from the vertical shall be sized in accordance with three listed conditions. In soil or waste stacks below the lowest horizontal branch, there shall be no change in diameter required if the offset is made at an angle not greater than 45 degrees (0.79 rad) from the vertical. If an offset greater than 45 degrees (0.79 rad) from the vertical is made, the offset and stack below it shall be sized as a building drain (See Table P3005.4.2).*

Discussion and Commentary: Depending on the conditions involved, there are various methods for sizing offsets in drainage piping. The most complicated of the methods involves horizontal offsets above the lowest branch.

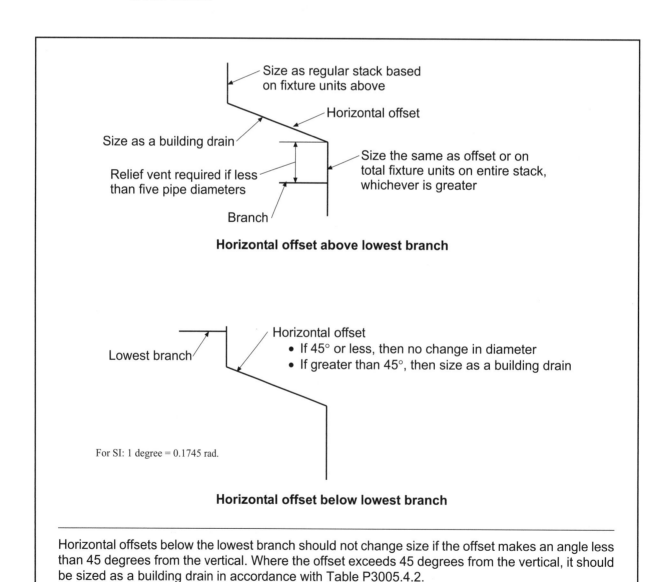

Horizontal offset above lowest branch

Horizontal offset below lowest branch

For SI: 1 degree = 0.1745 rad.

Horizontal offsets below the lowest branch should not change size if the offset makes an angle less than 45 degrees from the vertical. Where the offset exceeds 45 degrees from the vertical, it should be sized as a building drain in accordance with Table P3005.4.2.

Quiz

Study Session 15
IRC Chapters 28, 29 and 30

1. A required overflow pan for a water heater shall be a minimum of _____ in depth and of sufficient size and shape to collect all dripping and condensate.

    a. 1 inch

    b. $1^1/_2$ inches

    c. 2 inches

    d. 3 inches

    Reference _____

2. Where the termination of a drain pan for a water heater extends to the exterior of the building, it shall terminate a minimum of _____ and a maximum of _____ above the adjacent ground surface.

    a. 3 inches, 12 inches

    b. 3 inches, 24 inches

    c. 6 inches, 12 inches

    d. 6 inches, 24 inches

    Reference _____

3. Where installed in a garage, water heaters having an ignition source shall be elevated such that the source of ignition is a minimum of _____ above the garage floor unless the appliance is listed as flammable vapor ignition resistant.

    a. 6 inches

    b. 12 inches

    c. 18 inches

    d. 24 inches

    Reference _____

4. A water heater pressure-relief valve shall be set to open at not less than _____ above the system pressure but not over _____ .

     a. 15 psi, 125 psi   b. 15 psi, 150 psi

     c. 25 psi, 125 psi   d. 25 psi, 150 psi

Reference _____

5. A water heater temperature-relief valve shall be set to open at a maximum temperature of _____ .

     a. 150°F   b. 165°F

     c. 180°F   d. 210°F

Reference _____

6. Where located close to a wall, what is the minimum required air gap for a lavatory with an effective opening of $\frac{1}{2}$ inch?

     a. 1 inch   b. $1\frac{1}{2}$ inches

     c. 2 inches   d. 3 inches

Reference _____

7. A temperature-relief valve installed in a water heater shall be installed such that the temperature-sensing element monitors the water within the top _____ of the tank.

     a. one-sixth   b. one-fourth

     c. 6 inches   d. 12 inches

Reference _____

8. In determining the peak demand for the water service and water distribution service, a flow rate of _____ and a flow pressure of _____ shall be used for a bathtub at the point of outlet discharge.

     a. 3 gpm, 20 psi   b. 4 gpm, 20 psi

     c. 4 gpm, 8 psi   d. 5 gpm, 8 psi

Reference _____

9. A water closet shall have a maximum consumption of _____ gallons per flushing cycle.

    a. 1.0   b. 1.6

    c. 2.0   d. 2.3

Reference _____

10. A handheld shower spray shall have a maximum flow rate of _____ .

    a. 2.2 gpm at 60 psi   b. 2.2 gpm at 80 psi

    c. 2.5 gpm at 60 psi   d. 2.5 gpm at 80 psi

Reference _____

11. What are the minimum and maximum static pressures required for a water service?

    a. 40 psi, 80 psi   b. 40 psi, 100 psi

    c. 50 psi, 80 psi   d. 50 psi, 100 psi

Reference _____

12. When sizing a water distribution system, a tank-type water closet shall have a water-supply fixture unit value of _____ .

    a. 1.0   b. 1.4

    c. 2.2   d. 2.7

Reference _____

13. Where a fire sprinkler system is installed, sprinklers are not required in bathrooms having a maximum floor area of _____ square feet.

    a. 40   b. 50

    c. 55   d. 65

Reference _____

14. Pipe and fittings used in the water supply system shall have a maximum of _____ lead content.

    a. 1 percent   b. 2 percent

    c. 5 percent   d. 8 percent

Reference _____

15. Bends of polyethylene pipe shall have a minimum installed radius of pipe curvature greater than _____ pipe diameters.

    a. 15   b. 20

    c. 25   d. 30

Reference _____

16. Hot-water-distribution piping within dwelling units shall have a minimum pressure rating of _____ at 180°F.

    a. 80 psi   b. 100 psi

    c. 110 psi   d. 125 psi

Reference _____

17. What is the minimum required radius for bends in copper tubing used for water distribution?

    a. four tube diameters   b. six tube diameters

    c. 4 inches   d. 6 inches

Reference _____

18. What is the calculated load on DWV-system piping for two bath groups, a kitchen group, a laundry group and a clothes washer standpipe?

    a. 15 d.f.u.   b. 16 d.f.u.

    c. 18 d.f.u.   d. 20 d.f.u.

Reference _____

19. Drainage pipe cleanouts shall be installed a maximum of _____ apart in horizontal drainage lines.

    a. 40 feet   b. 60 feet

    c. 80 feet   d. 100 feet

Reference _____

20. The minimum size of a cleanout serving a 3-inch pipe shall be _____.

    a. 3 inches   b. $2^1/_2$ inches

    c. 2 inches   d. $1^1/_2$ inches

Reference _____

21. Where more than one change in direction occurs in a run of piping in a drainage system, only one cleanout is required in each _____ of developed length of the drainage piping.

    a. 25 feet   b. 40 feet

    c. 50 feet   d. 75 feet

Reference _____

22. What is the minimum required clearance in front of a cleanout for a 2-inch drainage pipe?

    a. 12 inches   b. 18 inches

    c. 24 inches   d. 30 inches

Reference _____

23. Three-inch horizontal drainage piping shall be installed at a minimum uniform slope of _____ per foot.

    a. $^1/_8$ inch   b. $^3/_{16}$ inch

    c. $^1/_4$ inch   d. $^5/_{16}$ inch

Reference _____

24. Below grade drain pipes shall be a minimum of _____ in diameter.

    a. $1^1/_2$ inches   b. $2^1/_2$ inches

    c. 3 inches   d. 4 inches

Reference _____

25. What is the maximum number of fixture units allowed to be connected to a 3-inch-diameter vertical drain stack?

    a. 10   b. 12

    c. 20   d. 48

Reference _____

26. A water heater discharge pipe serving a pressure-relief valve shall terminate atmospherically a maximum of _____ inch(es) above the floor or waste receptor.

    a. 1   b. 2

    c. 4   d. 6

Reference _____

27. A hose connection backflow preventer shall conform to which of the following applicable standards?

    a. ASSE 1012   b. ASSE 1020

    c. ASSE 1052   d. ASSE 1056

Reference _____

28. Water service pipe installed underground and outside of a structure shall have a minimum working pressure rating of _____ psi at 73°F.

    a. 100   b. 140

    c. 160   d. 175

Reference _____

29. Back-to-back water closet connections to double sanitary tee patterns are permitted provided the horizontal developed length between the outlet of the water closet and the connection to the double sanitary tee is a minimum of _____ inches.

    a. 18   b. 24

    c. 30   d. 36

Reference _____

30. The area of coverage of a single fire sprinkler is limited to a maximum of _____ square feet and shall be based on the sprinkler listing and the manufacturer's installation instructions.

    a. 225   b. 300

    c. 350   d. 400

Reference _____

31. Where chemicals are intended to be introduced into a lawn irrigation system, the potable water supply shall be protected against backflow by a(n) _____ .

    a. atmospheric-type vacuum breaker

    b. double check-valve assembly

    c. pressure-type vacuum breaker assembly

    d. reduced pressure principle backflow prevention assembly

Reference _____

32. Where polyethylene plastic pipe is used in water service piping, bends must occur a minimum of _____ pipe diameters from any fitting or valve.

    a. 4   b. 6

    c. 10   d. 12

Reference _____

33. Galvanized steel DWV pipe used within a building shall be maintained a minimum of _____ inches above ground.

    a. 2   b. 3

    c. 4   d. 6

Reference _____

34. A quarter-bend fitting is permitted for a vertical to horizontal change in direction in drainage piping provided the fixture drain is a maximum of _____ inches in diameter.

    a. $1^1/_4$ b. $1^1/_2$

    c. 2   d. 3

Reference _____

35. Where a fire sprinkler system is installed in a one-story dwelling with a floor area of 2,600 square feet, the water supply shall have the capacity to provide the required design flow rate for sprinklers for a minimum of _____ minutes.

    a. 7   b. 10

    c. 12   d. 15

Reference _____

2012 IRC Chapters 31, 32 and 33
Vents, Traps and Storm Drainage

OBJECTIVE: To provide an understanding of the installation requirements for piping, tubing and fittings for vent systems; vent sizes and connections; fixture traps; and storm drainage.

REFERENCE: Chapters 31, 32 and 33, 2012 *International Residential Code*

KEY POINTS:
- What is the purpose of a vent piping system?
- Where must the required vent pipe terminate? How must it be installed? What is its minimum required size?
- At what minimum height must an open vent pipe terminate above a roof? What if the roof is used for a purpose other than weather protection?
- What is the minimum required size for a vent extension through a roof or wall in areas subject to cold climates? Where must any increase in size occur?
- How must a vent terminal be located in relationship to a door, openable window or similar air intake? To an adjoining building?
- What is the maximum distance allowed between a fixture trap and its vent? The minimum distance required? What is the maximum fall due to pipe slope?
- How many traps or trapped fixtures may be vented by a common vent? How is the connection regulated where the drains connect at the same level? At different levels?
- Under what conditions is a wet vent permitted? What are the limitations of a wet vent? How is a vertical wet vent regulated?
- How is a waste stack to be sized?
- Where is circuit venting acceptable? Where shall the connection be located? What is the maximum slope of the vent section of the horizontal branch drain?
- Where may a combination waste and vent system be used? How shall it be installed? What is the minimum size of the vent pipe?
- What fixtures are permitted to be vented by an island fixture venting system? Where is the connection to be located?
- What is the minimum required diameter of various vents and venting methods?
- How are air admittance valves regulated? Where are such valves permitted? Where should they be located?

KEY POINTS:
(Cont'd)

- How shall traps be designed? How shall they be constructed? What types of traps are prohibited?
- What is the minimum required liquid seal for a trap? The maximum allowable trap seal? What special provisions apply to floor drains?
- Where shall a trap be located in relationship to the fixture outlet? What is the maximum vertical distance permitted between the fixture outlet and the trap weir? The maximum horizontal distance?
- What is the minimum trap size for various plumbing fixtures?
- How are subsoil drains to be installed? To where must they discharge?
- What are the requirements for a sump pump and sump pit?

Topic: Roof Extension

Category: Vents

Reference: IRC P3103.1

Subject: Vent Terminals

Code Text: *Open vent pipes that extend through a roof shall be terminated at least 6 inches (152 mm) above the roof or 6 inches (152 mm) above the anticipated snow accumulation, whichever is greater, except that where a roof is to be used for any purpose other than weather protection, the vent extensions shall be run at least 7 feet (2134 mm) above the roof.*

Discussion and Commentary: At a very minimum, open vent pipes shall extend at least 6 inches above the roof surface. It is anticipated that the roof will be unoccupied, so the only concern is that gases and odors will discharge well above the roof surface. If it is expected that snow will accumulate on the roof at various times, then the height must be increased so that the vent pipes will maintain a height of at least 6 inches above the snow level. This determination is to be made by the building official based on local information. However, if the roof is used as an occupiable space, such as an entertainment deck or observation platform, the termination point must be above the level of individuals occupying the roof.

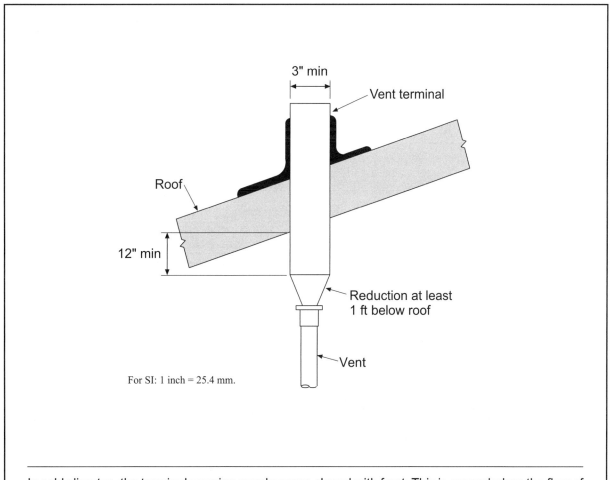

For SI: 1 inch = 25.4 mm.

In cold climates, the terminal opening may become closed with frost. This is caused when the flow of warm, moist air rising through the vent stack comes in contact with the frigid outside air, forming frost on the vent's interior. The chance of closure is reduced by increasing the vent size.

Study Session 16

401

Code Text: *An open vent terminal from a drainage system shall not be located less than 4 feet (1219 mm) directly beneath any door, openable window, or other air intake opening of the building or of an adjacent building, nor shall any such vent terminal be within 10 feet (3048 mm) horizontally of such an opening unless it is not less than 3 feet (914 mm) above the top of such opening.*

Discussion and Commentary: The vent terminal opening should not be located directly beneath any opening where sewer gas and odors can enter the building. Sewer gases are undesirable, unhealthy and potentially explosive. For the same reasons, the proximity of any vent terminal to openings within a 10-foot horizontal distance, whether in the same building or an adjacent building, are also regulated.

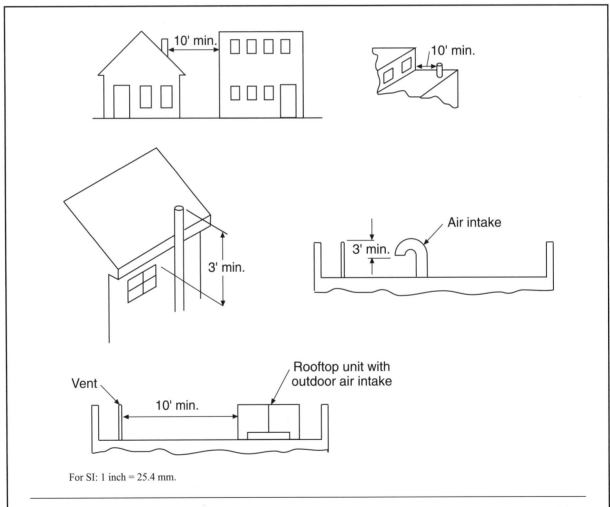

For SI: 1 inch = 25.4 mm.

Where a vent terminal extends through an exterior wall, the termination point must be at least 10 feet from any adjoining lot line. Vertically, the vent terminal must extend at least 10 feet above the highest grade located within 10 feet horizontally.

Code Text: *Each fixture trap shall have a protecting vent located so that the slope and the developed length in the fixture drain from the trap weir to the vent fitting are within the requirements set forth in Table P3105.1. See exception for self-siphoning fixtures. The total fall in a fixture drain resulting from pipe slope shall not exceed one pipe diameter, nor shall the vent pipe connection to a fixture drain, except for water closets, be below the weir of the trap. A vent shall not be installed within two pipe diameters of the trap weir.*

Discussion and Commentary: The length of each trap arm should not exceed that specified in Table P3105.1, measured from the weir of the trap to the vent along the centerline of the pipe. The measured length must include turns and offsets.

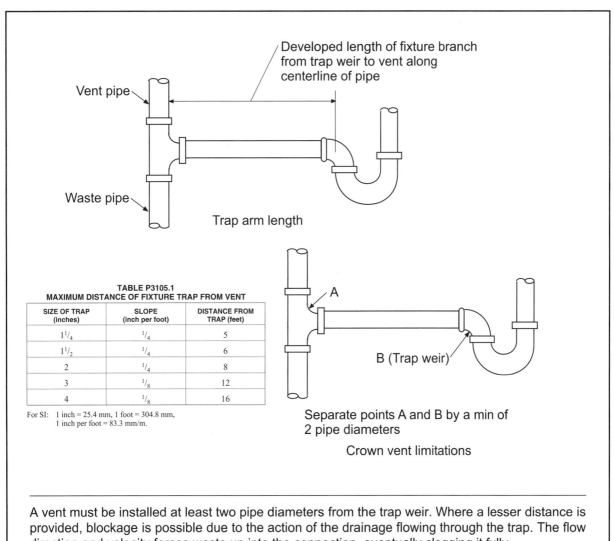

Developed length of fixture branch from trap weir to vent along centerline of pipe

Vent pipe

Waste pipe

Trap arm length

TABLE P3105.1
MAXIMUM DISTANCE OF FIXTURE TRAP FROM VENT

SIZE OF TRAP (inches)	SLOPE (inch per foot)	DISTANCE FROM TRAP (feet)
$1\frac{1}{4}$	$\frac{1}{4}$	5
$1\frac{1}{2}$	$\frac{1}{4}$	6
2	$\frac{1}{4}$	8
3	$\frac{1}{8}$	12
4	$\frac{1}{8}$	16

For SI: 1 inch = 25.4 mm, 1 foot = 304.8 mm, 1 inch per foot = 83.3 mm/m.

A

B (Trap weir)

Separate points A and B by a min of 2 pipe diameters

Crown vent limitations

A vent must be installed at least two pipe diameters from the trap weir. Where a lesser distance is provided, blockage is possible due to the action of the drainage flowing through the trap. The flow direction and velocity forces waste up into the connection, eventually clogging it fully.

Code Text: *Each trap and trapped fixture is permitted to be provided with an individual vent. The individual vent shall connect to the fixture drain of the trap or trapped fixture being vented.*

Discussion and Commentary: The simplest form of venting a trap or trapped fixture is an individual vent for each trap. A single vent pipe is connected between the trap of a fixture and the branch connection to the drainage system. With a properly installed individual vent, only the drainage of the fixture served is flowing past the vent.

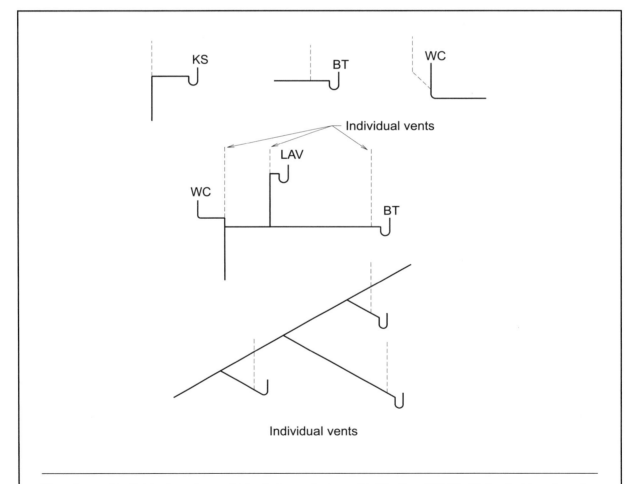

Individual vents

The sizes of individual vents are to be in accordance with Section P3113. Sizing is based on the minimum required diameter of the drain served and the total developed length. Any vent must be a minimum of one-half the diameter of the drain that it serves, with a minimum diameter of $1^1/_4$ inches.

Code Text: *An individual vent is permitted to vent two traps or trapped fixtures as a common vent. The traps or trapped fixtures being common vented shall be located on the same floor level. Where the fixture drains being common vented connect at the same level, the vent connection shall be at the interconnection of the fixture drains or downstream of the interconnection.*

Discussion and Commentary: A common vent installed vertically may be used to protect two fixture traps when the drains connect at the same level on a vertical drain. A typical installation is two horizontal drains connecting together to a vertical drain to a double-pattern fitting. The extension of the vertical pipe serves as the vent. The vent may connect at the interconnection of the fixture drains or downstream along the horizontal drain.

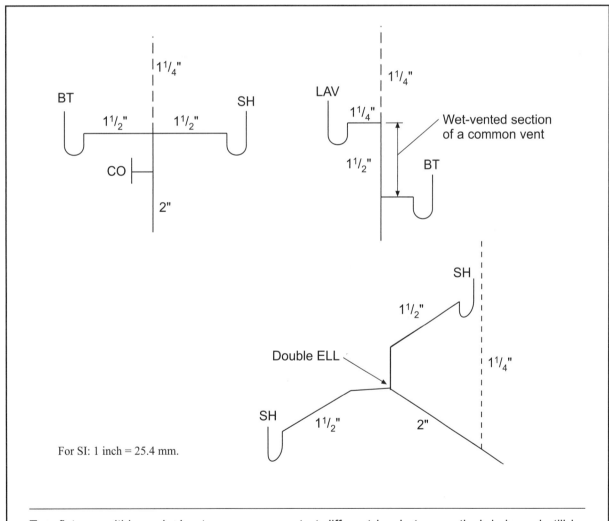

For SI: 1 inch = 25.4 mm.

Two fixtures within a single story may connect at different levels to a vertical drain and still be common-vented, provided the drain pipe between the upper and lower fixtures is oversized. Although drainage flows in the pipe between the two fixtures, the piping is not classified as a wet vent.

Code Text: *Any combination of fixtures within two bathroom groups located on the same floor level are permitted to be vented by a horizontal wet vent. The wet vent shall be considered the vent for the fixtures and shall extend from the connection of the dry vent along the direction of the flow in the drain pipe to the most downstream fixture drain connection.*

Discussion and A wet vent is a vent pipe that is wet because is also conveys drainage. The wet venting
Commentary: concept is based on employing oversized piping to allow for the flow of air above the waste flow. The low probability of simultaneous fixture discharges and the low-flow velocity that results from fixtures on the same floor level provide the necessary degree of reliability that there will always be adequate volume within the wet vent pipe to permit required airflow.

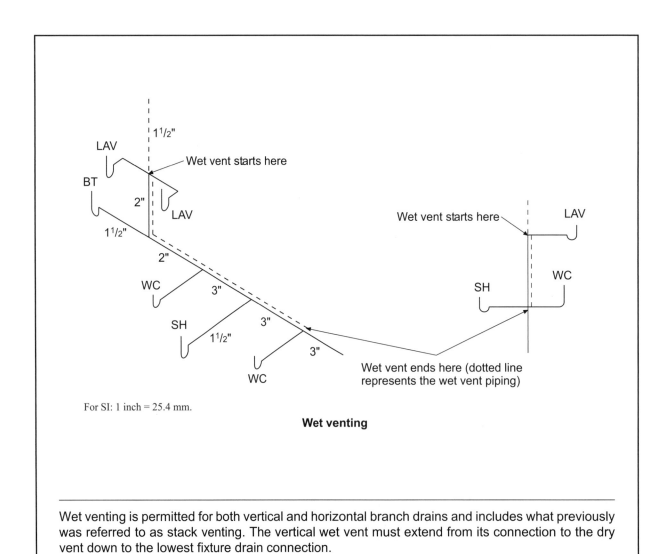

For SI: 1 inch = 25.4 mm.

Wet venting

Wet venting is permitted for both vertical and horizontal branch drains and includes what previously was referred to as stack venting. The vertical wet vent must extend from its connection to the dry vent down to the lowest fixture drain connection.

Topic: Installation
Reference: IRC P3109

Category: Vents
Subject: Waste Stack Vent

Code Text: *A waste stack shall be considered a vent for all of the fixtures discharging to the stack where installed in accordance with the requirements of* Section P3109. *The waste stack shall be vertical, and both horizontal and vertical offsets shall be prohibited between the lowest fixture drain connection and the highest fixture drain connection to the stack.. Every fixture drain shall connect separately to the waste stack. The stack shall not receive the discharge of water closets or urinals. A stack vent shall be installed for the waste stack. The size of the stack vent shall not be less than the size of the waste stack.*

Discussion and Commentary: A waste stack vent employs a waste stack as a vent for the fixtures. The principles of use are based on some of the original research that was done in plumbing. The system has been identified by a variety of names, including vertical wet vent and multi-floor stack venting.

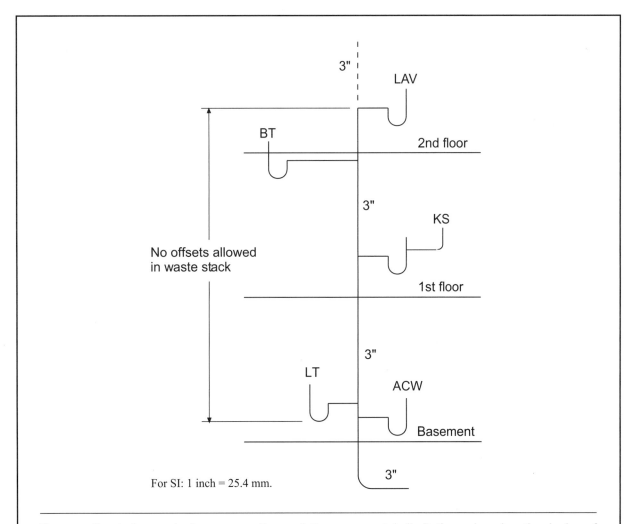

Because the drainage stack serves as the vent, there are certain limitations placed on the design of a waste stack vent. The system is specifically identified as a "waste" stack vent because it prohibits the connection of water closets and urinals. Only "waste" can discharge to the stack.

Code Text: *A maximum of eight fixtures connected to a horizontal branch drain shall be permitted to be circuit vented. Each fixture drain shall connect horizontally to the horizontal branch being circuit vented. The horizontal branch drain shall be classified as a vent from the most downstream fixture drain connection to the most upstream fixture drain connection to the horizontal branch.*

Discussion and Commentary: The principle of circuit venting is that the flow of drainage never exceeds a half-full flow condition. The air for venting the fixtures circulates in the top half of the horizontal branch drainpipe. The flow velocity in the horizontal branch is slow and nonturbulent, thereby preventing pressure differentials from affecting the connecting fixtures.

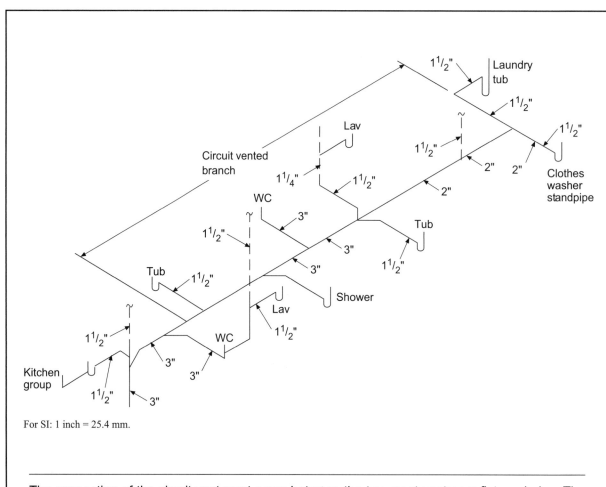

For SI: 1 inch = 25.4 mm.

The connection of the circuit vent must occur between the two most upstream fixture drains. The circuit vent pipe cannot receive the drainage of any soil or waste. In addition, the slope of the vent section of the horizontal branch drain is limited to 1:12 (8-percent slope).

Code Text: *The only vertical pipe of a combination waste and vent system shall be the connection between the fixture drain of a sink, lavatory or floor drain, and the horizontal combination waste and vent pipe. The vertical distance shall be not greater than 8 feet (2438 mm).*

Discussion and Commentary: A combination drain and vent system is a means of extending the distance from a trap to its vent for an unlimited distance because the vent and drain are literally one. The number of fixtures connecting to a combination drain and vent system is also unlimited, provided that the fixtures are floor drains, sinks or lavatories. The system is useful in areas that cannot accommodate vertical vent risers, such as where island sinks are installed or where floor drains are located in large open areas.

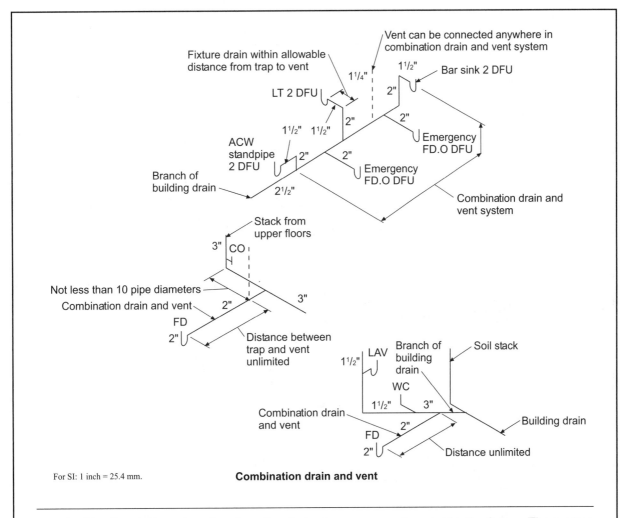

For SI: 1 inch = 25.4 mm.

Combination drain and vent

In a combination drain and vent system, the drain also serves as the vent for the fixture. The system is intended to be a horizontal piping system, with the only vertical piping (limited to 8 feet) being the connection to a sink, lavatory or floor drain located above the level of the vent.

Code Text: *Island fixture venting shall not be permitted for fixtures other than sinks and lavatories. The island fixture vent shall connect to the fixture drain as required for an individual or common vent. The vent shall rise vertically to above the drainage outlet of the fixture being vented before offsetting horizontally or vertically downward. The vent or branch vent for multiple island fixture vents shall extend to a minimum of 6 inches (152 mm) above the highest island fixture being vented before connecting to the outside vent terminal.*

Discussion and Commentary: It is not uncommon for one or more fixtures to be located without the nearby walls or partitions necessary for conventional venting methods to be used. Through the use of loop venting, the loop is typically able to be located below the counter height. The island fixture vent must rise vertically to a height above the fixture drain before it turns down.

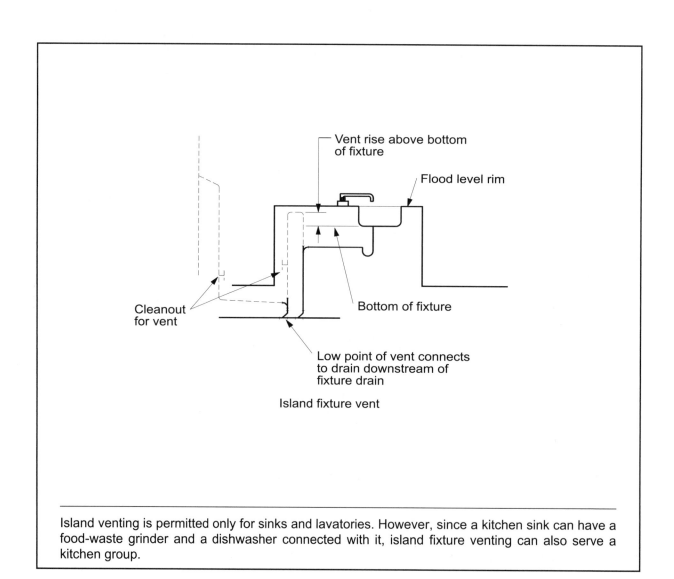

Island venting is permitted only for sinks and lavatories. However, since a kitchen sink can have a food-waste grinder and a dishwasher connected with it, island fixture venting can also serve a kitchen group.

Topic: Location

Category: Vents

Reference: IRC P3114.4

Subject: Air Admittance Valves

Code Text: *Individual and branch air admittance valves shall be located not less than 4 inches (102 mm) above the horizontal branch drain or fixture drain being vented. Stack-type air admittance valves shall be located not less than 6 inches (152 mm) above the flood level rim of the highest fixture being vented. The air admittance valve shall be located within the maximum developed length permitted for the vent. The air admittance valve shall be installed not less than 6 inches (152 mm) above insulation materials where installed in attics.*

Discussion and Commentary: Individual vents, branch vents, circuit vents and stack vents may all terminate with a connection to an air admittance valve rather than extend to the open air. Air enters the upper plumbing drainage system when negative pressures develop in the piping system. The device closes by gravity and seals the vent terminal when the internal pressure is equal to or greater than atmospheric pressure.

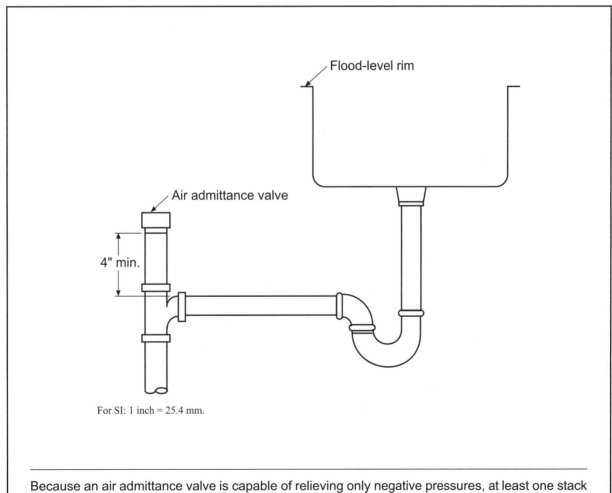

For SI: 1 inch = 25.4 mm.

Because an air admittance valve is capable of relieving only negative pressures, at least one stack vent or vent stack must extend to the outdoors. The vent to the open air serves as the positive pressure relief for the drainage system.

Code Text: *Traps shall have a liquid seal not less than 2 inches (51 mm) and not more than 4 inches (102 mm). Traps for floor drains shall be fitted with a trap primer or shall be of the deep seal design. Traps shall be set level with respect to their water seals and shall be protected from freezing. Trap seals shall be protected from siphonage, aspiration or back pressure by an approved system of venting.*

Discussion and Commentary: A trap is a simple method used to keep sewer gases from emanating out of the drainage system. The water seal prevents the sewer gases and aerosol-borne bacteria from entering the building space. The configuration of a trap interferes with the flow of the drainage; however, the interference is minimal because of the construction of the trap and the relatively high inlet velocity of the waste flow.

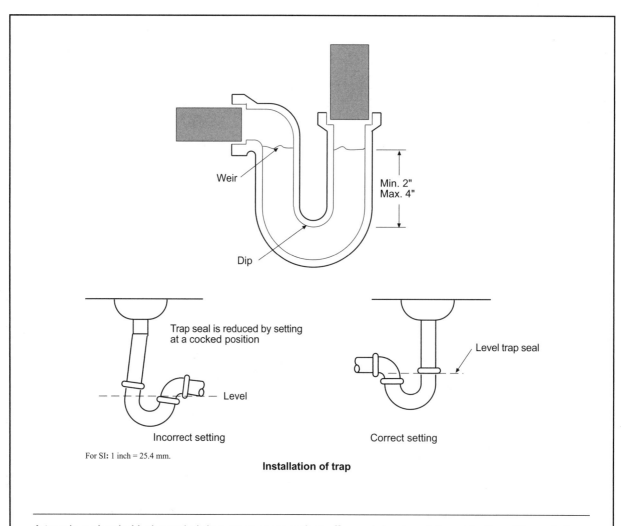

Weir

Min. 2"
Max. 4"

Dip

Trap seal is reduced by setting at a cocked position

Level

Level trap seal

Incorrect setting

Correct setting

For SI: 1 inch = 25.4 mm.

Installation of trap

A trap is a simple U-shaped piping arrangement that offers minimal resistance to flow. The only type of fixture trap permitted is the "P" trap. All other types, such as bell traps, drum traps, "S" traps, and traps with moving parts, have undesirable characteristics and are prohibited.

Code Text: *Each plumbing fixture shall be separately trapped by a water seal trap.* See three conditions where separate trapping is not required. *The vertical distance from the fixture outlet to the trap weir shall not exceed 24 inches (610 mm) and the horizontal distance shall not exceed 30 inches (762 mm) measured from the centerline of the fixture outlet to the centerline of the inlet of the trap. Fixtures shall not be double trapped.*

Discussion and Commentary: The limit on trap location is based on multiple criteria. It is desirable to locate the trap as close as possible to the fixture to minimize the amount of drainpipe on the inlet side of the trap. Buildup on the wall of the fixture outlet pipe will breed bacteria and cause odors to develop. The vertical distance is also limited to control the velocity of the drainage flow. If the trap has too great a vertical separation from the fixture, the velocity of flow at the trap inlet may create self-siphoning of the water in the trap. The horizontal dimension is believed to be the maximum length through which the flow energy is able to move waste.

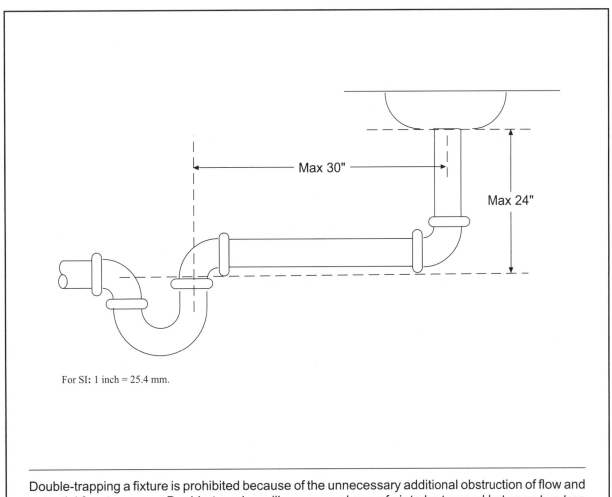

For SI: 1 inch = 25.4 mm.

Double-trapping a fixture is prohibited because of the unnecessary additional obstruction of flow and potential for stoppages. Double-trapping will cause a volume of air to be trapped between two trap seals, and the "air-bound" drain will impede the flow.

Topic: Size of Fixture Traps

Category: Traps

Reference: IRC P3201.7

Subject: Fixture Traps

Code Text: *Fixture trap size shall be sufficient to drain the fixture rapidly and not less than the size indicated in Table P3201.7. A trap shall not be larger than the drainage pipe into which the trap discharges.*

Discussion and Commentary: Where a fixture trap is properly sized, it allows the fixture to drain rapidly and function as intended. Although the code does not regulate for maximum size, an oversized trap will not scour itself and will be prone to clogging. If the trap is larger than the drainage pipe into which it discharges, there is a greater potential for clogging because the reduction does not allow a water velocity that is necessary to scour the trap.

TABLE P3201.7
SIZE OF TRAPS AND TRAP ARMS FOR PLUMBING FIXTURES

PLUMBING FIXTURE	TRAP SIZE MINIMUM (inches)
Bathtub (with or without shower head and/or whirlpool attachments)	$1^1/_2$
Bidet	$1^1/_4$
Clothes washer standpipe	2
Dishwasher (on separate trap)	$1^1/_2$
Floor drain	2
Kitchen sink (one or two traps, with or without dishwasher and garbage grinder)	$1^1/_2$
Laundry tub (one or more compartments)	$1^1/_2$
Lavatory	$1^1/_4$

Shower (based on the total flow rate through showerheads and bodysprays)
Flow rate:
   5.7 gpm and less
   More than 5.7 gpm up to 12.3 gpm
   More than 12.3 gpm up to 25.8 gpm
   More than 25.8 gpm up to 55.6 gpm
	
	$1^1/_2$
	2
	3
	4

Water closet	Note a

For SI: 1 inch = 25.4 mm.

a. Consult fixture standards for trap dimensions of specific bowls.

In Section P3201.6, the provisions of Exception 2 allow a set of not more than three similar fixtures (lavatories, kitchen sinks or laundry tubs) to be drained by one trap. The trap size need not be increased because, in most instances, side-by-side fixtures are either served by a single faucet or do not experience heavy use simultaneously.

Code Text: *The sump pit shall not be less than 18 inches (457 mm) in diameter and 24 inches (610 mm) deep, unless otherwise approved. The pit shall be accessible and located so that all drainage flows into the pit by gravity. The sump pit shall be constructed of tile, steel, plastic, cast-iron, concrete or other approved material, with a removable cover adequate to support anticipated loads in the area of use. The pit floor shall be solid and provide permanent support for the pump.*

Discussion and Commentary: The pit size for a sump is important in the design of the storm drainage system because it determines the amount of water that is stored in the system prior to discharge. It also determines the operation time of the pump for each "on" cycle. The pit must have a removable cover to allow access to the pump and related items such as the check valve.

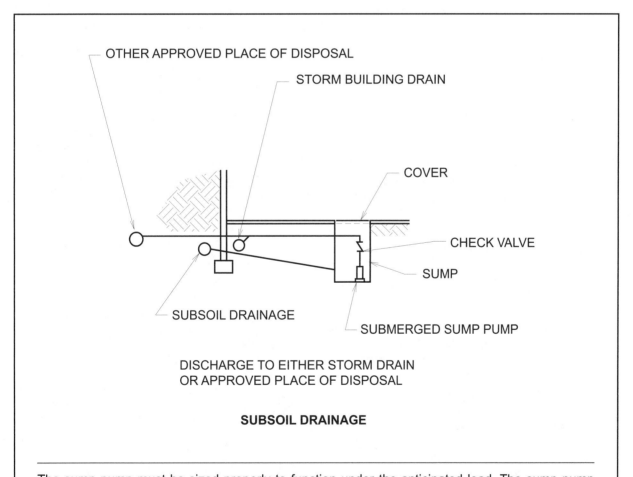

DISCHARGE TO EITHER STORM DRAIN
OR APPROVED PLACE OF DISPOSAL

SUBSOIL DRAINAGE

The sump pump must be sized properly to function under the anticipated load. The sump pump system must be designed to remove the subsoil drainage at the anticipated inflow rate under a worst-case condition. The sizing is determined by evaluating the local ground conditions, the length of piping and the weather in the area.

Study Session 16
IRC Chapters 31, 32 and 33

1. Where the accumulation of snow is not a concern, open vent pipes shall extend a minimum of _____ inches above a roof used solely for weather protection.

    a. 6   b. 12

    c. 18   d. 24

    Reference _____

2. Where a vent extension through a roof is increased in size due to cold climatic conditions, the size increase shall occur inside the structure a minimum of _____ inches below the roof.

    a. 6   b. 12

    c. 24   d. 36

    Reference _____

3. Horizontal vent pipes forming branch vents shall be a minimum of _____ inches above the flood level rim of the highest fixture served.

    a. 3   b. 4

    c. 6   d. 12

    Reference _____

4. Where a roof is to be used for any purpose other than weather protection, the vent extension shall run a minimum of _____ above the roof.

    a. 2 feet   b. 5 feet

    c. 7 feet   d. 10 feet

    Reference _____

5. Unless located a minimum of _____ above the top of the opening, an open vent terminal from a drainage system shall be located at least 10 feet horizontally from a door, openable window or other air intake opening of the building.

    a. 12 inches   b. 24 inches

    c. 36 inches   d. 60 inches

    Reference _____

6. Every dry vent shall rise vertically a minimum of _____ above the flood level rim of the highest trap or trapped fixture being vented.

    a. 1 inch   b. 2 inches

    c. 4 inches   d. 6 inches

    Reference _____

7. What is the maximum permitted developed length of the fixture drain from the trap weir to the vent fitting for a self-siphoning fixture such as a water closet?

    a. 6 feet   b. 8 feet

    c. 12 feet   d. no limit

    Reference _____

8. A maximum distance of _____ is permitted between a $1^1/_2$-inch fixture trap and the vent fitting.

    a. 4 feet   b. 5 feet

    c. 6 feet   d. 8 feet

    Reference _____

9. The total fall in a fixture drain due to pipe slope is limited to a maximum of
_____ .

    a. 1 inch   b. one pipe diameter

    c. 2 inches   d. two pipe diameters

Reference _____

10. A vent shall be installed a minimum of _____ from the trap weir.

    a. 1 inch   b. one pipe diameter

    c. 2 inches   d. two pipe diameters

Reference _____

11. An individual vent may vent a maximum of _____ trap(s) or trap fixture(s) as a common vent.

    a. one   b. two

    c. three   d. four

Reference _____

12. Where fixture drains connect at different levels and the vent connects as a vertical extension of the vertical drain, the vertical drain pipe connecting the two fixture drains shall be considered the vent for the lower fixture drain. If the vent pipe size is 2 inches, what is the maximum permitted discharge from the upper fixture drain?

    a. 1 d.f.u.   b. 4 d.f.u.

    c. 6 d.f.u.   d. 8 d.f.u.

Reference _____

13. A vertical wet vent shall have a minimum pipe size of _____ where serving a total fixture unit load of 9 d.f.u.

    a. 2 inches   b. $2^{1}/_{2}$ inches

    c. 3 inches   d. 4 inches

Reference _____

14. A 3-inch waste stack vent may be used for a maximum total discharge to the stack of
_____ .

      a. 8 d.f.u.   b. 12 d.f.u.

      c. 24 d.f.u.   d. 32 d.f.u.

Reference _____

15. What is the maximum number of fixtures connected to a horizontal branch drain that are permitted to be circuit vented?

      a. two   b. four

      c. six   d. eight

Reference _____

16. Which one of the following fixtures shall not be served by a combination waste and vent system?

      a. sink   b. floor drain

      c. standpipe   d. lavatory

Reference _____

17. The maximum permitted slope of a horizontal combination waste and vent pipe shall be _____ unit vertical in 12 units horizontal.

      a. one-eighth   b. one-fourth

      c. one-half   d. one

Reference _____

18. A $2^1/_2$-inch diameter pipe used as a combination waste and vent shall serve a maximum of _____ fixture units where connected to a horizontal branch.

      a. 3   b. 6

      c. 12   d. 26

Reference _____

19. Vents having a minimum developed length exceeding _____ shall be increased by one nominal pipe size for the entire developed length of the vent pipe.

    a. 25   b. 40

    c. 50   d. 60

Reference _____

20. Where used in a vent system, an individual air admittance valve shall be located a minimum of _____ above the horizontal branch drain or fixture drain being vented.

    a. 1 inch   b. 2 inches

    c. 4 inches   d. 6 inches

Reference _____

21. Traps shall have a minimum liquid seal of _____ inches and a maximum seal of _____ inches.

    a. 2, 4   b. 2, 6

    c. 3, 4   d. 3, 6

Reference _____

22. Where a fixture is separately trapped, what is the maximum vertical distance permitted from the fixture outlet to the trap weir?

    a. 6 inches   b. 12 inches

    c. 18 inches   d. 24 inches

Reference _____

23. Where common trapped fixture outlets are permitted, they shall be located a maximum of _____ apart.

    a. 18 inches   b. 24 inches

    c. 30 inches   d. 36 inches

Reference _____

24. Unless otherwise approved, a sump pit shall be a minimum of _____ inches in diameter and _____ inches deep.

      a. 12, 15   b. 15, 18

      c. 18, 24   d. 24, 36

Reference _____

25. The minimum required size for a trap arm serving a clothes washer standpipe shall be _____ .

      a. $1^1/_4$ inches   b. $1^1/_2$ inches

      c. 2 inches   d. $2^1/_2$ inches

Reference _____

26. An open vent terminal extending through an exterior wall shall terminate a minimum of _____ feet from the lot line.

      a. 4   b. 5

      c. 6   d. 10

Reference _____

27. In circuit venting, the maximum slope of the vent section of the horizontal branch drain shall be _____ unit(s) vertical in 12 units horizontal.

      a. $^1/_4$ b. $^1/_2$

      c. 1   d. 2

Reference _____

28. What is the maximum vertical distance between the fixture drain of a sink and a horizontal combination waste and vent pipe?

      a. 30 inches   b. 4 feet

      c 6 feet   d. 8 feet

Reference _____

29. The vent or branch vent for multiple island fixture vents shall extend a minimum of _____ inch(es) above the highest island fixture being vented before connecting to the outside vent terminal.

    a. 1   b. 2

    c. 4   d. 6

Reference _____

30. Stack-type air admittance valves shall be located a minimum of _____ inches above the flood level rim of the highest fixture being vented.

    a. 4   b. 6

    c. 12   d. 15

Reference _____

31. Offsets in a stack vent installed for the waste stack shall be located a minimum of _____ inches above the flood level of the highest fixture.

    a. 2   b. 4

    c. 6   d. 8

Reference _____

32. Vent pipes shall be a minimum of _____ inches in diameter.

    a. $1^{1}/_{4}$ b. $1^{1}/_{2}$

    c. 2   d. $2^{1}/_{2}$

Reference _____

33. The air pressure relief pipe from a pneumatic sewage ejector shall be a minimum of _____ inch(es) in size.

    a. $^{3}/_{4}$ b. 1

    c. $1^{1}/_{4}$ d. $1^{1}/_{2}$

Reference _____

34. Where the developed length of a sump vent is 110 feet, what is the minimum required vent size for a sump with a sewage pump having a discharge capacity of 60 gallons per minute?

   a. $1^1/_4$ inches

   b. $1^1/_2$ inches

   c. 2 inches

   d. $2^1/_2$ inches

   Reference _____

35. The horizontal distance from a fixture outlet to the trap weir shall be a maximum of _____ inches measured from the centerline of the fixture outlet to the centerline of the trap inlet.

   a. 24

   b. 30

   c. 36

   d. 48

   Reference _____

Study Session

17

2012 IRC Chapters 34, 35, 36 and 37

General Electrical Requirements, Definitions, Services, and Branch Circuits and Feeder Requirements

OBJECTIVE: To obtain an understanding of the general requirements for electrical systems, equipment and components, including provisions addressing service conductors, branch circuits and feeders.

REFERENCE: Chapters 34, 35, 36 and 37, 2012 *International Residential Code*

KEY POINTS:
- What minimum working clearances are required at energized equipment and panelboards? How is clearance above a panelboard regulated? What is the minimum required headroom?
- What materials are permitted for use as conductors? What is the minimum permitted size? How are stranded conductors regulated? Conductors in parallel? How are conductors to be identified?
- What is the difference between "accessible" and "readily accessible"? How do "damp," "dry," and "wet" locations differ? How is "rainproof" different from "rain tight"?
- What are the requirements for "labeled" and "listed" equipment or materials? How do these terms related to that of "identified"?
- Where must the service disconnecting means be located? What is the maximum number of disconnects?
- How is the minimum load for ungrounded service conductors to be calculated?
- What minimum clearances are mandated for overhead service conductors from doors, porches, decks, stairs and balconies? Above roofs? Above pedestrian areas and sidewalks? Over residential property and driveways? Over public streets and alleys or parking areas subject to truck traffic?
- How are connections at service heads to be installed?
- What is the basic method for creating a grounding electrode system? When is the use of a metal underground water pipe permitted for such a such system? A concrete-encased electrode? Ground rings? Plate electrodes?
- What is a "made" electrode? When are such electrodes permitted? What minimum sizes are required for rod and pipe electrodes? How are they to be installed? What is the minimum required resistance to ground? How may this be achieved?

KEY POINTS:
(Cont'd)

- Where is bonding required to ensure electrical continuity? What methods of bonding are permitted?

- What are the rating requirements for branch circuits serving lighting units and general utilization equipment? Fixed utilization equipment? Motors? Ranges and cooking appliances? Heating loads? Air-conditioning and heat pumps?

- How many branch circuits are required to serve the kitchen and dining area? The laundry area? Bathrooms?

- How is the size of feeders to be determined? Conductors?

Code Text: *Wood-framed structural members shall not be drilled, notched or altered in any manner except as provided for in the IRC. Electrical installations in hollow spaces, vertical shafts, and ventilation or air-handling ducts shall be made so that the possible spread of fire or products of combustion will not be substantially increased. Electrical penetrations into or through fire-resistance-rated walls, partitions, floors or ceilings shall be protected by approved methods to maintain the fire-resistance rating of the element penetrated. Penetrations through fireblocking and draftstopping shall be protected in an approved manner to maintain the integrity of the element penetrated.*

Discussion and Commentary: Through penetrations and membrane penetrations of fire-resistive elements must be protected in accordance with the requirements of Section R302.4. Where fireblocking and draftstopping is mandated by Section R302.11 and Section R302.12, respectively, the penetration by electrical installations must also be done in an appropriate manner.

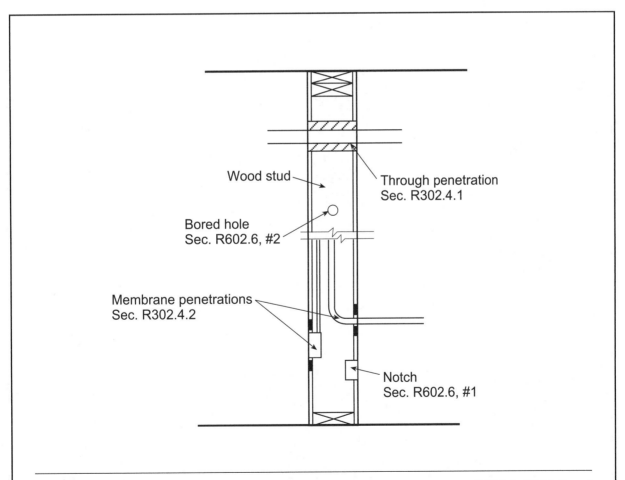

Wood stud

Through penetration
Sec. R302.4.1

Bored hole
Sec. R602.6, #2

Membrane penetrations
Sec. R302.4.2

Notch
Sec. R602.6, #1

For the notching and boring of structural wood framing members, Sections R502.8, R602.6 and R802.7 must be followed. These requirements are in addition to those mandated in Table E3802.1 for the protection of the electrical cables and conductors installed in such framing.

Category: Electrical Requirements
Subject: Equipment Location and Clearances

Code Text: *Except as otherwise specified in Chapters 34 through 43, the dimension of the working space in the direction of access to panelboards and live parts likely to require examination, adjustment, servicing or maintenance while energized shall be not less than 36 inches (914 mm) in depth. In addition to the 36-inch dimension (914 mm), the work space shall not be less than 30 inches (762 mm) wide in front of the electrical equipment and not less than the width of such equipment. The work space shall be clear and shall extend from the floor or platform to a height of 6.5 feet (1981 mm) or the height of the equipment, whichever is greater.*

Discussion and Commentary: A minimum depth of 3 feet in front of the access point to an electrical panelboard is needed to allow for adequate clearance for a person from any live parts. It provides enough working depth to both maintain a safe distance and allow free movement when a person is servicing or maintaining the equipment.

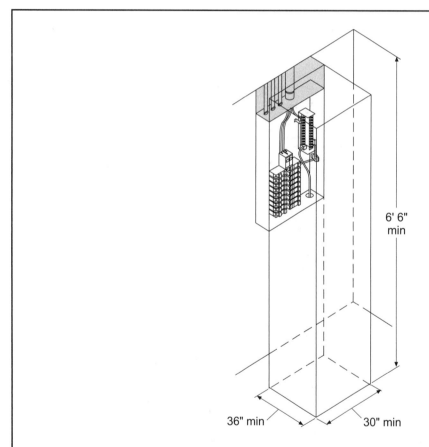

6' 6"
min

36" min   30" min

For SI: 1 inch = 25.4 mm, 1 foot = 304.8 mm.

Regardless of the size of the panelboard, a minimum clear width of 30 inches is required for working space. It need not be centered in front of the equipment but should be located for safe access. If the panelboard exceeds 30 inches in width, the clear space must extend the full width.

Code Text: *The space equal to the width and depth of the panelboard and extending from the floor to a height of 6 feet (1829 mm) above the panelboard, or to the structural ceiling, whichever is lower, shall be dedicated to the electrical installation. Piping, ducts, leak protection apparatus and other equipment foreign to the electrical installation shall not be installed in such dedicated space.*

Discussion and Commentary: In addition to the working space that must extend vertically in front of a panelboard, the area directly above the panelboard itself is regulated. This area must be free of any equipment or building features that would interfere with the direct run of cables and/or conduits to the equipment. The only exception allows for a suspended ceiling to encroach into the dedicated panelboard space, provided it has ceiling panels that are removable.

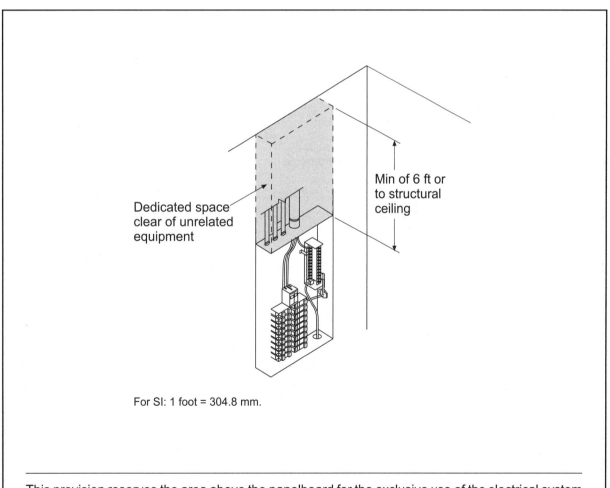

Dedicated space clear of unrelated equipment

Min of 6 ft or to structural ceiling

For SI: 1 foot = 304.8 mm.

This provision reserves the area above the panelboard for the exclusive use of the electrical system and its components. Plumbing and HVAC piping, ducts and equipment must be located outside the designated area.

Code Text: *Where conductors are to be spliced, terminated or connected to fixtures or devices, a minimum length of 6 inches (150 mm) of free conductor shall be provided at each outlet, junction or switch point. The required length shall be measured from the point in the box where the conductor emerges from its raceway or cable sheath. Where the opening to an outlet, junction, or switch point is less than 8 inches (200 mm) in any dimension, each conductor shall be long enough to extend at least 3 inches (75 mm) outside of such opening.*

Discussion and Commentary: To provide adequate conductor length for connection purposes, at least 6 inches of free conductor must be available. The conductor extension also ensures that undue stress will not be placed on the conductor during or after the splice or connection.

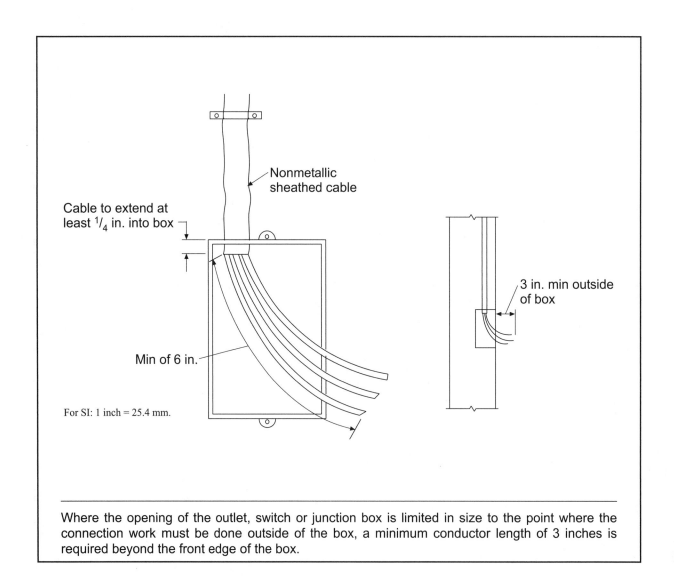

For SI: 1 inch = 25.4 mm.

Where the opening of the outlet, switch or junction box is limited in size to the point where the connection work must be done outside of the box, a minimum conductor length of 3 inches is required beyond the front edge of the box.

Code Text: *Insulated grounded conductors of sizes 6 AWG or smaller shall be identified by a continuous white or gray outer finish or by three continuous white stripes on other than green insulation along the entire length of the conductors. Equipment grounding conductors of sizes 6 AWG and smaller shall be identified by a continuous green color or a continuous green color with one or more yellow stripes on the insulation or covering, except where bare. Insulation on the ungrounded conductors shall be a continuous color other than white, gray and green. See exception for insulated ungrounded conductors that are part of a cable or flexible cord assembly.*

Discussion and Commentary: In addition to the identification of conductors, electrical devices common to residential construction, such as receptacles, that connect to both the ungrounded (hot) and grounded (neutral) conductors typically require properly identified terminals. The identification requirement does not apply to panelboards or devices rate over 30 amps.

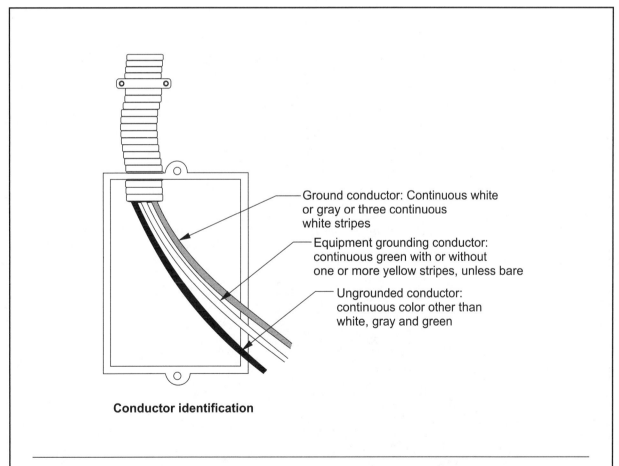

Ground conductor: Continuous white or gray or three continuous white stripes

Equipment grounding conductor: continuous green with or without one or more yellow stripes, unless bare

Ungrounded conductor: continuous color other than white, gray and green

Conductor identification

Grounded conductors larger than 6 AWG may have the same markings as those 6 AWG and smaller, but the code also permits other colors of insulation, typically black, provided the electrician applies white or gray tape or marking around the conductor at the terminal ends at the time of installation. A similar allowance is provided for grounding conductors larger than 6 AWG.

Topic: Service Load
Reference: IRC E3602.2

Category: Services
Subject: Service Size and Rating

Code Text: *The minimum load for ungrounded service conductors and service devices that serve 100 percent of the dwelling unit load shall be computed in accordance with Table E3602.2. Ungrounded service conductors and service devices that serve less than 100 percent of the dwelling unit load shall be computed as required for feeders in accordance with Chapter 37.*

Discussion and Commentary: Table E3602.2 sets forth a procedure for the calculation of the minimum service load. It is based on the square footage of the dwelling unit, the number of small appliance circuits and the rating of any fastened-in-place, permanently connected or dedicated circuit-supplied appliances such as ranges, ovens, clothes dryers and water heaters. A demand factor is applied to the calculated sum, to which is added a load for the air-conditioning and/or heating equipment.

TABLE E3602.2
MINIMUM SERVICE LOAD CALCULATION

LOADS AND PROCEDURE
3 volt-amperes per square foot of floor area for general lighting and general use receptacle outlets.
Plus
1,500 volt-amperes multiplied by total number of 20-ampere-rated small appliance and laundry circuits.
Plus
The nameplate volt-ampere rating of all fastened-in-place, permanently connected or dedicated circuit-supplied appliances such as ranges, ovens, cooking units, clothes dryers not connected to the laundry branch circuit and water heaters.
Apply the following demand factors to the above subtotal:
The minimum subtotal for the loads above shall be 100 percent of the first 10,000 volt-amperes of the sum of the above loads plus 40 percent of any portion of the sum that is in excess of 10,000 volt-amperes.
Plus the largest of the following:
One-hundred percent of the nameplate rating(s) of the air-conditioning and cooling equipment.
One hundred percent of the nameplate rating(s) of the heat pump where a heat pump is used without any supplemental electric heating.
One-hundred percent of the nameplate rating of the electric thermal storage and other heating systems where the usual load is expected to be continuous at the full nameplate value. Systems qualifying under this selection shall not be figured under any other category in this table.
One-hundred percent of nameplate rating of the heat pump compressor and sixty-five percent of the supplemental electric heating load for central electric space-heating systems. If the heat pump compressor is prevented from operating at the same time as the supplementary heat, the compressor load does not need to be added to the supplementary heat load for the total central electric space-heating load.
Sixty-five percent of nameplate rating(s) of electric space-heating units if less than four separately controlled units.
Forty percent of nameplate rating(s) of electric space-heating units of four or more separately controlled units.
The minimum total load in amperes shall be the volt-ampere sum calculated above divided by 240 volts.

Ungrounded service conductors must have an ampacity of no less than the load to be served. The minimum rating of the ungrounded service conductors for a single-family dwelling is 100 amperes, with an allowance of 60 amperes for other installations.

Code Text: *Conductors used as ungrounded service entrance conductors, service lateral conductors, and feeder conductors that serve as the main power feeder to a dwelling unit shall be those listed in Table E3603.1. Ungrounded service conductors shall have a minimum size in accordance with Table E3603.1. The grounded conductor size shall not be less than the maximum unbalance of the load and its size shall not be smaller than the required minimum grounding electrode conductor size specified in Table E3603.1.*

Discussion and Commentary: The minimum conductor size listed in Table E3603.1 is based on the maximum allowable load. Conductor size differs based on the allowable ampacity and the type of conductor material, either copper or aluminum. The same criteria are used for sizing the grounding electrode conductor.

TABLE E3603.1
SERVICE CONDUCTOR AND GROUNDING ELECTRODE CONDUCTOR SIZING

CONDUCTOR TYPES AND SIZES-THHN, THHW, THW, THWN, USE, RHH, RHW, XHHW, RHW-2, THW-2, THWN-2, XHHW-2, SE, USE-2 (Parallel sets of 1/0 and larger conductors are permitted in either a single raceway or in separate raceways)		SERVICE OR FEEDER RATING (AMPERES)	MINIMUM GROUNDING ELECTRODE CONDUCTOR SIZE[a]	
Copper (AWG)	Aluminum and copper-clad aluminum (AWG)	Maximum load (amps)	Copper (AWG)	Aluminum (AWG)
4	2	100	8[b]	6[c]
3	1	110	8[b]	6[c]
2	1/0	125	8[b]	6[c]
1	2/0	150	6[c]	4
1/0	3/0	175	6[c]	4
2/0	4/0 or two sets of 1/0	200	4[d]	2[d]
3/0	250 kcmil or two sets of 2/0	225	4[d]	2[d]
4/0 or two sets of 1/0	300 kcmil ortwo sets of 3/0	250	2[d]	1/0[d]
250 kcmil or two sets of 2/0	350 kcmil or two sets of 4/0	300	2[d]	1/0[d]
350 kcmil or two sets of 3/0	500 kcmil or two sets of 250 kcmil	350	2[d]	1/0[d]
400 kcmil or two sets of 4/0	600 kcmil or two sets of 300 kcmil	400	1/0[d]	3/0[d]

For SI: 1 inch = 25.4 mm.

a. Where protected by a ferrous metal raceway, grounding electrode conductors shall be electrically bonded to the ferrous metal raceway at both ends.

b. An 8 AWG grounding electrode conductor shall be protected with rigid metal conduit, intermediate metal conduit, rigid polyvinyl chloride (Type PVC) nonmetallic conduit, rigid thermosetting resin (Type RTRC) nonmetallic conduit, electrical metallic tubing or cable armor.

c. Where not protected, 6 AWG grounding electrode conductor shall closely follow a structural surface for physical protection. The supports shall be spaced not more than 24 inches on center and shall be within 12 inches of any enclosure or termination.

d. Where the sole grounding electrode system is a ground rod or pipe as covered in Section E3608.2, the grounding electrode conductor shall not be required to be larger than 6 AWG copper or 4 AWG aluminum. Where the sole grounding electrode system is the footing steel as covered in Section E3608.1.2, the grounding electrode conductor shall not be required to be larger than 4 AWG copper conductor.

An ungrounded service conductor that serves a workshop, detached garage or similar accessory structure need only have a minimum rating of 60 amperes. The sizing of such conductors shall be in accordance with Chapter 37 for feeders.

Code Text: *Open conductors and multiconductor cables without an overall outer jacket shall have a clearance of not less than 3 feet (914 mm) from the sides of doors, porches, decks, stairs, ladders, fire escapes, and balconies, and from the sides and bottom of windows that open.*

Discussion and Commentary: The clearances for overhead service drops are limited to conductors and cables that are not installed in a raceway or provided with an approved outer jacket. The requirement reduces the potential for damage to the service conductors and limits the risk of accidental contact. The clearance requirement from openable windows applies only to the sides and below the window opening. Service conductors and drip loops located just above an opening are deemed to be out of reach.

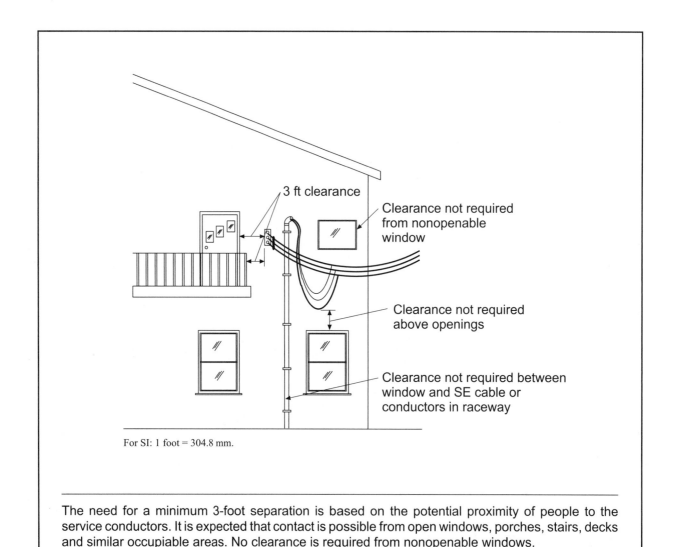

For SI: 1 foot = 304.8 mm.

The need for a minimum 3-foot separation is based on the potential proximity of people to the service conductors. It is expected that contact is possible from open windows, porches, stairs, decks and similar occupiable areas. No clearance is required from nonopenable windows.

Code Text: *Conductors shall have a vertical clearance of not less than 8 feet (2438 mm) above the roof surface. The vertical clearance above the roof level shall be maintained for a distance of not less than 3 feet (914 mm) in all directions from the edge of the roof.* See five exceptions allowing for reduced clearances.

Discussion and Commentary: The general provisions mandate at least an 8-foot clearance between a service-drop conductor and a roof. This clearance must not only occur directly above the roof surface but also extend horizontally at least 3 feet, except for the point where the service drop is attached to the side of the building. However, the separation may be reduced to 3 feet where the roof slope is 4:12 or greater. It is anticipated that travel across such a steeply-sloping roof will be minimal.

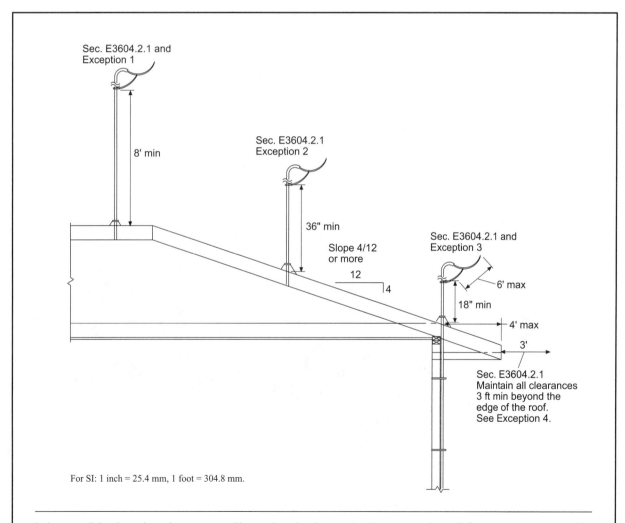

Sec. E3604.2.1 and Exception 1

8' min

Sec. E3604.2.1 Exception 2

36" min

Slope 4/12 or more

12
4

Sec. E3604.2.1 and Exception 3

6' max

18" min

4' max

3'

Sec. E3604.2.1 Maintain all clearances 3 ft min beyond the edge of the roof. See Exception 4.

For SI: 1 inch = 25.4 mm, 1 foot = 304.8 mm.

It is possible that the clearance will need to be increased to more than 8 feet where the roof is expected to have pedestrian access. This could occur where there is a rooftop court or sundeck or where the roof is also the upper deck of a parking garage.

Code Text: *Overhead service conductors shall have the following minimum clearances from final grade: 1) for conductors supported on and cabled together with a grounded bare messenger wire, the minimum vertical clearance shall be 10 feet (3048 mm) at the electric service entrance to buildings, at the lowest point of the drip loop of the building electric entrance, and above areas or sidewalks accessed by pedestrians only; 2) twelve feet (3658 mm)—over residential property and driveways; and 3) eighteen feet (5486 mm)—over public streets, alleys, roads or parking areas subject to truck traffic.*

Discussion and Commentary: The minimum clearances for overhead service conductors are based on the expected height of potential hazards. Where there is no anticipated vehicle traffic, at least 10 feet of vertical clearance is mandated. Where vehicles are present, higher clearances are necessary.

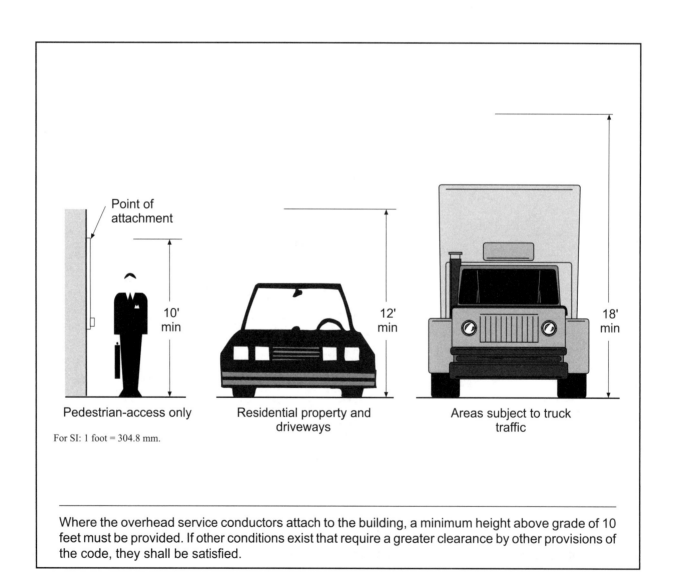

Point of attachment

10' min

Pedestrian-access only

For SI: 1 foot = 304.8 mm.

12' min

Residential property and driveways

18' min

Areas subject to truck traffic

Where the overhead service conductors attach to the building, a minimum height above grade of 10 feet must be provided. If other conditions exist that require a greater clearance by other provisions of the code, they shall be satisfied.

Code Text: *All electrodes specified in Sections E3608.1.1* (metal underground water pipe)*, E3608.1.2* (concrete-encased electrode)*, E3608.1.3* (ground rings)*, E3608.1.4* (rod and pipe electrodes)*, E3608.1.5* (plate electrodes) *and E3608.1.6* (other listed electrodes) *that are present at each building or structure served shall be bonded together to form the grounding electrode system. Where none of these electrodes are present, one or more of the electrodes specified in Sections E3608.1.3, E3608.1.4, E3608.1.5 and E3608.1.6 shall be installed and used.* See exception for concrete-encased electrodes of existing buildings.

Discussion and Commentary: One method of providing a grounding electrode is the use of a metal underground water pipe that is in direct contact with the earth for a minimum distance of 10 feet. Where this method is used, it is necessary to also provide a supplementary electrode, such as electrode of pipe, conduit or rod, a ground ring, or a plate electrode.

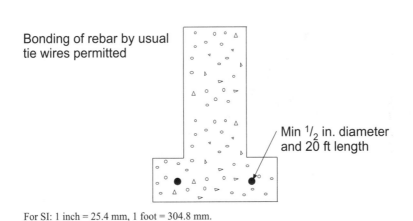

Bonding of rebar by usual tie wires permitted

Min $^1/_2$ in. diameter and 20 ft length

For SI: 1 inch = 25.4 mm, 1 foot = 304.8 mm.

Concrete-encased electrode

One or more complying reinforcing bars can be used as a grounding electrode when encased in concrete. The bar must be at least $^1/_2$ inch in diameter, be at least 20 feet in length, and be located within and near the bottom of the concrete foundation or footing.

Code Text: *Where more than one rod, pipe or plate electrode is used, each electrode of one grounding system shall be not less than 6 feet (1829 mm) from any other electrode of another grounding system. Two or more grounding electrodes that are effectively bonded together shall be considered as a single grounding electrode system.*

Discussion and Commentary: Electrodes of rods or pipes can be utilized in a grounding electrode system. The minimum length of such made electrodes shall be 8 feet, embedded below the permanent moisture level if possible. Pipes or conduit must be a minimum of $^3/_4$-inch diameter in trade size, while rod-type grounding electrodes of stainless steel and copper or zinc-coated steel are to be at least $^5/_8$-inch diameter. A plate electrode can be considered a grounding electrode, provided it exposes at least 2 square feet of surface to the exterior soil and is installed a minimum of 30 inches below the ground.

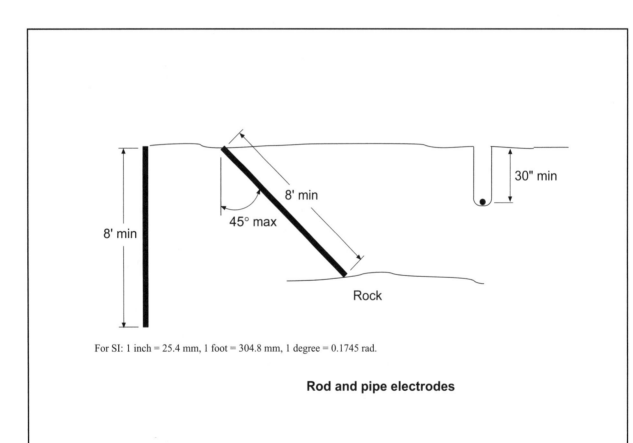

For SI: 1 inch = 25.4 mm, 1 foot = 304.8 mm, 1 degree = 0.1745 rad.

Rod and pipe electrodes

At least 8 feet of the electrode must be in contact with the soil. The electrode should be driven vertically unless soil conditions are prohibitive. In such cases, the electrode can be driven at an angle of up to 45 degrees or placed in a minimum 30-inch deep trench.

Code Text: *The requirements for circuits having two or more outlets, or receptacles, other than the receptacle circuits of Section E3703.2 (kitchen and dining area receptacles), E3703.3 (laundry circuit) and E3705.4 (bathroom circuits), are summarized in Table E3702.13. Branch circuits in dwelling units shall supply only loads within that dwelling unit or loads associated only with that dwelling unit. Branch circuits installed for the purpose of lighting, central alarm, signal, communications or other purposes for public or common areas of a two-family dwelling shall not be supplied from equipment that supplies an individual dwelling unit.*

Discussion and Commentary: Whereas a 20-ampere rated receptacle is not permitted to be installed on a 15-ampere rated circuit, it is acceptable for a 15-ampere rated receptacle to be located on a 20-ampere circuit. However, Section E4002.1.1 requires that a single receptacle on an individual branch be rated equal to the rating of the branch circuit.

TABLE E3702.13
BRANCH-CIRCUIT REQUIREMENTS—SUMMARY[a,b]

	CIRCUIT RATING		
	15 amp	**20 amp**	**30 amp**
Conductors: Minimum size (AWG) circuit conductors	14	12	10
Maximum overcurrent-protection device rating Ampere rating	15	20	30
Outlet devices: Lampholders permitted Receptacle rating (amperes)	Any type 15 maximum	Any type 15 or 20	N/A 30
Maximum load (amperes)	15	20	30

a. These gages are for copper conductors.
b. N/A means not allowed.

A two-family dwelling may have areas that are common to both residences, such as landscape lighting and exterior accent lighting. Under such conditions, the power for the common area loads must be supplied from a separate service and panelboard.

Code Text: *Central heating equipment other than fixed electric space heating shall be supplied by an individual branch circuit. A minimum of two 20-ampere-rated branch circuits shall be provided to serve all wall and floor receptacle outlets located in the kitchen, pantry, breakfast area, dining area or similar area of a dwelling. A minimum of one 20-ampere-rated branch circuit shall be provided for receptacles located in the laundry area and shall serve only receptacle outlets located in the laundry area. A minimum of one 20-ampere branch circuit shall be provided to supply the bathroom receptacle outlet(s). The minimum number of branch circuits shall be determined from the total calculated load and the size or rating of the circuits used.*

Discussion and Commentary: In addition to general lighting and convenience outlet circuits, a number of specific areas of a dwelling unit must be provided with their own dedicated branch circuits.

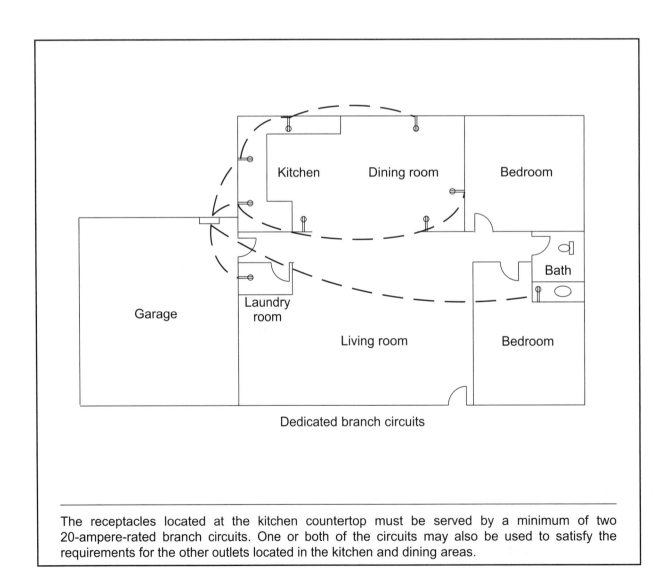

The receptacles located at the kitchen countertop must be served by a minimum of two 20-ampere-rated branch circuits. One or both of the circuits may also be used to satisfy the requirements for the other outlets located in the kitchen and dining areas.

Code Text: *All circuits and circuit modifications shall be legibly identified as to their clear, evident, and specific purpose or use. The identification shall include sufficient detail to allow each circuit to be distinguished from all others. Spare positions that contain unused overcurrent devices or switches shall be described accordingly. The identification shall be included in a circuit directory located on the face of the panelboard enclosure or inside the panel door. Circuits shall not be described in a manner that depends on transient conditions of occupancy.*

Discussion and Commentary: The importance of adequately identifying the various circuits cannot be overstated. The code mandates that all circuits be noted in a legible fashion, indicating their purpose in a clear, evident and specific manner. It is also important to adequately differentiate each circuit from all of the others.

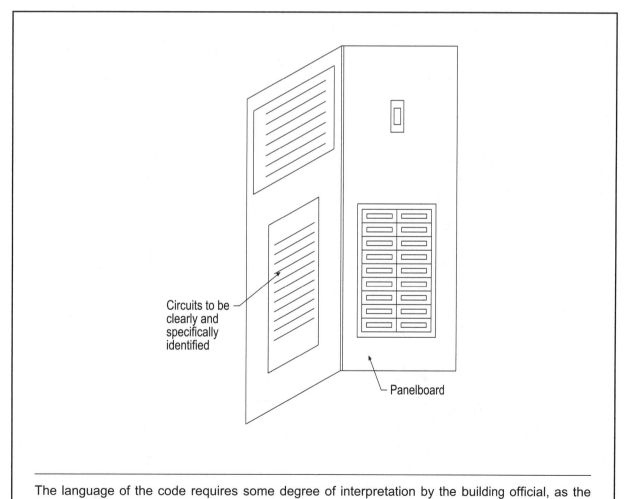

Circuits to be clearly and specifically identified

Panelboard

The language of the code requires some degree of interpretation by the building official, as the provision is primarily performance in nature. However, it is obvious that generic terminology such as "lighting" or "kitchen" is not acceptable.

Quiz

Study Session 17
IRC Chapters 34, 35, 36 and 37

1. Energized parts operating at a minimum of _____ shall be guarded against accidental contact by people through the use of approved enclosures.

    a. 50 volts

    b. 60 volts

    c. 90 volts

    d. 110 volts

    Reference _____

2. An electrical panelboard requiring access while energized shall be provided with a minimum of _____ in depth measured in the direction of access.

    a. 30 inches

    b. 36 inches

    c. 42 inches

    d. 48 inches

    Reference _____

3. Work space in front of an electrical panelboard shall be a minimum of _____ in width, but no less than the width of the panelboard.

    a. 30 inches

    b. 36 inches

    c. 42 inches

    d. 48 inches

    Reference _____

4. The dedicated space above an electrical panelboard shall be a minimum of _____ high or to the structural ceiling, whichever is lower.

    a. 4 feet   b. 5 feet

    c. 6 feet   d. 6.5 feet

Reference _____

5. The minimum size of electrical conductors for feeders and branch circuits, when of copper, shall be _____ AWG.

    a. No. 16   b. No. 14

    c. No. 12   d. No. 10

Reference _____

6. Where electrical conductors are to be spliced, terminated or connected to fixtures or devices, a minimum length of _____ of free conductor shall be provided at each outlet, junction or switch point.

    a. 4 inches   b. 6 inches

    c. 8 inches   d. 12 inches

Reference _____

7. Insulated grounded conductors of sizes 6 AWG and smaller may be identified by all but which one of the following methods?

    a. continuous white outer finish

    b. continuous natural gray outer finish

    c. continuous black outer finish

    d. three continuous white stripes (on other than green insulation)

Reference _____

8. Equipment grounding conductors of sizes 6 AWG and smaller may be identified by all but which one of the following methods?

    a. continuous white color

    b. continuous green color

    c. bare

    d. continuous green color with one or more yellow stripes

Reference _____

9. In general, which of the following continuous colors is permitted for the insulation on ungrounded conductors?

    a. gray   b. white

    c. green   d. red

Reference _____

10. What is the minimum permitted size of a grounded service conductor serving a maximum load of 150 amps?

    a. No. 1 aluminum   b. No. 1 copper

    c. No. 1/0 aluminum   d. No. 1/0 copper

Reference _____

11. An open service conductor without an overall outer jacket shall have a minimum clearance of _____ from the sides of doors, porches and openable windows.

    a. 3 feet   b. 4 feet

    c. 6 feet   d. 8 feet

Reference _____

12. An overhead service conductor shall have a minimum clearance above a 3:12 roof of _____ .

    a. 3 feet   b. 6 feet

    c. 7 feet   d. 8 feet

Reference _____

13. Overhead service conductors shall have a minimum clearance of _____ over residential property and driveways.

    a. 8 feet   b. 10 feet

    c. 12 feet   d. 15 feet

Reference _____

14. In all cases, the point of attachment of service-drop conductors to a building shall be a minimum of _____ above finished grade.

      a. 8 feet   b. 10 feet

      c. 12 feet   d. 14 feet

      Reference _____

15. A metal underground water pipe used as part of the grounding electrode system shall be in direct contact with the earth for a minimum of _____ .

      a. 8 feet   b. 10 feet

      c. 15 feet   d. 20 feet

      Reference _____

16. Where more than one rod, pipe or plate electrode is used in one grounding electrode system, each electrode shall be located a minimum of _____ from any other electrode of another grounding system.

      a. 5 feet   b. 6 feet

      c. 10 feet   d. 20 feet

      Reference _____

17. Where a grounding electrode is made of a stainless steel rod, the rod shall be a minimum of _____ in diameter and a minimum of _____ in length unless listed..

      a. $^{1}/_{2}$ inch, 8 feet   b. $^{1}/_{2}$ inch, 10 feet

      c. $^{5}/_{8}$ inch, 8 feet   d. $^{5}/_{8}$ inch, 10 feet

      Reference _____

18. Where used outside, an aluminum or copper-clad aluminum grounding electrode conductor shall be installed a minimum of _____ from the earth.

      a. 6 inches   b. 12 inches

      c. 18 inches   d. 30 inches

      Reference _____

19. The rating of any one cord- and plug-connected utilization equipment not fastened in place shall be a maximum of _____ of the branch-circuit ampere rating.

    a. 80 percent   b. 100 percent

    c. 110 percent   d. 125 percent

Reference _____

20. A minimum of _____ branch circuit(s) shall be provided to serve receptacles located in the kitchen, pantry, breakfast area and dining area.

    a. one 15-ampere-rated   b. one 20-ampere-rated

    c. two 15-ampere-rated   d. two 20-ampere-rated

Reference _____

21. A minimum of _____ branch circuit(s) shall be provided for receptacles located in the laundry room and a minimum of _____ branch circuit(s) shall be provided to supply the bathroom receptacle outlet(s).

    a. one 15-ampere-rated, one 15-ampere-rated

    b. one 20-ampere-rated, one 20-ampere-rated

    c. one 15-ampere-rated, two 15-ampere-rated

    d. one 20-ampere-rated, two 20 ampere-rated

Reference _____

22. In the calculation of electrical feeder loads, an electric clothes dryer load shall be considered _____ VA for each dryer circuit or the nameplate rating load of each dryer, whichever is greater.

    a. 4,000   b. 5,000

    c. 5,500   d. 6,000

Reference _____

23. When sizing feeder conductors, a minimum unit load of _____ shall constitute the minimum lighting and convenience receptacle load for each square foot of floor area.

    a. two volt-amperes   b. three volt-amperes

    c. five volt-amperes   d. six volt-amperes

Reference _____

24. What is the maximum overcurrent-protection-device rating permitted for No. 12 copper conductors?

    a. 15 amps   b. 20 amps

    c. 25 amps   d. 30 amps

Reference _____

25. Which one of the following ratings is not a standard ampere rating for fuses and inverse time circuit breakers?

    a. 25 amperes   b. 50 amperes

    c. 75 amperes   d. 100 amperes

Reference _____

26. Where metallic plugs or plates are used with nonmetallic electrical enclosures to close unused openings, they shall be recessed a minimum of _____ inch from the outer surface of the enclosure.

    a. $^1/_8$ b. $^1/_4$

    c. $^3/_8$ d. $^1/_2$

Reference _____

27. Electrical panelboards and overcurrent protection devices shall not be located in _____ .

    a. clothes closets   b. sleeping rooms

    c. garages   d. storage rooms

Reference _____

28. Where the opening to an outlet, junction, or switch point is less than 8 inches in any dimension, each electrical conductor shall be long enough to extend a minimum of _____ inches outside of such opening.

    a. 2   b. 3

    c. 4   d. 6

Reference _____

29. Electrical service disconnecting means shall not be installed in _____ .

    a. storage closets   b. laundry rooms

    c. sleeping rooms   d. bathrooms

Reference _____

30. In order to be used as a part of the grounding electrode system, interior metal water pipe shall be located a maximum of _____ feet from the entrance to the building.

    a.. 3   b. 5

    c. 6   d. 10

Reference _____

31. Electrical conductors located above a roof shall be installed with a minimum vertical clearance of _____ feet where the roof has a slope of 8:12.

    a. 3   b. 4

    c. 6   d. 7

Reference _____

32. Service-entrance cables shall be supported by straps or other approved means within _____ inches of every electrical service head and at maximum intervals of _____ inches.

    a. 8, 30   b. 8, 36

    c. 12, 30   d. 12, 36

Reference _____

33. A plate electrode installed as a part of a grounding electrode system shall be located a minimum of _____ inches below the surface of the earth.

    a. 12   b. 18

    c. 24   d. 30

Reference _____

34. Where an electrical branch circuit has multiple outlets, what is the minimum required size of circuit conductors for a circuit rated at 30 amps?

    a. 14 AWG   b. 12 AWG

    c. 10 AWG   d. 8 AWG

    Reference _____

35. An overcurrent protection device shall be installed so that the center of the switch or circuit breaker, when in its highest position, is located a maximum of _____ above the floor or working platform.

    a. 5 feet, 0 inches   b. 6 feet, 4 inches

    c. 6 feet, 7 inches   d. 7 feet, 0 inches

    Reference _____

Study Session

18

2012 IRC Chapters 38, 39, 40, 41 and 42

Lighting Distribution, Devices and Lighting Fixtures, Appliance Installation and Swimming Pools

OBJECTIVE: To gain an understanding of the provisions regulating wiring methods, power and lighting distribution, devices and lighting fixtures, appliance installation and swimming pools.

REFERENCE: Chapters 38, 39, 40, 41 and 42, 2012 *International Residential Code*

KEY POINTS:
- How are electrical cables to be installed where they are run in attics across the top of structural members? Where they are run parallel to framing members? How must they be protected from physical damage?
- What are the minimum cover requirements for direct buried cable or raceways installed underground? How are direct buried cables to be protected from damage where emerging from the ground?
- How must backfill be placed in an excavation containing electrical cables or raceways?
- What method is used to determine the minimum spacing requirements for convenience receptacles? What wall spaces are to be considered? How are floor receptacles regulated?
- Where are small appliance receptacles regulated? How many receptacles must be provided? How must they be distributed?
- In what manner are counter receptacles to be located? How do the provisions differ for island counter spaces? For peninsular counter spaces?
- How many wall receptacles are required in a bathroom? Outdoors? In laundry areas? In basements and garages? In hallways?
- Where is ground-fault circuit-interrupter protection mandated?
- In what locations are arc-fault circuit interrupter protection mandated?
- In what rooms and spaces must wall-switch-controlled lighting outlets be installed? Under what conditions are wall-switch-controlled receptacles permitted as an alternate to lighting outlets?
- How is nonmetallic-sheathed cable to be installed in nonmetallic boxes? What is the minimum required depth of such boxes? How is box volume and fill calculated? How are boxes to be installed?
- How are luminaires to be installed in wet or damp locations? In bathtub and shower areas? In clothes closets?

KEY POINTS:
(Cont'd)

- What are the limits for supporting luminaires by the screw shell of a lampholder? By an outlet box? What clearances are required between a recessed luminaire and thermal insulation? Other combustible material?

- How is track lighting to be installed? Where is track lighting prohibited?

- Where shall receptacle outlets be located in relationship to a pool, spa or hot tub? Which receptacles require ground-fault circuit-interrupter protection? How are luminaires, lighting outlets and ceiling-suspended paddle fans regulated?

- What parts of a swimming pool, spa or hot tub must be bonded? What are the acceptable bonding methods? Which equipment must be grounded? How is pool equipment regulated?

Code Text: *The allowable wiring methods for electrical installations shall be those listed in Table E3801.2. Single conductors shall be used only where part of one of the recognized wiring methods listed in Table E3801.2. As used in* the IRC, *abbreviations of the wiring-method types shall be as indicated in Table E3801.2.*

Discussion and Commentary: Single insulated conductors are not permitted as a wiring method without being part of a cable assembly or installed in tubing or conduit. In many older houses, a wiring method known as "knob-and-tube" was used, consisting of single insulated conductors supported in free air on porcelain insulators and, where run through a wood framing member, installed through insulating tubes. Where remodeling, retrofitting or repair is being done in such a house, the transition to new wiring should be to cable.

TABLE E3801.2
ALLOWABLE WIRING METHODS

ALLOWABLE WIRING METHOD	DESIGNATED ABBREVIATION
Armored cable	AC
Electrical metallic tubing	EMT
Electrical nonmetallic tubing	ENT
Flexible metal conduit	FMC
Intermediate metal conduit	IMC
Liquidtight flexible conduit	LFC
Metal-clad cable	MC
Nonmetallic sheathed cable	NM
Rigid polyvinyl chloride conduit (Type PVC)	RNC
Rigid metallic conduit	RMC
Service entrance cable	SE
Surface raceways	SR
Underground feeder cable	UF
Underground service cable	USE

In many older houses, the wiring is still sound and in good shape, but when additions or extensions are made, the possibility for arcing, bad connections, heating and overloading come into play. Thus, one of the allowable wiring methods must be utilized.

Code Text: *Wiring methods shall be installed and supported in accordance with Table E3802.1.*

Discussion and Commentary: A number of different installation conditions are addressed in an effort to protect and support various wiring methods permitted by the code. For NM cable and other materials subject to damage, the installation procedures address the common situation of cable within stud cavities. If the cable is run parallel along the side of the stud or furring strip, it must be located at least $1\frac{1}{4}$ inches from the stud edge or otherwise physically protected. The same minimum dimension is necessary where cable is run perpendicular to the vertical framing members through bored holes. Where the mandated depth is not provided, it is common to install a minimum $\frac{1}{16}$-inch-thick steel plate across the stud edge.

TABLE E3802.1
GENERAL INSTALLATION AND SUPPORT REQUIREMENTS FOR WIRING METHODS[a, b, c, d, e, f, g, h, i, j, k]

INSTALLATION REQUIREMENTS (Requirement applicable only to wiring methods marked "A")	AC MC	EMT IMC RMC	ENT	FMC LFC	NM UF	RNC	SE	SR[a]	USE
Where run parallel with the framing member or furring strip, the wiring shall be not less than $1\frac{1}{4}$ inches from the edge of a furring strip or a framing member such as a joist, rafter or stud or shall be physically protected.	A	—	A	A	A	—	A	—	—
Bored holes in framing members for wiring shall be located not less than $1\frac{1}{4}$ inches from the edge of the framing member or shall be protected with a minimum 0.0625-inch steel plate or sleeve, a listed steel plate or other physical protection.	A[k]	—	A[k]	A[k]	A[k]	—	A[k]	—	—
Where installed in grooves, to be covered by wallboard, siding, paneling, carpeting, or similar finish, wiring methods shall be protected by 0.0625-inch-thick steel plate, sleeve, or equivalent, a listed steel plate or by not less than $1\frac{1}{4}$-inch free space for the full length of the groove in which the cable or raceway is installed.	A	—	A	A	A	—	A	A	A
Securely fastened bushings or grommets shall be provided to protect wiring run through openings in metal framing members.	—	—	A[j]	—	A[j]	—	A[j]	—	—
The maximum number of 90-degree bends shall not exceed four between junction boxes.	—	A	A	A	—	A	—	—	—
Bushings shall be provided where entering a box, fitting or enclosure unless the box or fitting is designed to afford equivalent protection.	A	A	A	A	—	A	—	A	—
Ends of raceways shall be reamed to remove rough edges.	—	A	A	A	—	A	—	A	—
Maximum allowable on center support spacing for the wiring method in feet.	4.5[b, c]	10[i]	3[b]	4.5[b]	4.5[i]	3[d, l]	2.5[e]	—	2.5
Maximum support distance in inches from box or other terminations.	12[b, f]	36	36	12[b, g]	12[h, i]	36	12	—	—

For SI: 1 inch = 25.4 mm, 1 foot = 304.8 mm, 1 degree = 0.0175 rad.

a. Installed in accordance with listing requirements.
b. Supports not required in accessible ceiling spaces between light fixtures where lengths do not exceed 6 feet.
c. Six feet for MC cable.
d. Five feet for trade sizes greater than 1 inch.
e. Two and one-half feet where used for service or outdoor feeder and 4.5 feet where used for branch circuit or indoor feeder.
f. Twenty-four inches where flexibility is necessary.
g. Where flexibility after installation is necessary, lengths of flexible metal conduit and liquidtight flexible metal conduit measured from the last point where the raceway is securely fastened shall not exceed: 36 inches for trade sizes $\frac{1}{2}$ through $1\frac{1}{4}$, 48 inches for trade sizes $1\frac{1}{2}$ through 2 and 5 feet *for trade sizes $2\frac{1}{2}$ and larger.*
h. Within 8 inches of boxes without cable clamps.
i. Flat cables shall not be stapled on edge.
j. Bushings and grommets shall remain in place and shall be listed for the purpose of cable protection.
k. See Sections R502.8 and R802.7 for additional limitations on the location of bored holes in horizontal framing members.

A concern in accessible attics is the presence of cables that may be subject to contact and potential physical damage. The code requires protection for these cables, particularly where they are located within 6 feet of the attic access opening.

Code Text: *Direct buried cable or raceways shall be installed in accordance with the minimum cover requirements of Table E3803.1. Direct buried conductors and cables emerging from the ground shall be protected by enclosures or raceways extending from the minimum cover distance below grade required by Section 3803.1 to a point at least 8 feet (2438 mm) above finished grade. In no case shall the protection be required to exceed 18 inches (457 mm) below finished grade.*

Discussion and Commentary: To protect cables or raceways buried below ground level from damage, the code establishes a minimum depth for burial. The burial depth is dependent upon two factors: the location or method of burial, and the type of wiring method. Where direct buried conductors and cables are permitted, they must be protected from below ground level to a point high enough that physical damage is improbable.

TABLE E3803.1
MINIMUM COVER REQUIREMENTS, BURIAL IN INCHES[a, b, c, d, e]

LOCATION OF WIRING METHOD OR CIRCUIT	TYPE OF WIRING METHOD OR CIRCUIT				
	1 Direct burial cables or conductors	2 Rigid metal conduit or intermediate metal conduit	3 Nonmetallic raceways listed for direct burial without concrete encasement or other approved raceways	4 Residential branch circuits rated 120 volts or less with GFCI protection and maximum overcurrent protection of 20 amperes	5 Circuits for control of irrigation and landscape lighting limited to not more than 30 volts and installed with type UF or in other identified cable or raceway
All locations not specified below	24	6	18	12	6
In trench below 2-inch-thick concrete or equivalent	18	6	12	6	6
Under a building	0 (In raceway only or Type MC identified for direct burial)	0	0	0 (In raceway only or Type MC identified for direct burial)	0 (In raceway only or Type MC identified for direct burial)
Under minimum of 4-inch-thick concrete exterior slab with no vehicular traffic and the slab extending not less than 6 inches beyond the underground installation	18	4	4	6 (Direct burial) 4 (In raceway)	6 (Direct burial) 4 (In raceway)
Under streets, highways, roads, alleys, driveways and parking lots	24	24	24	24	24
One- and two-family dwelling driveways and outdoor parking areas, and used only for dwelling-related purposes	18	18	18	12	18
In solid rock where covered by minimum of 2 inches concrete extending down to rock	2 (In raceway only)	2	2	2 (In raceway only)	2 (In raceway only)

For SI: 1 inch = 25.4 mm.

a. Raceways approved for burial only where encased concrete shall require concrete envelope not less than 2 inches thick.

b. Lesser depths shall be permitted where cables and conductors rise for terminations or splices or where access is otherwise required.

c. Where one of the wiring method types listed in columns 1 to 3 is combined with one of the circuit types in columns 4 and 5, the shallower depth of burial shall be permitted.

d. Where solid rock prevents compliance with the cover depths specified in this table, the wiring shall be installed in metal or nonmetallic raceway permitted for direct burial. The raceways shall be covered by a minimum of 2 inches of concrete extending down to the rock.

e. Cover is defined as the shortest distance in inches (millimeters) measured between a point on the top surface of any direct-buried conductor, cable, conduit or other raceway and the top surface of finished grade, concrete, or similar cover.

The backfill used to cover cables or raceways must be placed in such a manner as to avoid damage to the wiring method. It may be necessary to use boards, sleeves, granular material or other suitable methods to protect the raceway or cable from physical damage.

Code Text: *In every kitchen, family room, dining room, living room, parlor, library, den, sun room, bedroom, recreation room, or similar room or area of dwelling units, receptacle outlets shall be installed . . . so that no point measured horizontally along the floor line of any wall space is more than 6 feet (1829 mm), from a receptacle outlet. A wall space shall include: 1) any space that is 2 feet (610 mm) or more in width (including space measured around corners), and that is unbroken along the floor line by doorways and similar openings, fireplaces and fixed cabinets; 2) the space occupied by fixed panels in exterior wall, excluding sliding panels; and 3) the space created by fixed room dividers such as railings and freestanding bar-type counters. Receptacle outlets in floors shall not be counted as part of the required number of receptacle outlets except where located within 18 inches (457 mm) of the wall.*

Discussion and Commentary: The maximum spacing between convenience receptacles is basically 12 feet, allowing flexibility in the placement of electric appliances without the need of any extension cord. Those spaces that are subject to the requirement tend to be habitable spaces, those areas of a dwelling used for living, sleeping, dining or cooking.

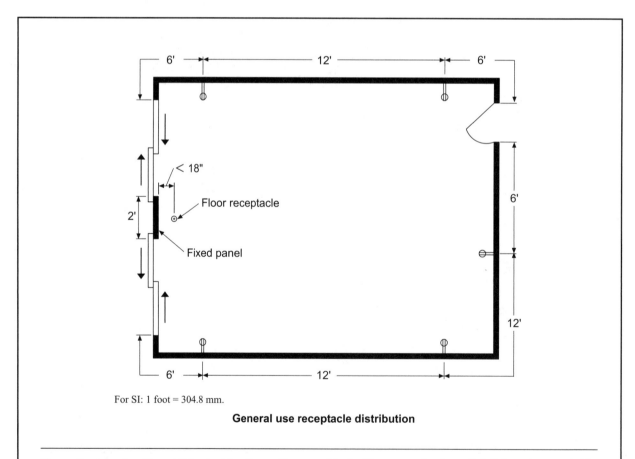

For SI: 1 foot = 304.8 mm.

General use receptacle distribution

If a wall space is at least 2 feet wide, it must be provided with a receptacle outlet. It is not acceptable to measure across a doorway or similar opening for compliance with the 6-foot rule. An extension cord across a door opening is a considerable hazard.

Topic: Countertop Receptacles
Reference: IRC E3901.4

Category: Power and Lighting Distribution
Subject: Receptacle Outlets

Code Text: *In kitchens, pantries, breakfast rooms, dining rooms and similar areas of dwelling units, receptacle outlets for countertop spaces shall be installed . . . at each wall counter space 12 inches (305 mm) or wider. Receptacle outlets shall be installed so that no point along the wall line is more than 24 inches (610 mm), measured horizontally from a receptacle outlet in that space. See exception for outlets on a wall directly behind a range or sink. Receptacle outlets shall be located not more than 20 inches (508 mm) above the countertop. Receptacle outlets shall not be installed in a face-up position in the work surfaces or countertops.*

Discussion and Commentary: Receptacle outlets are mandated in a repetitive fashion along a kitchen counter for both safety and convenience. With a number of appliances potentially in use at the same time, and because they are likely to be moved to a number of different positions, it is important that access to multiple receptacles be provided.

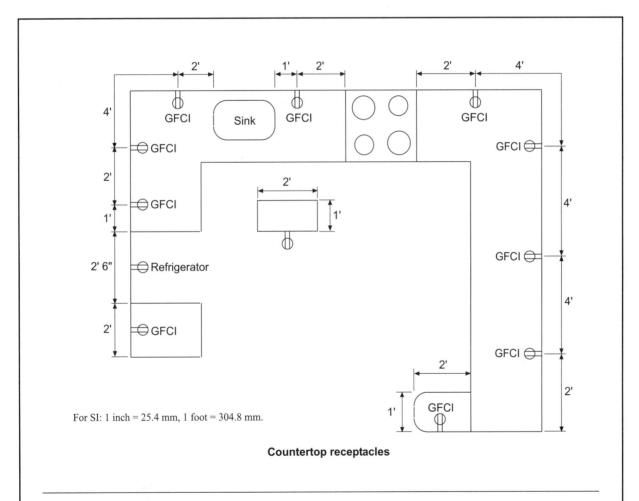

For SI: 1 inch = 25.4 mm, 1 foot = 304.8 mm.

Countertop receptacles

If an island counter space exceeds a specified size, it too must be provided with at least one receptacle outlet. The same holds true for peninsular counter spaces of similar dimensions. Appliance cords pose an obvious problem when extended through work areas or traffic paths.

Topic: Miscellaneous Receptacle Outlets **Category:** Power and Lighting Distribution
Reference: IRC E3901.6 – E3901.10 **Subject:** Receptacle Outlets

Code Text: *At least one wall receptacle outlet shall be installed in bathrooms and such outlet shall be located within 36 inches (914 mm) of the outside edge of each lavatory basin. At least one receptacle outlet that is accessible while standing at grade level and located not more than 6 feet 6 inches (1981 mm) above grade, shall be installed outdoors at the front and back of each dwelling unit having direct access to grade. At least one receptacle outlet shall be installed to serve laundry appliances. At least one receptacle outlet, in addition to any provided for specific equipment, shall be installed in each basement and in each attached garage, and in each detached garage or accessory building that is provided with electrical power. Hallways of 10 feet (3048 mm) or more in length shall have at least one receptacle outlet.*

Discussion and Commentary: In addition to habitable spaces, those spaces typically defined as not habitable also require a limited number of receptacle outlets. Although the number of outlets mandated may be fewer than required for habitable rooms, it is important that outlets be provided to serve any equipment or activity that is anticipated for the space.

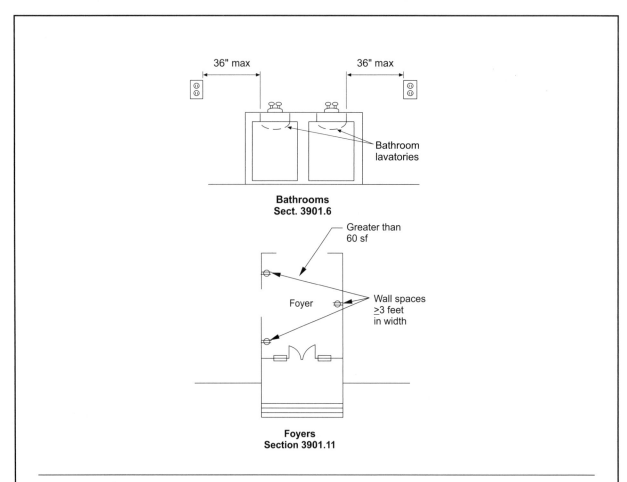

Heating, air-conditioning and refrigeration equipment, other than evaporative coolers, shall be provided with a convenience outlet installed within 25 feet of the equipment. To be used during maintenance or repair activities, the outlet must be located on the same level as the equipment.

| **Topic:** Required GFCI Protection | **Category:** Power and Lighting Distribution |
| **Reference:** IRC E3902.1 – E3902.7 | **Subject:** Ground-Fault Protection |

Code Text: *(Summary) All 125-volt, single-phase, 15- and 20-ampere receptacles installed in bathrooms; garages; grade-level portions of unfinished accessory buildings used for storage or work areas; outdoors; in a crawl space at or below grade; and unfinished basements shall have ground-fault circuit-interrupter protection. See exceptions. All 125-volt, single-phase, 15- and 20-ampere receptacles that serve kitchen countertop surfaces, and all 125-volt, single-phase, 15- and 20-ampere receptacles that are located within 6 feet (1829 mm) of the outside edge of a sink that is located in an area other than a kitchen, shall have ground-fault circuit-interrupter protection for personnel.*

Discussion and Commentary: Where there is an unbalance between the ungrounded "hot" conductor and the grounded "neutral" conductor, the purpose of a ground-fault circuit interrupter (GFCI) is to trip the circuit off. Where the unbalance is caused by human contact, this trip prevents serious injury or death.

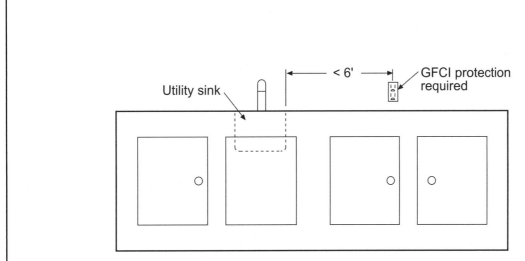

For SI: 1 foot = 304.8 mm.

The purpose of the GFCI is to recognize an unbalanced load and shut off the circuit, but it does not reduce the magnitude of the ground-fault current. A very strong, but brief, shock is still encountered by a person during the time period the GFCI device needs to recognize the unbalance and trip.

Topic: Nonmetallic Boxes	**Category:** Power and Lighting Distribution
Reference: IRC E3905.3	**Subject:** Boxes, Conduit Bodies and Fittings

Code Text: *Nonmetallic boxes shall be used only with cabled wiring methods with entirely nonmetallic sheaths, flexible cords and nonmetallic raceways. See exceptions for special bonding conditions. Where nonmetallic-sheathed cable is used, the cable assembly, including the sheath, shall extend into the box not less than $^1/_4$ inch (6.4 mm) through a nonmetallic-sheathed cable knockout opening. Where nonmetallic-sheathed cable is used with boxes not larger than a nominal size of $2^1/_4$ inches by 4 inches (57 mm by 102 mm) mounted in walls or ceilings, and where the cable is fastened within 8 inches (203 mm) of the box measured along the sheath, and where the sheath extends through a cable knockout not less than $^3/_4$ inch (6.4 mm), securing the cable to the box shall not be required.*

Discussion and Commentary: To provide adequate insulation for the conductors up and into the box, the sheathing must extend into the box a limited distance.

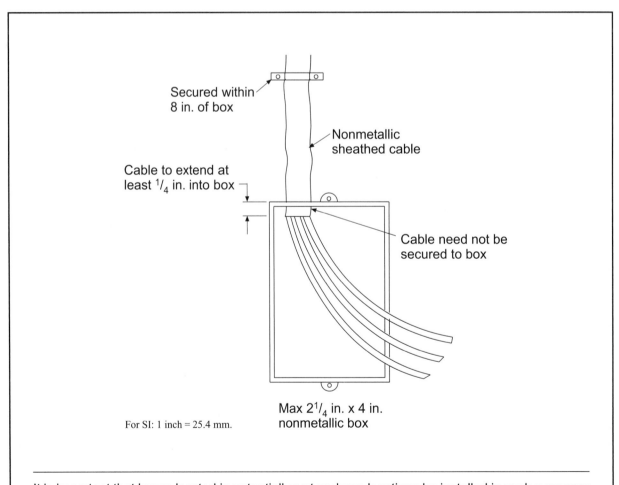

Secured within 8 in. of box

Nonmetallic sheathed cable

Cable to extend at least $^1/_4$ in. into box

Cable need not be secured to box

For SI: 1 inch = 25.4 mm.

Max $2^1/_4$ in. x 4 in. nonmetallic box

It is important that boxes located in potentially wet or damp locations be installed in such a manner as to prevent the entry and accumulation of moisture. Where installed in such areas, the boxes must be listed for use in wet locations.

Code Text: *Cord-connected luminaires, chain-, cable-, or cord-suspended luminaires, lighting track, pendants, and ceiling-suspended (paddle) fans shall not have any parts located within a zone measured 3 feet (914 mm) horizontally and 8 feet (2438 mm) vertically from the top of a bathtub rim or shower stall threshold. This zone is all encompassing and includes the zone directly over the tub or shower. Luminaires within the actual outside dimension of the bathtub or shower to a height of 8 feet (2438 mm) vertically from the top of the bathtub rim or shower threshold shall be marked for damp locations and where subject to shower spray, shall be marked for wet locations.*

Discussion and Commentary: In most homes, a bathroom ceiling is not high enough to permit any fixture other than a surface-mounted or recessed-type fixture above a bathtub or shower. Fixtures such as chain-hung or pendant types where a cord is exposed above a bathtub or shower could be a safety hazard, and there is more of a danger of electrocution where water is present.

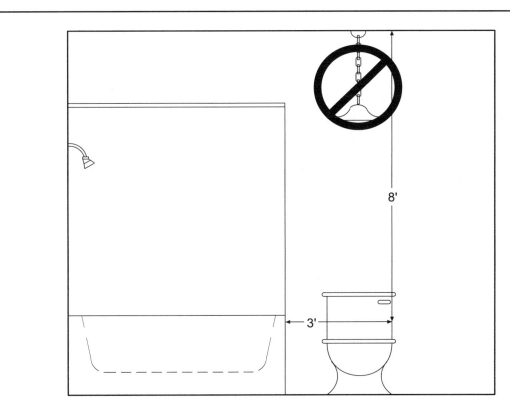

Luminaries in bathtub and shower

For SI: 1 foot = 304.8 mm

Although specific luminaires are prohibited in the tub and shower zone, other types, such as those that are surface-mounted, are acceptable when marked for damp locations. A listing marking the allowed use of the luminaire in a wet location is mandated where the luminaire will be subjected to shower spray.

Code Text: *The types of luminaires installed in clothes closets shall be limited to surface-mounted or recessed incandescent or LED luminaires with completely enclosed light sources, surface-mounted or recessed fluorescent luminaires, and surface-mounted fluorescent or LED luminaires identified as suitable for installation within the closet storage area. Incandescent luminaires with open or partially enclosed lamps and pendant luminaires or lamp-holders shall be prohibited. The minimum clearance between luminaires installed in clothes closets and the nearest point of a closet storage area shall be as follows:* See five methods for providing complying clearances.

Discussion and Commentary: It is quite probable that a clothes closet is the most common place in a house where a light source will be located adjacent to combustible materials. Therefore, various clearances are set forth based on the type of luminaire installed.

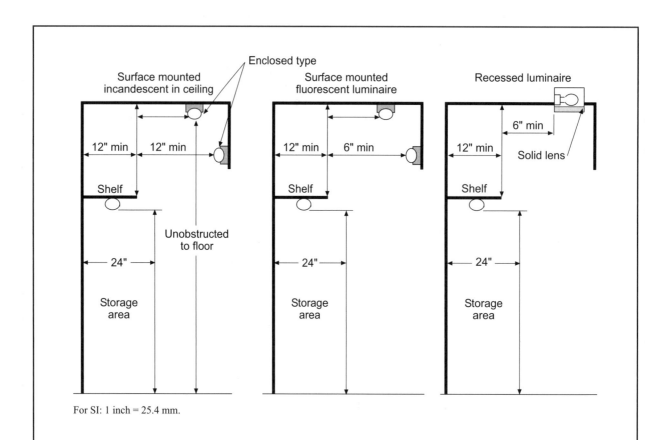

For SI: 1 inch = 25.4 mm.

It is impossible to fully regulate the use of a space once the dwelling unit is occupied; therefore, the code defines for its purposes the extent of a "storage space." It is bounded by the sides and back of the closet walls and extends to a height of 6 feet or the highest clothes-hanging rod.

Code Text: *Luminaires and lampholders shall be securely supported. A luminaire that weighs more than 6 pounds (2.72 kg) or exceeds 16 inches (406 mm) in any dimension shall not be supported by the screw shell of a lampholder. Per Section E3905.6.2, outlet boxes shall be capable of supporting a luminaire weighing up to 50 pounds (22.7 kg). A luminaire that weighs more than 50 pounds (22.7 kg) shall be supported independently of the outlet box unless the outlet box is listed and marked for the maximum weight to be supported. A recessed luminaire that is not identified for contact with insulation shall have all recessed parts spaced at least $^1/_2$ inch (12.7 mm) from combustible materials.*

Discussion and Commentary: Lampholders and luminaires should only be carried by the lampholder screw shell when they are of limited weight. The screw shell is not designed as a supporting element beyond supporting a bulb and simple shade or diffuser.

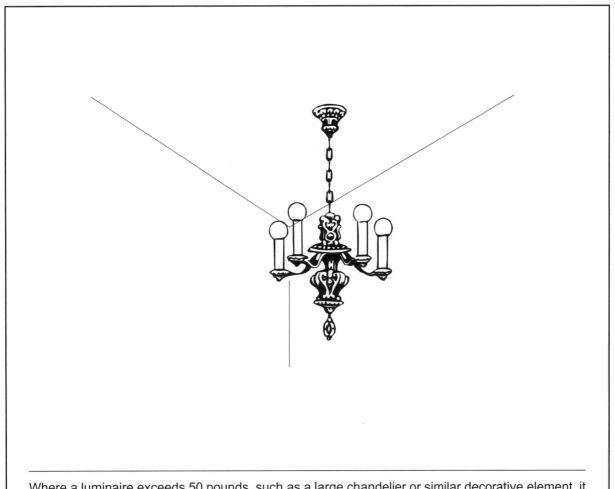

Where a luminaire exceeds 50 pounds, such as a large chandelier or similar decorative element, it cannot be adequately supported by an outlet box. Therefore, it is necessary to provide an independent means of support, typically addressed in the luminaire's installation instructions.

Code Text: *Ceiling-suspended fans (paddle) shall be supported independently of an outlet box or by a listed outlet box or outlet box system identified for the use and installed in accordance with Section E3905.8. Outlet boxes and outlet box systems used as the sole support of ceiling-suspended fans (paddle) shall be marked by their manufacturer as suitable for this purpose and shall not support ceiling-suspended fans (paddle) that weigh more than 70 pounds (31.8 kg). For outlet boxes and outlet box systems designed to support ceiling-suspended fans (paddle) that weigh more than 35 pounds (15.9 kg), the required marking shall include the maximum weight to be supported.*

Discussion and Commentary: Ceiling fans can produce a potentially hazardous condition if not installed correctly. The combination of fan weight and movement of the fan creates a condition that is specifically addressed by the code. It is important to consider both the weight of the fan and any light fixture or other accessory when determining the requirements.

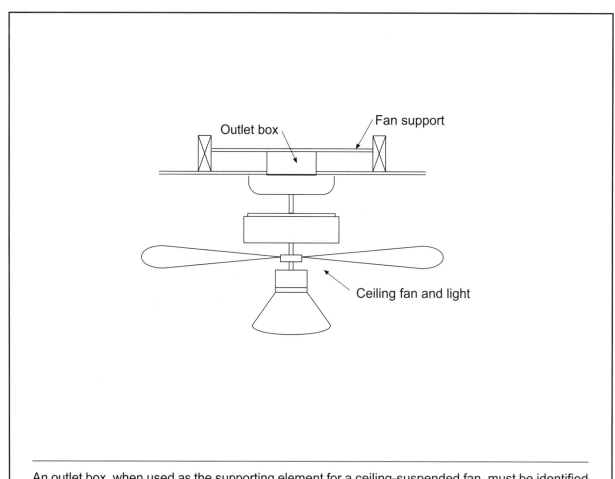

An outlet box, when used as the supporting element for a ceiling-suspended fan, must be identified in a manner that will indicate its acceptable use as a fan support. Although the fan weight is limited to 70 pounds, any fan that weighs over 35 pounds must utilize an outlet box that is specifically marked to indicate the maximum weight that it can support.

Code Text: *Receptacles that provide power for water-pump motors or other loads directly related to the circulation and sanitation system shall be permitted to be located between 6 feet and 10 feet (1829 mm and 3048 mm) from the inside walls of pools and outdoor spas and hot tubs. Other receptacles on the property shall be located not less than 6 feet (1829 mm) from the inside walls of pools and outdoor spas and hot tubs. At least one 125-volt 15- or 20-ampere receptacle supplied by a general-purpose branch circuit shall be located a minimum of 6 feet (1829 mm) from and not more than 20 feet (6096 mm) from the inside wall of pools and outdoor spas and hot tubs.*

Discussion and Commentary: Convenience receptacles must be located at least 6 feet from both indoor and outdoor swimming pools, as well as outdoor spas and hot tubs. At least one outlet shall be provided, but it cannot be located more than 20 feet from the pool edge. This required receptacle must not be located more than 78 inches above the floor, platform or grade level serving the pool, spa or hot tub.

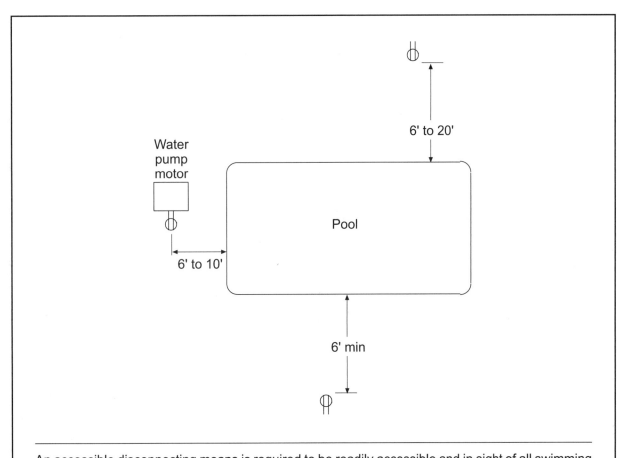

An accessible disconnecting means is required to be readily accessible and in sight of all swimming pools, spas and hot tubs.

Code Text: *Receptacles shall be located not less than 6 feet (1829 mm) from the inside walls of indoor spas and hot tubs. A minimum of one 125-volt receptacle shall be located between 6 feet (1829 mm) and 10 feet (3048 mm) from the inside walls of indoor spas or hot tubs. All 125 volt receptacles rated 30 amperes or less and located within 10 feet (3048 mm) of the inside walls of spas and hot tubs installed indoors shall be protected by ground-fault circuit-interrupters.*

Discussion and Commentary: The requirements applicable to receptacles adjacent to indoor hot tubs and spas are not as restrictive as those for both indoor and outdoor swimming pools, as well as outdoor hot tubs and spas. The limited space available in most indoor spa and hot tub areas would make it impractical to apply the provisions mandated for outdoor facilities.

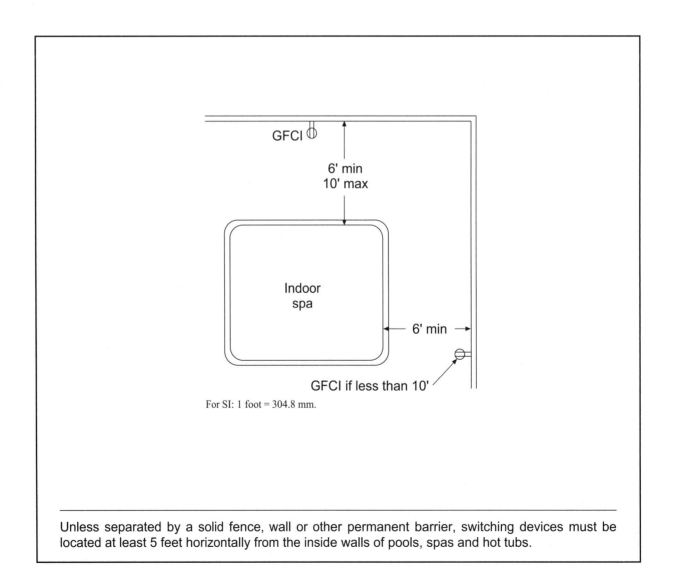

For SI: 1 foot = 304.8 mm.

Unless separated by a solid fence, wall or other permanent barrier, switching devices must be located at least 5 feet horizontally from the inside walls of pools, spas and hot tubs.

Code Text: *In outdoor pool, outdoor spas and outdoor hot tubs areas, luminaires, lighting outlets and ceiling-suspended paddle fans shall not be installed over the pool or over the area extending 5 feet (1524 mm) horizontally from the inside walls of a pool except where no part of the luminaires or ceiling-suspended paddle fan is less than 12 feet (3658 mm) above the maximum water level.*

Discussion and Commentary: The code mandates clearances for luminaires in connection with both outdoor and indoor swimming pools. Typically the provisions for outdoor pools are consistent with those located indoors; however, a reduction in the luminaires or paddle fan height requirements to 7 feet, 6 inches is permitted for indoor installations. To provide this reduced clearance, the luminaires must be of the totally-enclosed type, and a GFCI must be installed in the branch circuit serving the fan or luminaires.

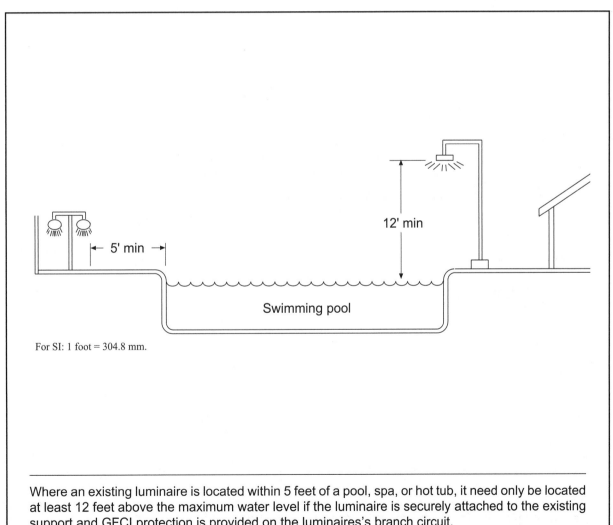

For SI: 1 foot = 304.8 mm.

Where an existing luminaire is located within 5 feet of a pool, spa, or hot tub, it need only be located at least 12 feet above the maximum water level if the luminaire is securely attached to the existing support and GFCI protection is provided on the luminaires's branch circuit.

Code Text: *Except where installed with the clearances specified in Table E4203.5, the following parts of pools and outdoor spas and hot tubs shall not be placed under existing service-drop conductors or any other open overhead wiring; nor shall such wiring be installed above the following: 1) pools and the areas extending 10 feet (3048 mm) horizontally from the inside of the walls of the pool, 2) diving structures, or 3) observation stands, towers, and platforms.*

Discussion and Commentary: When installing a new swimming pool, spa or hot tub, it is important to take into account the position of any overhead service-drop conductors or wiring. If the location of the pool, spa or hot tub brings it within the clearances required by the code, the conductors or wiring must be rerouted to bring the installation into compliance.

TABLE E4203.5
OVERHEAD CONDUCTOR CLEARANCES

	INSULATED SUPPLY OR SERVICE DROP CABLES, 0-750 VOLTS TO GROUND, SUPPORTED ON AND CABLED TOGETHER WITH AN EFFECTIVELY GROUNDED BARE MESSENGER OR EFFECTIVELY GROUNDED NEUTRAL CONDUCTOR (feet)	ALL OTHER SUPPLY OR SERVICE DROP CONDUCTORS (feet)	
		Voltage to ground	
		0-15 kV	Greater than 15 to 50 kV
A. Clearance in any direction to the water level, edge of water surface, base of diving platform, or permanently-anchored raft	22.5	25	27
B. Clearance in any direction to the diving platform	14.5	17	18

For SI: 1 foot = 304.8 mm.

A minimum 10-foot (3048 mm) clearance is required for overhead communications conductors, coaxial cables and their supports that are owned, operated and maintained by a utility. The separation must be maintained to pools, diving boards, observation stands and similar structures.

Quiz

Study Session 18
IRC Chapters 38, 39, 40, 41 and 42

1. Electrical nonmetallic tubing is an allowable wiring method in all but which one of the following applications?

    a. feeders

    b. branch circuits

    c. embedded in masonry

    d. direct burial

    Reference _____

2. Unless installed parallel to framing members, NM wiring run in an attic accessed by a portable ladder shall be protected from damage where located within _____ of the nearest edge of the attic entrance.

    a. 6 feet

    b. 7 feet

    c. 8 feet

    d. 10 feet

    Reference _____

3. Type NM wiring shall be supported a maximum of _____ on center.

    a. 3 feet

    b. $4^1/_2$ feet

    c. 6 feet

    d. 12 feet

    Reference _____

4. In general, direct burial cable installed below a driveway serving a one-family dwelling unit shall have a minimum burial depth of _____ .

    a. 6 inches   b. 12 inches

    c. 18 inches   d. 24 inches

    Reference _____

5. Direct buried conductors and cables emerging from the ground shall be protected by enclosures or raceways extending to a minimum height of _____ above finished grade or the point of entrance to the building.

    a. 7 feet   b. 8 feet

    c. 10 feet   d. 12 feet

    Reference _____

6. Receptacle outlets shall be installed so that all points along the floor line of any wall space are a maximum of _____ from an outlet in that space when measured horizontally.

    a. 4 feet   b. 6 feet

    c. 8 feet   d. 12 feet

    Reference _____

7. An unbroken wall space having a minimum width of _____ shall be considered a wall space for the purpose of determining receptacle outlet distribution.

    a. 12 inches   b. 18 inches

    c. 24 inches   d. 48 inches

    Reference _____

8. Floor receptacle outlets cannot be counted as part of the required number of receptacle outlets unless located within _____ of the wall.

    a. 4 inches   b. 6 inches

    c. 12 inches   d. 18 inches

    Reference _____

9. At a kitchen counter, countertop receptacle outlets shall be installed with a maximum horizontal distance of _____ from any point along the wall line to an outlet.

   a. 12 inches

   b. 18 inches

   c. 24 inches

   d. 36 inches

   Reference _____

10. In a bathroom, at least one receptacle outlet shall be located a maximum of _____ from the outside edge of each lavatory.

    a. 18 inches

    b. 24 inches

    c. 30 inches

    d. 36 inches

    Reference _____

11. A hallway a minimum of _____ in length shall be provided with at least one receptacle outlet.

    a. 6 feet

    b. 8 feet

    c. 10 feet

    d. 12 feet

    Reference _____

12. All 125-volt, single-phase, 20-ampere receptacles shall be provided with ground-fault circuit-interrupter protection where located within _____ of the outside edge of a laundry sink.

    a. 2 feet

    b. 3 feet

    c. 4 feet

    d. 6 feet

    Reference _____

13. Where nonmetallic-sheathed cable is used, the cable assembly including the sheath shall extend into the box a minimum of _____ through a nonmetallic-sheathed cable knockout opening.

    a. $^1/_4$ inch

    b. $^3/_8$ inch

    c. $^1/_2$ inch

    d. $^3/_4$ inch

    Reference _____

14. Electrical outlet boxes that do not enclose devices or utilization equipment shall have a minimum internal depth of _____ .

    a. $^1/_2$ inch   b $^5/_8$ inch

    c. $^{15}/_{16}$ inch   d.   1 inch

Reference _____

15. What is the maximum number of No. 12 conductors permitted for a 4-inch by $1^1/_2$-inch standard round metal box not marked with a cubic-inch capacity?

    a.   4   b.   6

    c.   9   d.   10

Reference _____

16. In a wall constructed of wood or other combustible material, outlet boxes shall be installed so that the front edge of the box is set back a maximum of _____ from the finished surface.

    a.   0 inches; (the box must be flush or project outward)

    b. $^1/_8$ inch

    c. $^1/_4$ inch

    d. $^1/_2$ inch

Reference _____

17. All switches shall be installed so that the center of the grip of the operating handle, in its highest position, shall be located a maximum of _____ above the floor or working platform.

    a.   6 feet, 0 inches   b.   6 feet, 6 inches

    c.   6 feet, 7 inches   d.   6 feet, 9 inches

Reference _____

18. What is the required receptacle rating for a 20-ampere branch circuit supplying two or more receptacles or outlets?

    a.   15 amperes only   b.   20 amperes only

    c.   30 amperes only   d.   15 amperes or 20 amperes

Reference _____

19. Cord-connected luminaires, lighting track and pendants shall be located a minimum of _____ horizontally and _____ vertically from the top of a bathtub rim or shower stall threshold.

    a. 2 feet, 7 feet   b. 2 feet, 8 feet

    c. 3 feet, 7 feet   d. 3 feet, 8 feet

Reference _____

20. Where installed in a storage closet, a recessed fluorescent luminaires shall be located so as to provide a minimum clearance of _____ between the fixture and the nearest point of the storage space.

    a. 3 inches   b. 6 inches

    c. 8 inches   d. 12 inches

Reference _____

21. A luminaires that weighs more than _____ or exceeds _____ in any dimension shall not be supported by the screw shell of a lampholder.

    a. 4 pounds, 14 inches   b. 4 pounds, 16 inches

    c. 6 pounds, 14 inches   d. 6 pounds, 16 inches

Reference _____

22. Ceiling-suspended paddle fans weighing a maximum of _____ are permitted to be supported by outlet boxes identified for such use.

    a. 25 pounds   b. 35 pounds

    c. 42 pounds   d. 70 pounds

Reference _____

23. 125-volt receptacles located a minimum of _____ from the inside walls of a swimming pool need not be protected by a ground-fault circuit-interrupter.

    a. 10 feet   b. 15 feet

    c. 20 feet   d. 30 feet

Reference _____

24. At least one 125-volt receptacle shall be located a minimum of _____ from the inside walls of indoor spas and hot tubs.

     a. 5 feet   b. 6 feet

     c. 10 feet   d. 12 feet

     Reference _____

25. Where underground wiring is installed within 5 feet of the inside walls of a swimming pool, a minimum burial depth of _____ is required for intermediate metal conduit.

     a. 6 inches   b. 12 inches

     c. 18 inches   d. 30 inches

     Reference _____

26. Where three or more conductors of minimum size _____ AWG are run at angles with joists in unfinished basements, additional protection is not required if the cable assembly is attached directly to the bottom of the joists.

     a. 6   b. 8

     c. 10   d. 12

     Reference _____

27. Underground service conductors not encased in concrete and buried 18 inches or more below grade shall have their location identified by a warning ribbon placed a minimum of _____ above the underground installation.

     a. 4   b. 6

     c. 8   d. 12

     Reference _____

28. In general, all branch circuits that supply 125-volt, single-phase, 15- and 20-ampere outlets installed in dwelling unit _____ shall be protected by a combination type arc-fault circuit interrupter.

     a bathrooms   b. bedrooms

     c. kitchens   d. garages

     Reference _____

29. Wall-mounted luminaires weighing a maximum of _____ pounds are permitted to be supported on boxes other than those specifically designed for luminaires, provided the luminaires are secured to the box with at least two No. 6 or larger screws.

    a. 6   b. 10

    c. 14   d. 18

Reference _____

30. Lighting track shall be located a minimum of _____ above the finished floor unless protected from physical damage or the track operates at less than 30 volts rms open-circuit voltage.

    a. 5 feet, 0 inches   b. 6 feet, 0 inches

    c. 6 feet, 4 inches   d. 6 feet, 8 inches

Reference _____

31. The required bathroom receptacle outlet is not required to be mounted on a wall if it is installed on the side or face of the lavatory basin cabinet a maximum of _____ inches below the countertop.

    a. 6   b. 8

    c. 12   d. 18

Reference _____

32. For other than evaporative coolers, a receptacle outlet shall be installed for the servicing of HVAC equipment a maximum of _____ feet from such equipment.

    a. 20   b. 25

    c. 40   d. 50

Reference _____

33. Unless the outlet box is listed for the weight to be supported, an outlet box in a wall may support a luminaire having a maximum weight of _____ pounds.

    a. 28   b. 35

    c. 50   d. 70

Reference _____

34. Noncombustible surfaces that are broken or incomplete shall be repaired so that gaps or open spaces at the edge of a panelboard employing a flush-type cover are a maximum of _____ inch in size.

    a. 0 (no gaps permitted)   b. $^1/_{16}$

    c. $^1/_8$ d. $^1/_4$

Reference _____

35. A flexible cord for a cord-and-plug-connected range hood shall have a minimum length of _____ inches and a maximum length of _____ inches.

    a. 18, 36   b. 18, 48

    c. 30, 36   d. 36, 48

Reference _____

Answer Keys

Study Session 1
2012 *International Residential Code*

1.	c	Sec. R101.2		31.	c	Sec. R102.4, Exception
2.	b	Sec. R102.1		32.	c	Sec. R104.7
3.	c	Sec. R102.5		33.	b	Sec. R105.6
4.	a	Sec. R103.1		34.	d	Sec. R107.1
5.	d	Sec. R104.1		35.	d	Sec. R110.1, Exception 2
6.	b	Sec. R104.9.1				
7.	b	Sec. R104.10				
8.	a	Sec. R104.10.1				
9.	c	Sec. R104.11.1				
10.	d	Sec. R105.2, #B1				
11.	b	Sec. R105.2, #B7				
12.	b	Sec. R105.5				
13.	a	Sec. R105.7				
14.	a	Sec. R106.3.1				
15.	b	Sec. R106.5				
16.	d	Sec. R109.1				
17.	d	Sec. R109.3				
18.	d	Sec. R110.3				
19.	d	Sec. R110.4				
20.	b	Sec. R112.2				
21.	c	Sec. R112.3				
22.	c	Sec. R105.2, #B2				
23.	c	Sec. R114.1				
24.	d	Chapter 44				
25.	a	Chapter 44				
26.	d	Sec. R101.3				
27.	d	Sec. R105.2, #B9				
28.	d	Sec. R105.2, #M7				
29.	a	Sec. R109.3				
30.	b	Sec. R112.1				

Study Session 2
2012 *International Residential Code*

1.	a	Figure R301.2(3)	31.	c	Table R301.2.1.2, Note c
2.	d	Figure R301.2(4)A	32.	c	Table R301.7, Note b
3.	d	Figure R301.2(5)	33.	c	Table R301.5
4.	c	Figure R301.2(6)	34.	b	Table R302.1(2)
5.	c	Figure R301.2(7)	35.	b	Table R302.1(1)
6.	a	Table R301.2.1.2			
7.	a	Table R301.2.1.3			
8.	b	Sec. R301.2.1.4, #2			
9.	b	Table R301.2.2.1.1			
10.	c	Table 301.5			
11.	c	Secs. R301.2.2.3; R301.2.2.2.1, #5			
12.	d	Sec. R301.2.2.2.5, #4			
13.	c	Sec. R301.2.3			
14.	c	Table R301.5			
15.	c	Table R301.5, Note b			
16.	c	Table R301.5			
17.	c	Table R301.6			
18.	d	Table R301.7			
19.	b	Table R301.7			
20.	c	Table R302.1(1)			
21.	c	Tables R302.1(1), R302.1(2)			
22.	d	Sec. R302.1, Exception 3; R105.2, #B1			
23.	d	Tables R302.1(1), R302.1(2)			
24.	d	Sec. R302.4.1.2			
25.	a	Sec. R302.1, Exception 4			
26.	d	Sec. R301.2.1.4, #4			
27.	b	Sec. R301.2.2, #2			
28.	a	Sec. R302.5.1			
29.	a	Sec. R302.2, Exception			
30.	a	Table R301.5, Note f			

Study Session 3

2012 *International Residential Code*

1.	c	Sec. R303.1		31.	a	Sec. R310.2.2, Exeption
2.	a	Sec. R303.1		32.	d	Sec. R303.7.1
3.	a	Sec. R309.5		33.	a	Table R308.3.1(1)
4.	c	Sec. R303.1, Exception 2		34.	c	Sec. R310.1, Exception
5.	b	Sec. R304.1		35.	b	Sec. R310.5
6.	d	Sec. R304.4				
7.	b	Sec. R304.3				
8.	b	Sec. R305.1				
9.	a	Sec. R305.1.1, Exception				
10.	a	Figure R307.1				
11.	a	Figure R307.1				
12.	b	Sec. R307.2				
13.	a	Sec. R303.9				
14.	c	Sec. R308.4.5				
15.	a	Sec. R308.6.1				
16.	b	Sec. R308.6.8				
17.	c	Sec. R303.6				
18.	a	Figure R307.1				
19.	d	Sec. R308.2				
20.	c	Sec. R310.1				
21.	c	Sec. R310.1.1, Exception				
22.	a	Sec. R310.1.2, R310.1.3				
23.	d	Sec. R310.2				
24.	b	Sec. R309.2, Exception				
25.	c	Sec. R308.6.2, #1				
26.	c	Sec. R303.5.1				
27	a	Sec. R303.7				
28.	c	Sec. R305.1, Exception 2				
29.	c	Sec. R308.4.2, Exception 4				
30.	a	Sec. R308.4.1, Exception 1				

Study Session 4
2012 *International Residential Code*

1.	b	Sec. R311.6	31.	c	Sec. R311.3.1, Exception
2.	c	Sec. R311.7.6	32.	b	Sec. R311.7.5.3
3.	d	Sec. R311.8.1	33.	a	Sec. R319.1
4.	c	Sec. R311.8.3	34.	c	Sec. R312.1.3, Exception 2
5.	b	Sec. R311.8.2, #2	35.	a	Sec. R323.1
6.	c	Sec. R311.7.1			
7.	c	Sec. R311.7.1			
8.	c	Sec. R311.7.5.1, R311.7.5.2			
9.	b	Sec. R311.7.5.1			
10.	c	Sec. R311.7.5.3, Exception			
11.	b	Sec. R311.7.5.1			
12.	a	Sec. R322.3			
13.	d	Sec. R311.7.10.1			
14.	c	Sec. R311.7.8.1			
15.	b	Sec. R311.7.8.2			
16.	a	Sec. R311.7.8.3, #1			
17.	c	Sec. R312.1.1			
18.	b	Sec. R312.1.3			
19.	b	Sec. R314.3			
20.	b	Sec. R316.4			
21.	c	Sec. R315.1			
22.	a	Sec. R313.1			
23.	d	Sec. R315.1			
24.	b	Sec. R317.1, #2			
25.	d	Sec. R322.1.4, #1			
26.	d	Sec. R311.7.5.1			
27.	c	Sec. R316.5.11, #2			
28.	d	Sec. R311.7.8.3, #2			
29.	c	Sec. R311.7.8			
30.	a	Sec. R311.2			

Study Session 5

2012 *International Residential Code*

1.	b	Sec. R401.3	31.	a.	Sec. R403.1.6, Exception 2
2.	b	Table R401.4.1	32.	d	Sec. R404.1.2.3.4
3.	b	Table R402.2	33.	a	Sec. R406.1, #3
4.	c	Table R402.2, Note d	34.	a	Sec. R407.3
5.	c	Table R403.1	35.	c	Sec. R408.3, #1
6.	b	Sec. R403.1.1			
7.	b	Sec. R403.1.1			
8.	b	Figure R403.1(2)			
9.	a	Sec. R403.1.3.1			
10.	d	Sec. R403.1.4			
11.	b	Sec. R403.1.5			
12.	c	Sec. R403.1.6			
13.	a	Sec. R403.1.6			
14.	b	Table R405.1			
15.	b	Sec. R403.3			
16.	c	Table R403.3(1)			
17.	d	Table R404.1.1(1)			
18.	d	Table R404.1.2(4)			
19.	a	Sec. R404.1.6			
20.	b	Sec. R404.1.7, Exception			
21.	d	Sec. R404.2.2			
22.	b	Sec. R404.2.3			
23.	d	Sec. R405.1			
24.	c	Sec. R408.1			
25.	b	Sec. R408.4			
26.	d	Sec. R403.1.4.1, Exception 1			
27.	c	Sec. R403.2			
28.	a	Sec. R403.3.2			
29.	c	Sec. R406.2, #5			
30.	b	Sec. R408.2			

Study Session 6
2012 *International Residential Code*

1.	c	Sec. R502.3.1	31.	c	Sec. R502.6
2.	b	Table R502.3.1(1)	32.	d	Sec. R502.8.1
3.	a	Table R502.3.1(2)	33.	b	Sec. R502.8.1
4.	d	Table R502.5(1)	34.	d	Table R503.2.1.1(1), Note i
5.	b	Table R502.5(1)	35.	c	Sec. R506.2.4
6.	b	Table R502.5(2)			
7.	c	Sec. R502.4			
8.	b	Sec. R502.6			
9.	a	Sec. R502.6.1			
10.	d	Sec. R502.7.1			
11.	c	Sec. R502.8.1			
12.	d	Sec. R502.8.1			
13.	d	Sec. R502.10			
14.	c	Secs. R502.12, R302.12			
15.	b	Secs. R502.12, R302.12.1			
16.	d	Table R503.2.1.1(1), Note h			
17.	b	Table R503.2.1.1(2)			
18.	d	Sec. R505.1.1			
19.	c	Sec. R505.2.4			
20.	b	Table R505.3.1(2)			
21.	c	Table R505.3.2(1)			
22.	b	Sec. R502.6			
23.	c	Sec. R506.2.1			
24.	a	Sec. R506.2.2, Exception; Table R405.1			
25.	c	Sec. R506.2.3			
26.	a	Table R502.3.3(1)			
27.	c	Table R502.3.3(2)			
28.	d	Sec. R502.4			
29.	b	Sec. R502.10			
30.	b	Sec. R506.2.2			

Study Session 7

2012 *International Residential Code*

1.	c	Table R602.3(5)		31.	d	Sec. R602.6.1
2.	b	Table R602.3.1		32.	c	Sec. R602.11.1
3.	a	Sec. R602.3.2		33.	c	Sec. R613.7
4.	b	Sec. R602.3.2, Exception		34.	d	Sec. R602.1.1
5.	b	Table R602.3(3)		35.	d	Sec. R703.6.3
6.	c	Sec. R602.6, #1				
7.	c	Sec. R602.6, #2				
8.	d	Sec. R602.6, #2				
9.	d	Table R602.7.2				
10.	c	Sec. R602.8, R302.11.1				
11.	a	Sec. R602.8, R302.11.1.2				
12.	d	Sec. R602.9				
13.	b	Sec. R602.10.2.2				
14.	c	Sec. R602.10.2.2				
15.	b	Sec. R603.1.2, #1				
16.	d	Sec. R603.2.5.1, #6				
17.	b	Sec. R602.10.1.2				
18.	a	Figure R606.11(1)				
19.	d	Table R702.3.5				
20.	d	Sec. R702.3.6				
21.	a	Sec. R702.3.8				
22.	d	Table R703.4				
23.	a	Sec. R703.7.4.1, Exception				
24.	c	Sec. R703.9.4.1				
25.	c	Table R703.7.3.1				
26.	a	Sec. R602.3.1, Exception 1				
27.	d	Table R602.3(1), Item 14				
28.	b	Table R602.3(2)				
29.	d	Sec. R602.6.1				
30.	b	Sec. R703.6.2.1				

Study Session 8
2012 *International Residential Code*

1.	b	Sec. R801.3	31.	a	Sec. R802.3
2.	a	Sec. R802.3	32.	c	Sec. R806.2
3.	b	Sec. R802.3.1	33.	a	Sec. R806.5, #3
4.	b	Sec. R802.3.2	34.	d	Sec. R905.2.7.1
5.	c	Table R802.4(1)	35.	a	Sec. R905.2.8.5
6.	a	Table R802.5.1(4)			
7.	a	Sec. R802.5.1			
8.	a	Sec. R802.6			
9.	c	Sec. R802.7.1, R502.8.1			
10.	a	Sec. R802.9			
11.	c	Sec. R803.2.2, Table R503.2.1.1(1)			
12.	b	Tables R804.3.2.1(3), R804.3.2.1(1)			
13.	c	Sec. R804.3.2.1.1			
14.	a	Sec. R806.1			
15.	c	Sec. R807.1			
16.	d	Sec. R806.2, Exception, #1			
17.	a	Sec. R802.1.3			
18.	d	Sec. R905.2.6			
19.	d	Sec. R905.2.7			
20.	b	Sec. R905.2.8.3			
21.	c	Sec. R905.3.6			
22.	b	Sec. R802.7.1.1			
23.	b	Table R905.7.5			
24.	d	Sec. R905.8.6			
25.	c	Sec. R905.8.8			
26.	d	Secs. R802.7.1, R502.8.1			
27.	b	Sec. R802.9			
28.	b	Sec. R806.3			
29.	a	Sec. R807.1			
30.	c	Sec. R903.4.1			

Study Session 9

2012 *International Residential Code*

1.	d	Sec. R1001.2	31.	c	Sec. R1001.5.1
2.	a	Sec. R1003.6	32.	b	Sec. R1001.6, Exception
3.	d	Sec. R1003.9	33.	b	Sec. R1001.9.2
4.	c	Sec. R1003.12	34.	a	Sec. R1006.3
5.	b	Sec. R1003.13	35.	c	Sec. N1101.12.1.1
6.	a	Sec. R1003.17			
7.	c	Sec. R1003.18			
8.	b	Sec. R1001.2			
9.	d	Sec. R1001.3.1			
10.	c	Sec. R1001.5			
11.	b	Sec. R1001.5			
12.	c	Sec. R1001.7			
13.	d	Sec. R1001.9.1			
14.	d	Sec. R1001.10			
15.	b	Sec. R1001.9.2, Exception			
16.	d	Sec. R1001.11			
17.	c	Sec. R1003.18, Exception 2			
18.	c	Sec. R1001.11, Exception 4			
19.	d	Sec. R1004.1			
20.	b	Sec. R1006.4			
21.	a	Sec. N1101.12.1			
22.	a	Sec. N1101.13.1			
23.	b	Table N1102.1.1			
24.	c	Sec. N1102.4.3			
25.	b	Sec. N1103.2.1			
26.	b	Sec. R1003.2			
27.	d	Sec. R1003.9.2, #1			
28.	c	Table R1003.20			
29.	c	Sec. R1001.6			
30.	d	Sec. R1002.5			

1.	b	Sec. M1305.1	31.	c	Sec. M1305.1.3, Exception 2
2.	b	Sec. M1305.1.1	32.	a	Sec. M1307.4.1.1
3.	a	Sec. M1305.1.1	33.	a	Sec. M1411.5
4.	a	Sec. M1305.1.2	34.	d	Sec. M1408.3, #4
5.	a	Sec. M1305.1.3	35.	b	Sec. M1411.3.1
6.	c	Sec. M1305.1.4.1			
7.	c	Sec. M1305.1.4.2			
8.	b	Sec. M1306.2			
9.	b	Table M1306.2			
10.	d	Sec. M1306.2.1, Table M1306.2			
11.	b	Sec. M1307.3			
12.	b	Sec. M1402.3			
13.	a	Sec. M1403.1			
14.	b	Sec. M1403.2			
15.	a	Sec. M1406.3, #2			
16.	d	Sec. M1407.3			
17.	c	Sec. M1408.3, #3			
18.	a	Sec. M1408.4			
19.	a	Sec. M1408.5, #3			
20.	b	Sec. M1409.2, #2			
21.	c	Sec. M1411.2			
22.	b	Sec. M1411.3.2			
23.	b	Sec. M1411.3.1, #1			
24.	a	Sec. M1413.1, #2, M1305.1.4.1			
25.	c	Sec. M1414.2			
26.	a	Sec. M1305.1.4			
27.	b	Sec. M1307.2			
28.	c	Sec. M1307.4.1			
29.	d	Sec. M1308.2			
30.	b	Sec. M1410.2			

Study Session 11

2012 *International Residential Code*

1.	b	Sec. M1502.4.2	31.	b	Sec. M1502.3
2.	d	Sec. M1502.4.4.1	32.	d	Sec. M1502.4.3
3.	c	Table M1502.4.4.1	33.	a	Table M1601.1.1(1)
4.	d	Sec. M1503.1	34.	b	Sec. M1601.4.1
5.	a	Sec. M1503.2, Exception, #4	35.	b	Sec. M1601.4.7
6.	b	Sec. M1505.1			
7.	a	Sec. M1505.1			
8.	d	Sec. M1601.1.1, #1			
9.	c	Sec. M1601.1.1, #2			
10.	b	Table M1601.1.1(2)			
11.	a	Sec. M1601.1.1, #5			
12.	c	Sec. M1601.1.1, #6			
13.	b	Sec. M1601.1.2			
14.	b	Sec. M1601.1.2			
15.	a	Sec. M1601.3, #1			
16.	a	Sec. M1601.3, #3			
17.	b	Sec. M1601.2.1			
18.	c	Sec. M1601.4.3			
19.	d	Sec. M1601.5.2			
20.	c	Sec. M1601.5.3			
21.	d	Sec. M1602.2, #1			
22.	a	Sec. M1602.2, #4			
23.	a	Sec. M1506.2			
24.	b	Sec. M1901.1			
25.	d	Sec. M1902.4			
26.	b	Table M1507.4			
27.	c	Table M1507.4			
28.	a	Sec. M1601.5.4			
29.	b	Sec. M1601.5.5			
30.	c	Sec. M1902.4			

Study Session 12
2012 *International Residential Code*

1.	b	Sec. M1701.1
2.	a	Sec. M1802.2.1
3.	b	Sec. M1803.3.1, #2
4.	d	Sec. M1804.2.6, #4
5.	c	Sec. M1804.3.1
6.	a	Sec. M1805.2
7.	b	Sec. M2003.1
8.	b	Sec. M2103.2.1
9.	c	Sec. M2103.2.2
10.	d	Sec. M2105.1
11.	a	Sec. M2201.2.1
12.	b	Sec. M1803.3.2
13.	a	Sec. M1804.2.6, #5
14.	a	Sec. M2203.5
15.	b	Sec. M2104.4.2
16.	c	Sec. M1805.3.1
17.	b	Sec. M1803.3
18.	c	Sec. M1803.3.2
19.	c	Table M1803.3.4
20.	d	Sec. M1804.2.3
21.	a	Sec. M1805.3
22.	c	Sec. M2003.1.1
23.	d	Sec. M2103.4
24.	b	Sec. M2203.3
25.	c	Sec. M2301.2.9
26.	d	Sec. M2005.2
27.	a	Table M2101.9
28.	d	Sec. M2101.10
29.	b	Sec. M2201.2
30.	a	Sec. M2204.3

31.	c	Table M2003.2
32.	b	Sec. M1804.2.3
33.	d	Sec. M2002.4
34.	a	Sec. M2201.2.2
35.	c	Sec. M2203.4

Study Session 13

2012 *International Residential Code*

1.	d	Sec. G2406.2	31.	c	Table G2413.2
2.	c	Sec. G2408.2	32.	b	Sec. G2415.16
3.	b	Sec. G2408.3	33.	b	Sec. G2415.12.1
4.	d	Table G2409.2	34.	c	Sec. G2422.1.4
5.	b	Table G2409.2	35.	d	Sec. G2444.4
6.	a	Sec. G2412.5			
7.	a	Sec. G2413.6			
8.	a	Sec. G2413.6.1			
9.	c	Sec. G2414.5.2			
10.	b	Table G2414.9.2			
11.	d	Sec. G2415.3			
12.	d	Sec. G2415.7			
13.	b	Sec. G2415.12			
14.	a	Sec. G2415.16			
15.	c	Sec. G2415.17.3			
16.	a	Sec. G2416.2, #5			
17.	b	Sec. G2417.4.1			
18.	a	Sec. G2407.6			
19.	d	Sec. G2419.1			
20.	c	Sec. G2420.5			
21.	c	Sec. G2422.1.2.1			
22.	d	Sec. G2416.3, #3			
23.	a	Sec. G2425.9			
24.	a	Sec. G2440.7			
25.	b	Sec. G2445.3			
26.	c	Sec. G2408.4			
27.	c	Sec. G2415.9			
28.	d	Sec. G2417.2			
29.	c	Table G2424.1			
30.	a	Sec. G2450.3			

Study Session 14
2012 *International Residential Code*

1.	d	Sec. P2503.4		31.	a	Sec. P2708.1.1
2.	b	Sec. P2503.5.1, #2		32.	c	Sec. P2713.3
3.	d	Sec. P2503.7		33.	a	Sec. P2717.2
4.	d	Sec. P2603.2.1		34.	b	Sec. P2720.1
5.	c	Sec. P2603.2.1		35.	c	Sec. P2721.2
6.	d	Sec. P2603.4				
7.	b	Sec. P2603.5				
8.	c	Sec. P2603.5				
9.	d	Sec. P2604.1				
10.	b	Sec. P2604.3				
11.	d	Table P2605.1				
12.	a	Table P2605.1				
13.	c	Sec. P2703.1				
14.	a	Sec. P2704.1				
15.	b	Sec. P2706.2				
16.	b	Sec. P2706.2.1				
17.	a	Sec. P2708.1				
18.	a	Sec. P2705.1, #5				
19.	b	Sec. P2711.3				
20.	b	Sec. P2712.4				
21.	c	Sec. P2713.1				
22.	b	Sec. P2714.1				
23.	b	Sec. P2716.1				
24.	c	Sec. P2719.1				
25.	a	Sec. P2723.2				
26.	b	Sec. P2503.5.2, #2.1				
27.	a	Sec. P2503.9				
28.	b	Sec. P2604.4				
29.	c	Sec. P2715.1				
30.	b	Sec. P2705.1, #5				

Study Session 15
2012 *International Residential Code*

1.	b	Sec. P2801.5.1	31.	d	Sec. P2902.5.3
2.	d	Sec. P2801.5.2	32.	c	Sec. P2905.3
3.	c	Sec. P2801.6	33.	d	Sec. P3002.1
4.	d	Sec. P2803.3	34.	c	Table P3005.1, Note a
5.	d	Sec. P2803.4	35.	b	Sec. P2904.5.2, #2
6.	b	Table P2902.3.1			
7.	c	Sec. P2803.4			
8.	b	Table P2903.1			
9.	b	Table P2903.2			
10.	d	Table P2903.2, Note a			
11.	a	Sec. P2903.3, P2903.3.1			
12.	c	Table P2903.6			
13.	c	Sec. P2904.1.1, Exception 3			
14.	d	Sec. P2905.2			
15.	d	Sec. P2905.3			
16.	b	Sec. P2905.5			
17.	a	Sec. P2906.1			
18.	a	Table P3004.1			
19.	d	Sec. P3005.2.2			
20.	a	Sec. P3005.2.9			
21.	b	Sec. P3005.2.4			
22.	a	Sec. P3005.2.5			
23.	a	Sec. P3005.3			
24.	a	Sec. P3005.4.1			
25.	d	Table P3005.4.1			
26.	d	Sec. P2803.6.1, #10			
27.	c	Table P2902.3			
28.	c	Sec. P2905.4			
29.	a	Sec. P3005.1.1, Exception			
30.	d	Sec. P2904.2.4.1			

Study Session 16
2012 *International Residential Code*

1.	a	Sec. P3103.1	31.	c	Sec. P3109.3
2.	b	Sec. P3103.2	32.	a	Sec. P3113.1
3.	c	Sec. P3104.5	33.	c	Sec. P3113.4.2
4.	c	Sec. P3103.1	34.	c	Table P3113.4.1
5.	c	Sec. P3103.5	35.	b	Sec. P3201.6
6.	d	Sec. P3104.4			
7.	d	Sec. P3105.1, Exception			
8.	c	Table P3105.1			
9.	b	Sec. P3105.2			
10.	d	Sec. P3105.3			
11.	b	Sec. P3107.1			
12.	b	Table P3107.3			
13.	c	Table P3108.3			
14.	c	Table P3109.4			
15.	d	Sec. P3110.1			
16.	c	Sec. P3111.1			
17.	c	Sec. P3111.2.1			
18.	b	Table P3111.3			
19.	b	Sec. P3113.1			
20.	c	Sec. P3114.4			
21.	a	Sec. P3201.2			
22.	d	Sec. P3201.6			
23.	c	Sec. P3201.6, #2			
24.	c	Sec. P3303.1.2			
25.	c	Table P3201.7			
26.	d	Sec. P3103.6			
27.	c	Sec. P3110.3			
28.	d	Sec. P3111.2			
29.	d	Sec. P3112.2			
30.	b	Sec. P3114.4			

Study Session 17
2012 *International Residential Code*

1.	a	Sec. E3404.9	31.	a	Sec. E3604.2.1, Exception 2
2.	b	Sec. E3405.2	32.	c	Sec. E3605.7
3.	a	Sec. E3405.2	33.	d	Sec. E3608.1.5
4.	c	Sec. E3405.3	34.	c	Table E3702.13
5.	b	Sec. E3406.3	35.	c	Sec. E3705.7, #6
6.	b	Sec. E3406.11.3			
7.	c	Sec. E3407.1			
8.	a	Sec. E3407.2			
9.	d	Sec. E3407.3			
10.	b	Table E3603.1			
11.	a	Sec. E3604.1			
12.	d	Sec. E3604.2.1			
13.	c	Sec. E3604.2.2, #2			
14.	b	Sec. E3604.3			
15.	b	Sec. E3608.1.1			
16.	b	Sec. E3608.3			
17.	c	Sec. E3608.1.4, #2			
18.	c	Sec. E3610.2			
19.	a	Sec. E3702.3			
20.	d	Sec. E3703.2			
21.	b	Sec. E3703.3, E3703.4			
22.	b	Table E3704.2(1)			
23.	b	Sec. E3704.4			
24.	b	Table E3705.5.3			
25.	c	Sec. E3705.6			
26.	b	Sec. E3404.6			
27.	a	Sec. E3405.4			
28.	b	Sec. E3406.11.3			
29.	d	Sec. E3601.6.2			
30.	b	Sec. E3608.1.1			

Study Session 18

2012 *International Residential Code*

1.	d	Tables E3801.2, E3801.4	31.	c	Sec. E3901.6, Exception
2.	a	Sec. E3802.2.1	32.	b	Sec. E3901.11
3.	b	Table E3802.1	33.	c	Sec. E3905.6.1
4.	c	Table E3803.1	34.	c	Sec. E3907.4
5.	b	Sec. E3803.3	35.	a	Table E4101.3
6.	b	Sec. E3901.2.1			
7.	c	Sec. E3901.2.2, #1			
8.	d	Sec. E3901.2.3			
9.	c	Sec. E3901.4.1			
10.	d	Sec. E3901.6			
11.	c	Sec. E3901.10			
12.	d	Sec. E3902.7			
13.	a	Sec. E3905.3.1			
14.	a	Sec. E3905.4.1			
15.	b	Table E3905.12.1			
16.	a	Sec. E3906.5			
17.	c	Sec. E4001.6			
18.	d	Table E4002.1.2			
19.	d	Sec. E4003.11			
20.	b	Sec. E4003.12, #4			
21.	d	Sec. E4004.4			
22.	d	Sec. E4101.6, E3905.8			
23.	c	Sec. E4203.1.3			
24.	b	Sec. E4203.1.4			
25.	a	Table E4203.7			
26.	b	Sec. E3802.4			
27.	d	Sec. E3803.2			
28.	b	Sec. E3902.11			
29.	a	Sec. E3905.6.1, Exception			
30.	a	Sec. E4005.4, #8